BIOSTATISTICS

THE UNIVERSITY OF WESTERN ONTARIO SERIES IN PHILOSOPHY OF SCIENCE

A SERIES OF BOOKS
IN PHILOSOPHY OF SCIENCE, METHODOLOGY,
EPISTEMOLOGY, LOGIC, HISTORY OF SCIENCE
AND RELATED FIELDS

VOLUME 38

ADVANCES IN THE STATISTICAL SCIENCES

Festschrift in Honor of Professor V. M. Joshi's 70th Birthday

VOLUME V

BIOSTATISTICS

Edited by

IAN B. MacNEILL and GARY J. UMPHREY

Department of Statistical and Actuarial Sciences,
The University of Western Ontario

Associate editors:

ALLAN DONNER

Department of Epidemiology and Biostatistics,
The University of Western Ontario

V. KRISHNA JANDHYALA

Department of Statistical and Actuarial Sciences,
The University of Western Ontario

D. REIDEL PUBLISHING COMPANY

A MEMBER OF THE KLUWER ACADEMIC PUBLISHERS GROUP

DORDRECHT / BOSTON / LANCASTER / TOKYO

Library of Congress Cataloging in Publication Data

CIP

Biostatistics.

(Advances in the statistical sciences; v. 5)
(The University of Western Ontario series in philosophy of science; v. 38)
1. Biometry—Congresses. I. MacNeill, Ian B., 1931– . II. Umphrey, Gary J., 1953– . III. Series. IV. Series: University of Western Ontario series in philosophy of science; v. 38.
QA276.A1A39 vol. 5 519.5 s 86–29675
[QH323.5] [574'.072]

ISBN-13: 978-94-010-8626-4 e-ISBN-13: 978-94-009-4794-8
DOI: 10.1007/978-94-009-4794-8

Published by D. Reidel Publishing Company,
P.O. Box 17, 3300 AA Dordrecht, Holland.

Sold and distributed in the U.S.A. and Canada
by Kluwer Academic Publishers,
101 Philip Drive, Assinippi Park, Norwell, MA 02061, U.S.A.

In all other countries, sold and distributed
by Kluwer Academic Publishers Group,
P.O. Box 322, 3300 AH Dordrecht, Holland.

TABLE OF CONTENTS

CONTENTS OF THE OTHER VOLUMES OF THE JOSHI FESTSCHRIFT

VOLUME I

Applied Probability, Stochastic Processes, and Sampling Theory

VOLUME II

Foundations of Statistical Inference

VOLUME III

Time Series and Econometric Modelling

VOLUME IV

Stochastic Hydrology

VOLUME VI

Foundations of Actuarial Science

PREFACE

On May 27–31, 1985, a series of symposia was held at The University of Western Ontario, London, Canada, to celebrate the 70th birthday of Professor V. M. Joshi. These symposia were chosen to reflect Professor Joshi's research interests as well as areas of expertise in statistical science among faculty in the Departments of Statistical and Actuarial Sciences, Economics, Epidemiology and Biostatistics, and Philosophy.

From these symposia, the six volumes which comprise the "Joshi Festschrift" have arisen. The 117 articles in this work reflect the broad interests and high quality of research of those who attended our conference. We would like to thank all of the contributors for their superb cooperation in helping us to complete this project.

Our deepest gratitude must go to the three people who have spent so much of their time in the past year typing these volumes: Jackie Bell, Lise Constant, and Sandy Tamowski. This work has been printed from "camera ready" copy produced by our Vax 785 computer and QMS Lasergraphix printers, using the text processing software TEX. At the initiation of this project, we were neophytes in the use of this system. Thank you, Jackie, Lise, and Sandy, for having the persistence and dedication needed to complete this undertaking.

We would also like to thank Maria Hlawka-Lavdas, our systems analyst, for her aid in the layout design of the papers and for resolving the many difficult technical problems which were encountered. Nancy Nuzum and Elly Pakalnis have also provided much needed aid in the conference arrangements and in handling the correspondence for the Festschrift.

Professor Robert Butts, the Managing Editor of *The University of Western Ontario Series in Philosophy of Science* has provided us with his advice and encouragement. We are confident that the high calibre of the papers in these volumes justifies his faith in our project.

In a Festschrift of this size, a large number of referees were needed. Rather than trying to list all of the individuals involved, we will simply say "thank you" to the many people who undertook this very necessary task for us. Your contributions are greatly appreciated.

Financial support for the symposia and Festschrift was provided by The University of Western Ontario Foundation, Inc., The University of Western Ontario and its Faculties of Arts, Science, and Social Science, The UWO Statistical Laboratory, and a conference grant from the Natural Sciences and Engineering Research Council of Canada. Their support is gratefully acknowledged.

Finally, we would like to thank Professor Joshi for allowing us to hold the conference and produce this Festschrift in his honor. Professor Joshi is a very modest man who has never sought the limelight. However, his substantial contributions to statistics merit notice (see Volume I for a bibliography of his papers and a very spiffy photo). We hope he will accept this as a tribute to a man of the highest integrity.

INTRODUCTION TO VOLUME V

The discipline of biostatistics has rapidly expanded over the last fifteen years, a period which has seen many new methods developed as well as increasing insight gained into older methods. The sixteen papers in this volume reflect both of these trends very well.

The lead paper by Lawless and Singhal discusses and evaluates the rise of regression methodology, including proportional hazards and accelerated failure time models, to explore large medical data bases. An example is given of the application of these methods to a data base containing information on ovarian carcinoma patients.

The paper by Ciampi, Chang, Hogg and McKinney also deals with exploratory data analysis, discussing the application of Recursive Partition, a clustering method, to the determination of prognostic strata, propensity strata and treatment response strata in clinical and epidemiological studies. They also tackle the general problem of evaluating the statistics of exploratory analyses.

The third paper in this volume, by Chapman, Etezadi-Amoli, Selby, Boyd and Dalley, reports on the use of linear analogue scales in assessing the quality of life of cancer patients. The report is based on the results of a study in which the properties of these scales were investigated after administration to a large sample of breast cancer patients.

Krewski, Smythe and Colin discuss the use of historical control information to help increase the sensitivity of tests for increasing trend in tumor occurrence with increasing dose. They propose a two-stage procedure in which the historical control data is used only if the concurrent control response rate falls within a suitable tolerance interval defined in terms of the historical control distributions.

The next group of four papers deals with inferences concerning odds ratios in one or more sets of 2×2 contingency tables. Walter considers and evaluates the performance of alternative point estimators of a single odds ratio and its logarithm in 2×2 tables with small cell frequencies, while Wells and Donner develop formulas for the bias and mean square error of the logit estimator from a single 2×2 table. The two papers by Hauck and by Donald and Donner deal with multiple 2×2 contingency tables, focussing on the estimation of a common odds ratio. Hauck provides a review of various point and interval estimators for the situation, while Donald and Donner discuss the effect of clustering on the analyses of multiple contingency tables.

Mantel and Paul also focus on the analyses of categorical data, critiquing the use of standard goodness-of-fit tests in toxicological experiments involving litters of varying age.

The papers by Bull and Donner and by Koval and Donner deal with logistic regression analyses, a methodology which has become extremely popular among biostatisticians in recent years. Bull and Donner derive the large sample relative efficiency of multinomial logistic regression, with an arbitrary number of response groups, to multiple group discriminant analyses. Koval and Donner propose a model to incorporate clustering, as found in family data, for example, into logistic estimation.

Konishi and Gupta discuss inferences concerning interclass and intraclass correlation from familial data. They review the existing literature and propose a new procedure for testing the equality of intraclass correlations in two multivariate normal populations.

The next two papers focus on the topic of estimation in bioassay. Viveros and Sprott present an alternative to the Fieller method for obtaining approximate confidence intervals for LD50 and LD90; their method is based on transforming the parameters of interest. Walsh, Hubert and Carter continue their recent work in multivariate bioassay. They also propose a new test for "parallelism" for univariate symmetric parabolic bioassays.

Shoukri and Consul present an overview of the generalized Poisson distribution, and illustrate the generation of this model with a number of examples from biology and medicine.

The final paper ends this volume on a high note. Federer and Murty discuss the application of multivariate analyses for intercropping experiments. Dr. Federer is one of the world's leading experts in the field of experimental design, and we wish to congratulate him on the recent Festschrift for his retirement. However, he tells us that he will still be very active in research, and we look forward to his further contributions to statistical science.

J. F. Lawless and K. Singhal [1]

REGRESSION METHODS AND THE EXPLORATION OF LARGE MEDICAL DATA BASES

1. INTRODUCTION

Many moderate to large size data bases are assembled in connection with the study and treatment of chronic diseases. Examples are data bases related to the diagnosis and treatment of patients with various forms of cancer (e.g., Dembo *et al.*, 1982; Ciampi *et al.*, 1981) or coronary heart disease (e.g., Mock *et al.*, 1982), or those related to prospective studies, for example on the risk of coronary heart disease (Keys *et al.*, 1972). Such data bases typically contain survival or other event-history information about individuals, along with information on other pertinent variables. Our objective in this paper is to review the use of regression methods in exploring them.

We consider problems in which the relationship of some primary variable to other explanatory variables is of interest. Following usual regression terminology, we will refer to the primary variable as the response variable and the others as explanatory variables or covariates. We assume that the data base may have information on a fairly large number of individuals (somewhere between a hundred and many thousand is common) and that there may be a large number of explanatory variables possibly related to a given primary variable. Although sometimes all or part of the data may have arisen from controlled randomized trials, the data are often observational. Consequently we focus on rather general statistical objectives related to exploring and describing the structure of the data. These include

(i) assessment of which explanatory variables appear to be related to the response variable, and description of the nature of the relationship,

(ii) summarization and description of salient features of the data.

These are of course very broad objectives, about which we say more in following sections.

---

[1] Department of Statistics and Actuarial Science, University of Waterloo, Waterloo, Ontario N2L 3G1 (both authors)

I. B. MacNeill and G. J. Umphrey (eds.), Biostatistics, 1–22.

To use regression methods it is necessary to adopt models or paradigms to guide the analysis. In Section 2 we outline strategies which we have found useful when the response variable is a (continuous) survival or response time. We mention, in Section 6, situations where the response variable is categorical, but to conserve space do not discuss them at length. We do note that a large proportion of practical problems can be addressed using either continuous or discrete-response regression models.

Section 3 reviews methods for fitting, screening and criticising models; both important recent developments and gaps in current knowledge are noted. We discuss software in Section 4, including a package written by K. Singhal (1985) which provides flexible methodology for the regression analysis of lifetime and categorical response data. In Section 5 we illustrate points made earlier by examining data on the survival of patients with ovarian carcinoma. Section 6 concludes with a few additional remarks.

2. SURVIVAL TIME REGRESSION MODELS

We consider regression models for which the response variable Y is a survival time and $\mathbf{x} = (x_1, \ldots, x_k)$, is a vector of explanatory variables. The conditional survivor function (SF), probability density function (PDF), and hazard function (HF) of Y given $\mathbf{x}$ respectively are denoted by

$$S(y \mid \mathbf{x}) = \Pr(Y \geq y \mid \mathbf{x}),$$
$$f(y \mid \mathbf{x}) = -dS(y \mid \mathbf{x})/dy,$$

and

$$h(y \mid \mathbf{x}) = f(y \mid \mathbf{x}) \,/S(y \mid \mathbf{x}).$$

It is easily verified that $S(y \mid \mathbf{x}) = \exp(-\int_0^y h(u \mid \mathbf{x})du)$, which we note for future reference.

In the regression analysis of survival time data two main paradigms are used often: these are the so-called proportional hazards (PH) and accelerated failure time (AFT) families of models. We next outline these and comment on their features.

2.1 Proportional Hazards and Accelerated Failure Time Models

Proportional hazards models assume that $\mathbf{x}$ affects the distribution of Y by having a multiplicative effect on the HF. In other words, $h(y \mid \mathbf{x})$ can be expressed as $h_o(y)g(\mathbf{x})$, where $g(\mathbf{x})$ is a positive-valued function and $h_o(y)$ is a baseline HF corresponding to an $\mathbf{x}$ for which $g(\mathbf{x}) = 1$. Equivalently, the

SF of Y given $\mathbf{x}$ is expressible as

$$S(y \mid \mathbf{x}) = S_o(y)^{g(\mathbf{x})}, \tag{2.1}$$

where $S_o(y) = \exp(-\int_0^y h_o(s)ds)$ is the baseline SF for an individual with $g(\mathbf{x}) = 1$. In general one or both of $h_o(y)$ and $g(\mathbf{x})$ may involve unknown parameters, and there are two main approaches to statistical analysis with PH models. One due to Cox (1972) is semi-parametric: it does not assume any particular form for $h_o(y)$ or $S_o(y)$. It yields parametric inference procedures for parameters in $g(\mathbf{x})$ and nonparametric estimates of $h_o(y)$ or $S_o(y)$. Kalbfleisch and Prentice (1980, ch. 4), Lawless (1982, ch. 7) and Cox and Oakes (1984) give detailed discussions. The other approach is to assume a parametric form for both $g(\mathbf{x})$ and $h_o(y)$ and to use standard parametric inference methods.

In contrast with PH models, AFT models suppose that $\mathbf{x}$ affects the distribution of Y by acting multiplicatively on the time scale. That is, $S(y \mid \mathbf{x})$ can be expressed in the form

$$S(y \mid \mathbf{x}) = S_o(g(\mathbf{x})y), \tag{2.2}$$

where, as before, $g(\mathbf{x})$ is a positive-valued function and $S_o(y)$ is a baseline SF. Accelerated failure time models are location-scale models on the log time scale: if $W = \log\ Y$ then the distribution of W given $\mathbf{x}$ can be expressed as

$$W = -\log\ g(\mathbf{x}) + \varepsilon, \tag{2.3}$$

where ε is a random variable whose distribution does not depend on $\mathbf{x}$. The common approach to AFT analysis is to use a parametric model where $S_o(y)$ or in other words ε in (2.3) is taken to have a given form. The lognormal and Weibull models, wherein ε has a normal and an extreme value distribution, respectively, are two often-used models. Semi-parametric analyses which do not make parametric assumptions about $S_o(y)$ or ε are also possible (e.g., Kalbfleisch and Prentice 1980, ch. 6; Louis, 1981) but are computationally forbidding and not widely used at present.

2.2 Some Parametric Models

An approach we find useful is to use a fairly flexible class of parametric AFT models along with the Cox semi-parametric PH methods. In so doing we tend to work with the convenient and widely used form $g(\mathbf{x}) = \exp(\mathbf{x}'\boldsymbol{\beta})$ in (2.1) and (2.2), where $\boldsymbol{\beta} = (\beta_1, \ldots, \beta_k)$, is a vector of regression coefficients. Methods discussed below also apply to other parametric forms for $g(\mathbf{x})$ but for convenience we shall restrict discussion to this one.

Several families of AFT models possess similar properties and tend to fit survival data, and which ones to work with is, in part, a matter of taste. We prefer Weibull and log logistic models, and a Burr model that includes both of them as special cases, but we also use other models on occasion. The *Weibull model* (e.g., Lawless, 1982, ch. 6) can be expressed as

$$S(y \mid \mathbf{x}) = \exp[-\theta(\mathbf{x})y^p], \quad y > 0, \tag{2.4}$$

where $p > 0$ and $\theta(\mathbf{x}) = \exp(\alpha + \mathbf{x}'\boldsymbol{\beta})$. The corresponding HF is $h(y \mid \mathbf{x}) = py^{p-1}\theta(\mathbf{x})$, and we observe that (2.4) is both an AFT and a PH model. It is in fact the unique model with this property. Note also that the distribution of $W = \log Y$ can be expressed as

$$W = -\alpha\sigma - \mathbf{x}'\boldsymbol{\beta}\sigma + \sigma\varepsilon, \qquad -\infty < \varepsilon < \infty,$$

where $\sigma = p^{-1}$ and ε has a standard extreme value distribution with PDF $\exp(\varepsilon - e^{\varepsilon})$.

The *log logistic AFT model* (e.g., Bennett, 1983a) has SF

$$S(y \mid \mathbf{x}) = [1 + \theta(\mathbf{x})y^p]^{-1}, \qquad y > 0, \tag{2.5}$$

where $p > 0$ and $\theta(\mathbf{x}) = \exp(\alpha + \mathbf{x}'\boldsymbol{\beta})$. It is also expressible as

$$W = -\alpha\sigma - \mathbf{x}'\boldsymbol{\beta}\sigma + \sigma\varepsilon, \qquad -\infty < \varepsilon < \infty,$$

where $W = \log Y$, $\sigma = p^{-1}$ and ε has a standard logistic distribution with PDF $e^{\varepsilon}/(1 + e^{\varepsilon})^2$. The HF for this model is

$$h(y \mid \mathbf{x}) = \frac{py^{p-1}\theta(\mathbf{x})}{1 + \theta(\mathbf{x})y^p}$$

and is nonmonotonic in y for $p > 1$: it is easily seen that $h(y \mid \mathbf{x})$ increases up to $[(p-1)/\theta(\mathbf{x})]^{1/p}$ then decreases thereafter. This value of y is the $(p-1)/p$th quantile, independent of $\mathbf{x}$; for medical survival data p is often between one and two, so the decrease starts to the left of the median. For $p < 1$, $h(y \mid \mathbf{x})$ is decreasing for all $y > 0$.

A third distribution we find useful is the Burr XII model (e.g., Tadikamalla, 1980) with SF

$$S(y \mid \mathbf{x}) = [1 + \lambda y^p \theta(\mathbf{x})]^{\frac{-1}{\lambda}-1}, \qquad y > 0, \tag{2.6}$$

where $p > 0, \lambda > 0$ and $\theta(\mathbf{x}) = \exp(\alpha + \mathbf{x}'\boldsymbol{\beta})$. This family includes the log logistic ($\lambda = 1$) and Weibull (limit as $\lambda \to 0$) as special cases. The additional parameter λ provides extra flexibilty and the distribution can also be used to discriminate parametrically between the Weibull and log logistic distributions.

2.3 Discussion

The PH and AFT paradigms are flexible and convenient for survival data analysis, and in many situations allow a succinct representation of the approximate effect of explanatory variables on the distribution of survival time. In one (AFT) the ratio of the pth quantiles of survival time for two individuals with arbitrary covariate vectors $\mathbf{x}_1$ and $\mathbf{x}_2$ is constant for all p $(0 < p < 1)$, and in the other (PH) the ratio of the hazard functions $h(y \mid \mathbf{x}_1)$ and $h(y \mid \mathbf{x}_2)$ is constant for all $y > 0$. The PH model has been found realistic and useful in many areas, for example in clinical trial work and in epidemiological investigations of disease incidence and mortality. For the kinds of survival times associated with many chronic disease data bases, we often find the PH model less compelling. When the population is heterogeneous and there is considerable variation in survival times, the assumption of constant proportional hazards at all times y often seems unreasonable, and AFT or modified PH-type models often fit the data better. Gore *et al.* (1984) have given an excellent discussion of this and related points in connection with breast cancer survival, and we discuss in Section 5 some experiences with ovarian cancer survival.

We usually fit both PH and AFT models in a given situation, as a first step. With the PH family we use the semi-parametric Cox analysis partly because of its flexibilty in allowing HF's of different shapes; we deem this important because in large heterogeneous data sets there is often evidence of non-monotonic HF's. With the AFT family we tend to fit members of the Burr distribution (2.6), including the Weibull and log logistic models. The Weibull is of course also a PH model with, however, monotonic hazard functions. The Burr family is flexible and easily handled statistically and computationally. For $\lambda > 0$ it allows a wide range of hazard functions that increase to a point and then decrease, and for the Weibull case $\lambda = 0$ it gives monotonic hazard functions. For the PH model the flexibility and simplicity of the Cox analysis make it less urgent to develop more flexible fully parametric models, and we have not systematically explored this direction.

The PH and AFT models make strong assumptions about the effect of $\mathbf{x}$ and of course there are situations were neither model is reasonable. One type of departure from each model deserves special scrutiny. For AFT models it is assumed that the shape of the distribution of $\log Y$ does not change with $\mathbf{x}$, only the location. In particular, the dispersion is a constant and one should thus be on the lookout for heteroscedasticity in $\log Y$ (e.g., Cook and Weisberg, 1983). For the PH model the hazard function ratios are constant for any two $\mathbf{x}$'s, and one should watch for departures from this (e.g, Andersen, 1982). When departures of the kinds noted are serious, extensions to the PH or AFT models frequently provide a satisfactory representation of the data.

Other families of models can sometimes be used to good effect. For example, Bennett (1983b) has discussed the ***proportional odds*** (PO) family, in which log $\{S(y \mid \mathbf{x})/[1 - S(y \mid \mathbf{x})]\}$ is taken to be a linear function of $x_1, \ldots, x_k$. (Another nice feature of the parametric Burr family is that the special-case log logistic is the unique distribution which is both a PO and an AFT model.) Usually we find, however, that the PH and AFT models between them provide sufficient flexibility to deal with most situations, provided care is taken to accommodate extensions such as the possible dependence of p in (2.4), (2.5) or (2.6) on $\mathbf{x}$. At the same time, we have not found it particularly useful to adopt more refined parametric distributions within either the PH or AFT models. For example, the log F (e.g., Kalbfleisch and Prentice, 1980, ch. 3) is sometimes put forward as an "error" distribution within the AFT model (see the form (2.3)). This has a second shape parameter and includes the Burr family as a special case. We have found that, in view of the uncertainty surrounding an appropriate specification for $\theta(\mathbf{x})$ and the assumption of a constant p parameter, additional parametric flexibility in the baseline distribution beyond that provided in the Burr family is usually not essential. Indeed, with substantial censoring in the right-hand tail of the distribution it becomes difficult to discriminate well among baseline distributions anyway. In addition, the picture that emerges concerning the relative importance of explanatory variables is often fairly insensitive to the baseline distributed used. Of course, in situations with only a small number of explanatory variables, a more detailed look at the baseline distributions involved is feasible.

We next discuss some specific statistical considerations in the use of the regression models described here.

3. SOME STATISTICAL CONSIDERATIONS

To facilitate modelling and the exploration of data we need the means to manipulate and portray the data and fitted models (graphically and otherwise), to fit models, and to assess their adequacy. We discuss aspects of a few important procedures here, and consider associated software in Section 4.

For convenient preliminary examination of a large data base, good graphics and data manipulation and summarization methods are desirable. This is not discussed here except in the context of an example later. We turn immediately to model fitting and related matters.

3.1 Model Fitting and Screening

Relatively little needs to be said about the estimation of parameters, maximum likelihood estimation for models discussed in Section 2 being straightforward. The survival times in most data bases are subject to right censoring, so we consider the data on an individual to be $(y_i, \delta_i, \mathbf{x}_i)$, where y_i is either an observed survival time or a censoring time, δ_i is an indicator which takes the value 1 if y_i is a survival time and 0 if it is a censoring time, and $\mathbf{x}_i$ is the covariate vector. Under quite general conditions (e.g., Kalbfleisch and Prentice, 1980, ch. 5) the likelihood function from data on n independent individuals is then

$$L = \prod_{i=1}^{n} f(y_i \mid \mathbf{x}_i)^{\delta_i} S(y_i \mid \mathbf{x}_i)^{1-\delta_i}. \tag{3.1}$$

Maximum likelihood estimates (MLE's) for the parametric models of Section 2 are readily obtained by maximizing (3.1), and interval estimates and significance tests may be based on the usual large sample methods. Lawless (1982, ch. 6), Kalbfleisch and Prentice (1980, ch. 3) and Bennett (1983a) provide details for the Weibull, log logistic, Burr, and other models. For the Cox PH analysis a partial likelihood for $\boldsymbol{\beta}$ in $g(\mathbf{x}) = \exp(\mathbf{x}'\boldsymbol{\beta})$ of (2.1) is generally used for inference; see for example Kalbfleisch and Prentice (1980, ch. 4) for details.

Since there are often many potential explanatory variables present in the data, "all-subsets" regression procedures are useful for model screening purposes. In particular, with k explanatory variables where are 2^k regression models that can be obtained by including any subset of $x_1, \ldots, x_k$ in the model and excluding the rest. Stepwise regression techniques are often used to determine a "good" model, but much more information about the relationship between explanatory and response variables is provided by determining all of the better-fitting models and not just the few discovered by stepwise fitting. For normal linear models all-subsets methods are well known and implemented, for example, in SAS (PROC RSQUARE) and BMDP (9R). Lawless and Singhal (1978, 1985) discuss all-subsets procedures for may generalized linear models, including those of Section 2. These operate as follows: the models in Section 2 all have regression coefficient parameters $\beta_1, \ldots, \beta_k$ plus some additional parameters $\boldsymbol{\psi}$. The 2^k "subset" models are obtained by letting different subsets of $\beta_1, \ldots, \beta_k$ have value 0. The parameters in $\boldsymbol{\psi}$ are assumed to be always included in the model, the constant term α in such models as (2.4), (2.5) and (2.6) being one of these. For the Weibull model (2.4), for example, $\boldsymbol{\psi} = (\alpha, p)'$. Good subset models are defined in terms of the likelihood ratio statistic for testing the subset model against the full model. In particular, suppose $\boldsymbol{\beta} = (\boldsymbol{\beta}_1, \boldsymbol{\beta}_2)$ and consider the subset model corresponding to $\boldsymbol{\beta}_2 = \mathbf{0}$. The likelihood ratio statistic for testing $H_0 : \boldsymbol{\beta}_2 = \mathbf{0}$; $\boldsymbol{\beta}_1, \boldsymbol{\psi}$ unrestricted vs. $H_1 : \boldsymbol{\beta}_2, \boldsymbol{\beta}_1, \boldsymbol{\psi}$ unrestricted is

$$R = 2 \log L(\hat{\boldsymbol{\beta}}_1, \hat{\boldsymbol{\beta}}_2, \hat{\boldsymbol{\psi}}) - 2 \log L(\tilde{\boldsymbol{\beta}}_1, \mathbf{0}, \tilde{\boldsymbol{\psi}}), \tag{3.2}$$

where $(\hat{\boldsymbol{\beta}}_1, \hat{\boldsymbol{\beta}}_2, \hat{\boldsymbol{\psi}})$ is the unrestricted MLE of $(\boldsymbol{\beta}_1, \boldsymbol{\beta}_2, \boldsymbol{\psi})$ and $(\tilde{\boldsymbol{\beta}}_1, \mathbf{0}, \tilde{\boldsymbol{\psi}})$ is the MLE with $\boldsymbol{\beta}_2$ constrained to be $\mathbf{0}$. Large values of R provide evidence against the subset model, and under H_0, R is distributed approximately as $\chi^2_{(k-p)}$ in large samples, where $p = \dim(\boldsymbol{\beta}_1)$. Our approach is thus to determine those subset models for which R is small. To compare models with different numbers of covariates we may use the Akaike information criterion (AIC), defined as

$$\text{AIC} = -2 \log L(\tilde{\boldsymbol{\beta}}_1, \mathbf{0}, \tilde{\boldsymbol{\psi}}) + 2(p+m), \tag{3.3}$$

where $m = \dim(\boldsymbol{\psi})$. This is analogous to the use of Mallows' C_p statistic with normal linear models. Software for carrying out all-subset searches is described in Section 4.

3.2 Methods for Model/Data Criticism

It is essential that we be able to assess the adequacy of fitted models, and it is also important to identify unusual or otherwise interesting features of the data. There is a large recent literature on model/data criticism (e.g., see Cook and Weisberg, 1982), but for the models of Section 2 in conjunction with censored data, relatively little has been done. We will outline briefly our current approach, which is simply to determine fitted residuals and leverage values which can be examined in various ways.

We use the Cox-Snell (1968) type of residuals, which are discussed, for example, by Kalbfleisch and Prentice (1980, ch. 4), Kay (1977), Lawless (1982, chs. 6, 7). For the Burr models (2.6) these can take the form

$$r_i = y_i^{\hat{p}} \hat{\theta}(\mathbf{x}_i) = y_i^{\hat{p}} \exp(\hat{\alpha} + \mathbf{x}_i' \hat{\boldsymbol{\beta}}), \tag{3.4}$$

where $\hat{p}, \hat{\alpha}, \hat{\boldsymbol{\beta}}$ are MLE's. Since y_i may be either a survival or a censoring time, if (2.6) is appropriate the r_i's should look roughly like a censored sample from the corresponding standard distribution having $p = 1, \theta(\mathbf{x}) \equiv 1$, and the value of λ in question. In particular, for the Weibull ($\lambda = 0$) and log logistic ($\lambda = 1$) models the r_i's should look roughly like censored standard Weibull and log logistic samples, respectively. (It should perhaps be said that our approach is to fit members of the Burr family with fixed values of λ. Although we often determine the value of λ which maximizes the likelihood, we compute asymptotic variance estimates, confidence intervals and the like for $\boldsymbol{\beta}$, with λ treated as fixed.)

For the Cox analysis of the PH model we define residuals as in Kay (1977),

$$r_i = [-\log \hat{S}_o(y_i)]e^{x_i'\hat{\boldsymbol{\beta}}}, \tag{3.5}$$

where $\hat{\boldsymbol{\beta}}$ is the partial likelihood MLE of $\boldsymbol{\beta}$ and $\hat{S}_o(y)$ is the Nelson-Altschuler nonparametric estimator of $S_o(y)$ given by (7.2.34) of Lawless (1982). If the PH model is appropriate these r_i's should look roughly like a censored sample from the standard exponential ($\equiv$ standard Weibull) distribution.

For plots of residuals against covariates or other factors we adjust residuals r_i corresponding to censoring times upward by an amount equal to the median of ε_i, given $\varepsilon_i \geq r_i$, where ε_i has the standard distribution used to assess the r_i's. For example, for the Weibull model (2.4) and residuals (3.4), ε_i has a standard exponential distribution with survivor function $S(\varepsilon) = \exp(-\varepsilon)$. It is easily shown that

$$\text{median}(\varepsilon_i \mid \varepsilon_i \geq r_i) = r_i + \log 2,$$

so we define adjusted residuals to be $r_i + \log 2$ when y_i is a censoring time. In plots, residuals corresponding to uncensored and censored observations should still, however, be labelled differently.

Probability plots of residuals from the fully parametric models can also be useful. Since the r_i's are censored we form a product-limit survivor function estimate from the (unadjusted) r_i's assumed here to be distinct, as

$$\hat{S}(r) = \prod_{i:r_i<r} \left(\frac{n_i - 1}{n_i}\right)^{\delta_i}, \tag{3.6}$$

where n_i is the number of residuals $\geq r_i$. Then $\hat{S}(r)$ can be compared with what would be expected from the distribution of the corresponding ε. For the Weibull model $S(\varepsilon) = \exp(-\varepsilon)$ so a plot of $\log \hat{S}(r_i)$ vs. r_i should be approximately linear with slope one. An alternative is to plot $-\log[-\log \hat{S}(r_i)]$ vs. $\log r_i$, which reveals more detail near $r = 0$. For the log logistic model the corresponding plot is of $\log[(1 - \hat{S}(r_i))/\hat{S}(r_i)]$ vs. $\log r_i$.

On a cautionary note, it should be said that it is not clear how good residual probability plots are at detecting model inadequacy. It is recognized that for the Cox PH analysis, probability plots of the r_i's defined in (3.5) may not indicate model inadequacy at all in some instances (e.g., Crowley and Storer, 1983). Residual plots for the fully parametric models appear more useful, but systematic study of their properties is needed.

It is also of interest to identify observations which exert a strong influence on parameter estimates, and observations which are unusual in their y and/or $\mathbf{x}$ values. A standard way of assessing the influence of an observation is to consider the effect on estimation of dropping it (e.g., Cook and Weisberg, 1982). If a model has parameter $\boldsymbol{\theta}$ and likelihood function $L(\boldsymbol{\theta})$ let

$$\hat{\boldsymbol{\theta}}_{(i)} = \text{MLE obtained by using all the data except the } i\text{th observation}$$

and

$$LD_i = 2 \log L(\hat{\boldsymbol{\theta}}) - 2 \log L(\hat{\boldsymbol{\theta}}_{(i)}).$$

Several authors have considered approximations to $\hat{\boldsymbol{\theta}}_{(i)}$ and LD_i that avoid intensive computation needed to compute the exact quantities for each of n observations. No thorough comparison of approaches has been made for the models of Section 2, but see, for example, Hall *et al.* (1982) for discussion pertaining to the parametric models and Cain and Lange (1984), Moolgavkar *et al.* (1984), and Storer and Crowley (1985) for the Cox PH analysis. The various approximate influence measures are in a rough sense functions of some kind of generalized residual that measures agreement between y_i and the fitted model, and some measure of the distance from $\mathbf{x}_i$ to the centroid of $\mathbf{x}_1, \ldots, \mathbf{x}_n$. We are investigating the use of various approximations but given our present state of knowledge and the relatively large computational effort large data bases and some of the methods entail, our current approach is simply to examine the residuals r_i described above, along with the leverage values $h_{ii} = \mathbf{x}_i'(X'X)^{-1}\mathbf{x}_i$ used with the normal linear model. The h_{ii}'s help locate observations that are unusual in their $\mathbf{x}_i$ values, whereas the r_i's help identify interesting features in the fit of Y_i on $\mathbf{x}_i$. In addition, the most influential observations tend to be those with large r_i and/or h_{ii} values.

4. SOFTWARE

For data manipulation and portrayal, major packages such as SAS and BMDP provide good software. Good and widely accessible software for fitting and exploration of common survival analysis models was for a long time unavailable, but the situation is now improving. Both SAS and BMDP have Cox PH analysis programs, and the latest version of SAS has a program that handles censored data from the Weibull, log logistic, log normal and other regression models. Some other packages also provide a fairly wide range of model fitting capabilities, but are not as accessible or as viable with large data sets. Wagner and Meeker (1985) provide a very good survey of software in this area, and it appears that even better systems wll be available soon.

To provide a comprehensive package that handles both the important continuous response regression models and discrete-response models used in survival analysis, we have designed, and K. Singhal has written, a system called ISMOD. A more complete description of ISMOD is given by Lawless and Singhal (1985) and by Singhal (l985). For our purposes here, we note that the package can fit most of the common survival time regression models, including the exponential, Weibull, log logistic, Burr, log normal and Cox PH models. In fitting a single model its output includes parameter estimates, asymptotic covariance matrices, and residuals. It handles censored data and at present can accommodate up to 20 explanatory variables and large numbers of observations. It also has an all-subset feature and will find, according to likelihood ratio or approximate likelihood ratio criteria, the best-fitting models. In particular, it will find the "best m" models of each of sizes $1, 2, \ldots, k$ for $m \leq 10$, or the best m models overall in terms of the AIC criterion (3.3). This is done using efficient algorithms that keep computation to a minimum.

ISMOD contains facilities such as data transformation and data selection capabilities, so that one can transform variables or fit models to parts of the data. It also has the ability to generate two-way tables of summary statistics for pairs of variables: this is useful both in preliminary data exploration and in summarizing the data. A decision was taken not to include graphics capabilities or extensive diagnostic facilities inside ISMOD at present, because of our feeling that this area still required more development. However, ISMOD produces listing files of residuals and other quantities which can be plotted and manipulated using other software. We give an example in the next section.

5. AN EXAMPLE: OVARIAN CARCINOMA

We consider a data base that contains information on 704 ovarian carcinoma patients diagnosed during the period 1971–79 and registered at the Princess Margaret Hospital in Toronto. The response variable discussed is survival time, measured from diagnosis. Of the 704 patients, 341 had died by the time of the current data file update and hence there are 364 censored observations. After discussions with medical people and preliminary examination of the data it was decided to consider the explanatory variables $x_1, \ldots, x_{12}$ shown in Table 1. Except for age, all explanatory variables are categorical and we have used dummy covariates to represent them. The "histology" and "grade" covariates refer to pathological characteristics of the carcinoma: the five histology categories are serous or mixed (SER or MIX), endometroid (END), mucinous (MUC), clear cell (CLR) and unclassified (UNCL). Grade categories are moderately or well differentiated (MD or WD) and poorly differentiated (PD). The "stage" covariates refer to Figo staging and are related to the extent of the disease; prognosis tends to be poorer the higher the stage number. "Surgery" refers to whether a complete or incomplete Bilateral Salpingo Oopherectomy Hysterectomy (BSOH) was performed on the patient, and "residuum" refers to the amount of diseased tissue remaining after surgery. Clinical background on these data and an ovarian cancer in general can be found in Dembo *et al.* (1982), Dembo (1984) and other articles cited in these papers.

Table 1. *Ovarian Cancer—Initial Variables*

Variable	*Description*	*# Cases*	*# Uncensored*
Y	Time (days) from diagnosis to death	704	341
δ	1 – time uncensored, 0 – censored	704	341
x_1	age at diagnosis (years): mean 54	704	341
x_2	Histology 1: 1–SER or MIX, 0–Other	345	176
x_3	Histology 2: 1–END, 0–Other	143	57
x_4	Histology 3: 1–MUC, 0–Other	79	20
x_5	Histology 4: 1–CLR, 0–Other	28	14
	(Histology baseline is UNCL)	109	74
x_6	Grade: 1–MD or WD, 0–PD	160,544	37,304
x_y	Stage 1: 1–Stage I, 0–Other	117	19
x_8	Stage 2: 1–Stage II, 0–Other	217	78
x_9	Stage 3: 1–Stage III, 0–Other	275	172
	(Stage baseline is IV)	95	72
x_{10}	Surgery: 1–Complete, 0–Inc.	396,308	128,213
x_{11}	Residuum 1: 1–Small, 0–Other	110	54
x_{12}	Residuum 2: 1–Large, 0–Other	296	198
	(Residuum baseline is none)	298	89

The analysis which follows was carried out using the package ISMOD described in Section 4, along with the Integrated Software Systems Corp. graphics package DISSPLA. The data set is complex, and the analysis is meant to illustrate various points rather than to give an exhaustive discussion of its features.

We note to start that medical background provides a lot of information about the qualitative nature of the relationships between covariates and survival time, and among the covariates. For example, surgery, residuum and stage covariates are related, since the location and extent of the disease at diagnosis affects all three. It is also clear that the prognosis tends to be poorer for patients in higher stages, for patients with large residuum rather than none or small, and so on. In preliminary examination of the data we generated various plots and tables such as Tables 2 and 3, which show summary data for cross-classifications of explanatory variables. Table 2 and 3 give, for each pair of categories in the table, the total number of observations (n), the number of uncensored observations (n_u), and an estimated mean survival time $\bar{y}$ computed as the total of survival and censoring time divided by n_u. This estimate, which is the MLE of mean survival time when times follow an exponential distribution, provides a crude picture of the effects of covariates. It should be noted that the estimate may be very unreliable when n_u is small, and when censoring is very heavy. Note also that in considering the possibility of covariate effects under the models described in this paper, ratios of these means, rather than differences, are of interest.

Plots of the data are also useful. Figure 1, for example, shows product-limit survivor function estimates $\hat{S}(y)$ for individuals in each of Stages I to IV (unadjusted for any other covariates). Widely varying survival experience in the different stages is apparent.

As a first pass at model fitting we consider AFT and PH models with covariates $x_1, \ldots, x_{12}$ present. The Weibull, Cox PH and log logistic models turn out to give essentially the same picture regarding covariate effects, although they do not fit the data equally well. Table 4 shows MLE's and z values (= MLE $\div$ estimated standard deviation) under the three models. It should be remarked that the meaning of a regression coefficient β_i is not exactly the same under the three models. For the Cox PH and Weibull models, the β_i's have the same interpretation within the PH family, and direct comparison of values is relevant. Within the AFT family, values of $\hat{\beta}_i/\hat{p}$ should be compared under the Weibull and log logistic models.

From an examination of the fitted models and the data, it is clear that the stage covariates and the grade covariate are very important, with age, surgery, and large residuum also having some effect. Although the different models agree broadly as to the importance of the covariates, they do not fit the data equally well: the log logistic fits somewhat better than the Weibull and there is a clear indication that the Weibull model, at least with the given set of covariates, is unable to account for all of the heterogeneity in the data. For example, Figure 2 shows a residual probability plot for the

Table 2. *Summary Data for Stage x Surgery Categories*

		Stage				
		I	*II*	*III*	*IV*	*Total*
Surgery	*Complete*	$n = 111$ $n_u = 19$ $\overline{y} = 7219$	$n = 173$ $n_u = 48$ $\overline{y} = 4852$	$n = 93$ $n_u = 47$ $\overline{y} = 1515$	$n = 19$ $n_u = 14$ $\overline{y} = 2326$	$n = 396$ $n_u = 128$ $\overline{y} = 3701$
	Incomplete	$n = 6$ $n_u = 0$ $\overline{y} =$ undef.	$n = 44$ $n_u = 30$ $\overline{y} = 1445$	$n = 182$ $n_u = 125$ $\overline{y} = 790$	$n = 76$ $n_u = 58$ $\overline{y} = 200$	$n = 308$ $n_u = 213$ $\overline{y} = 839$
	Total	$n = 117$ $n_u = 19$ $\overline{y} = 7833$	$n = 217$ $n_u = 78$ $\overline{y} = 3542$	$n = 275$ $n_u = 172$ $\overline{y} = 988$	$n = 95$ $n_u = 72$ $\overline{y} = 260$	

Weibull model with all 12 covariates present. The plot is not close to linear and in fact consists very roughly of two segments, which is an indication that there is heterogeneity related to longer and shorter term survivors that is not accounted for by the Weibull model. Other evidence is also available: for example, within the Burr family (2.6) the maximized log likelihoods for the Weibull and log logistic models are -692.5 and -679.2, respectively, indicating strong support for the log logistic over the Weibull. In addition, plots of the nonparametric estimate $\hat{S}_o(y)$ (see (2.1)) obtained from the Cox PH analysis indicate a shape not consistent with the Weibull model and there is a suggestion of a non-monotonic hazard functon. The log logistic model fits the data somewhat better, though systematic departures from it are also apparent.

At this point, the evidence suggests that stage and grade are particularly important covariates, that several others are of some importance, and after adjusting for main effects of these covariates, there is an indication of a non-monotonic hazard function that first increases and then decreases. To look for further explanations of the wide variability in survival times, we decided next to examine the data separately within each stage. There are several reasons why this is a good idea, including the possibility of stage by other

Table 3. *Summary Data for Stage x Grade Categories*

		Stage				
		I	*II*	*III*	*IV*	*Total*
Grade	*Moderately or Well Diff.*	$n = 48$ $n_u = 2$ $\overline{y} = 35,664$	$n = 65$ $n_u = 14$ $\overline{y} = 7129$	$n = 38$ $n_u = 13$ $\overline{y} = 3038$	$n = 9$ $n_u = 8$ $\overline{y} = 479$	$n = 160$ $n_u = 37$ $\overline{y} = 5796$
	Poorly Diff.	$n = 69$ $n_u = 17$ $\overline{y} = 4558$	$n = 152$ $n_u = 64$ $\overline{y} = 2757$	$n = 237$ $n_u = 159$ $\overline{y} = 820$	$n = 86$ $n_u = 64$ $\overline{y} = 449$	$n = 544$ $n_u = 304$ $\overline{y} = 1359$
	Total	$n = 117$ $n_u = 19$ $\overline{y} = 7833$	$n = 217$ $n_u = 78$ $\overline{y} = 3542$	$n = 275$ $n_u = 172$ $\overline{y} = 988$	$n = 95$ $n_u = 72$ $\overline{y} = 260$	

covariate interactions, the fact that the data are highly unbalanced with regard to certain combinations of stage and other covariates, and the fact that the data set is very heavily censored and that the degree of censoring varies widely across stages. It is also clear that for stages I and II and to a lesser extent stage III there are substantial proportions of long term survivors; see Figure 1. With the heavy censoring in these stages it is difficult to say whether this can be explained completely in terms of the covariates under study, but it does not appear that it can. The picture here is rather different than for stage IV where there are many fewer long term survivors.

There was doubt that Weibull or log logistic models could hope adequately to fit the data for stages I to III, because of the appearance that there are individuals whose long term survival is not fully explainable by these models and the covariates that are present. Nevertheless we fitted these models in order to get an idea of the importance of covariates within stages. This prompts the following remarks.

1. The importance of the other covariates depends to a large extent on the stage. Table 5 shows the factors that appear important, for each stage. We find, for example, that the effect of age is more pronounced in stages I and IV than in II and III. Note also that Grade has an important effect

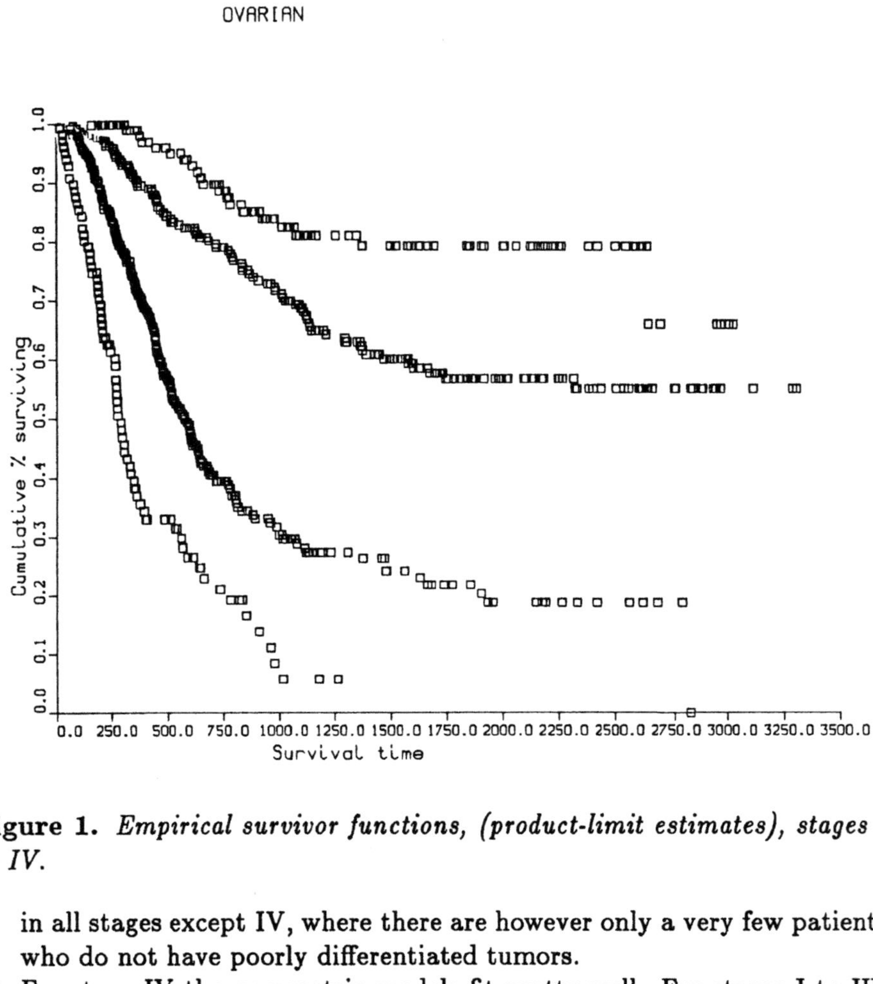

Figure 1. *Empirical survivor functions, (product-limit estimates), stages I to IV.*

in all stages except IV, where there are however only a very few patients who do not have poorly differentiated tumors.

2. For stage IV the parametric models fit pretty well. For stages I to III, however, the log logistic and Weibull models both overestimate the number of short to medium term survivors and underestimate the number of long term survivors, as one might expect from the discussion in 1. This is less pronounced in stage III than in I and II. Figures 3 and 4 show raw product-limit estimates $\hat{S}(y)$ computed from the lifetimes in stages II and III with no covariate adjustment, along with estimated survivor functions

$$\hat{S}(y) = \frac{1}{n_s} \sum_{i=1}^{n_s} [1 + \hat{\theta}(\mathbf{x}_i) y^{\hat{p}}]^{-1}$$

computed under the log logistic model with the covariates indicated on the graph. Here, the sum is over the individuals in the stage and n_s is

Table 4. *MLE's $\hat{\beta}_i$ Under Three Regression Models, for Full Data Set*

	Covariate	*Cox PH*	*Weibull*	*Log logistic*
1.	Age	.022(3.8)*	.024(4.1)	.026(3.3)
2.	Histology 1	–.073(–.5)	–.10(–.7)	–.23(–1.1)
3.	Histology 2	–.042(–.2)	–.11(–.6)	–.078(–.3)
4.	Histology 3	–.27(–1.0)	–.26(–1.0)	–.23(–.6)
5.	Histology 4	.49(1.6)	.55(1.8)	.56(1.3)
6.	Grade	–.70(–3.7)	–.83(–4.4)	–1.12(–4.6)
6.	Stage 1	–2.04(–6.5)	–2.21(–7.1)	–2.99(–7.6)
8.	Stage 2	–1.54(–7.5)	–1.72(–8.3)	–2.39(–8.5)
9.	Stage 3	–.74(–4.8)	–.74(–4.8)	–1.19(–5.3)
10.	Surgery	–.50(–2.8)	–.50(–2.8)	–.62(–2.5)
11.	Residuum 1	.29(1.4)	.26(1.3)	.41(1.6)
12.	Residuum 2	.29(1.6)	.30(1.7)	.59(2.4)
	MLE $\hat{p}$ (s.e.)		1.23(.052)	1.67(.075)

* *Z*-values are in brackets.

Table 5. *Important Factors Within Stages*

Stage			
I	*II*	*III*	*IV*
Age Grade	Grade Surgery Residuum	Histology MUC Grade	Age Surgery Residuum

the number of individuals in the stage. These plots portray the over- and underestimation mentioned above.

3. It is of interest to fit models that explicitly recognize the possibility of long term survivors. This can be done, though the heavy censoring present makes clear assessments difficult.

The data were also examined with regard to influential observations and in a number of other ways. In the interest of economy and brevity we will only remark that influence analysis did not yield any really striking results.

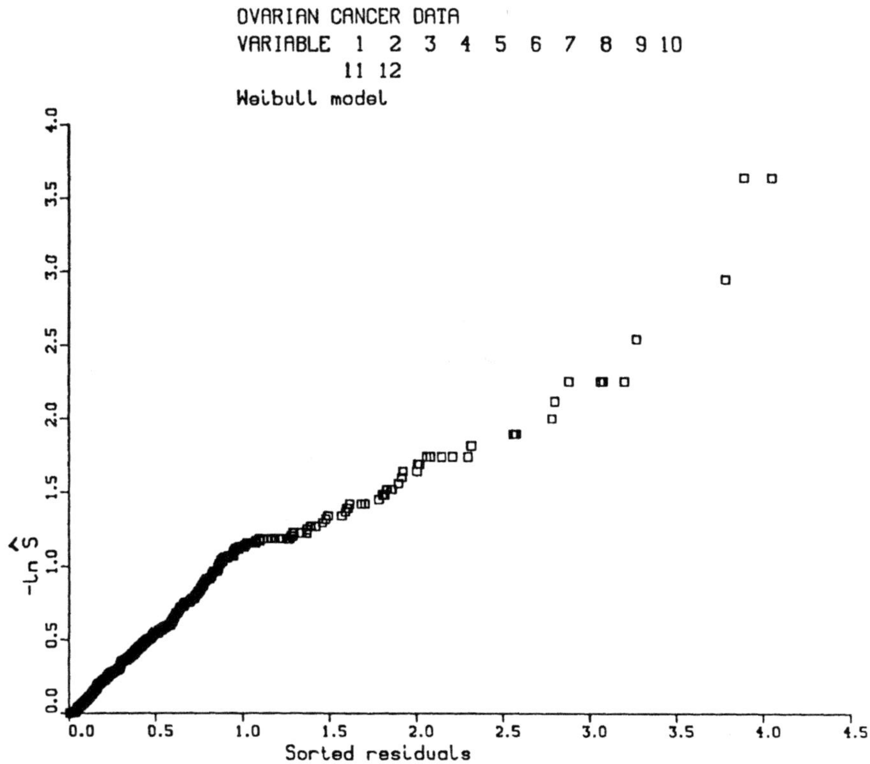

Figure 2. *Residual probability plot, Weibull model.*

The size of the data set and the heavy right censoring no doubt have someting to do with this. Of course, progressively more refined models can also be fitted to the data, but we will not discuss this here.

6. CONCLUDING REMARKS

Regression methodology provides useful tools for investigating large medical data bases. It is our experience that with software of the type discussed in Section 4 it is feasible to explore the data and to fit and assess regression models with moderate amounts of computer time. Models such as the Weibull, log logistic and Cox PH models discussed in this paper are easily handled and provide between them sufficient flexibility to determine many features of the data. Although in large data bases they invariably will not give a completely satisfactory description of the data, they provide convenient and useful baselines against which to look for other interesting

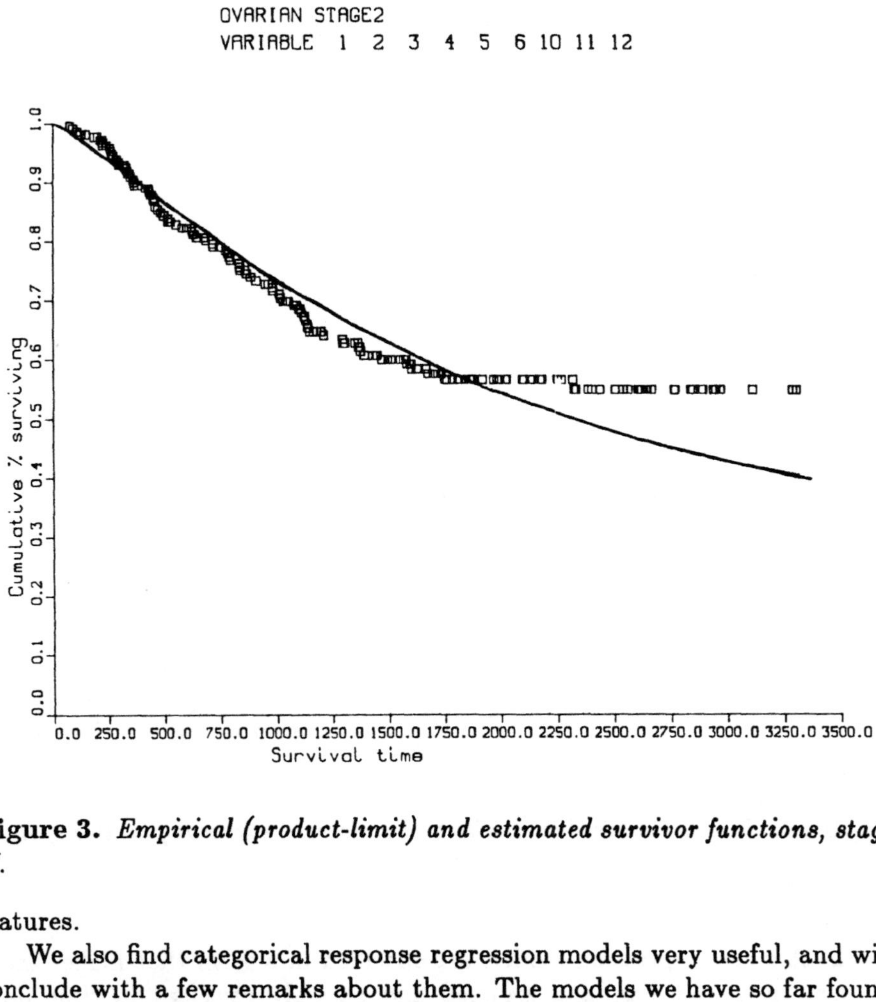

Figure 3. *Empirical (product-limit) and estimated survivor functions, stage II.*

features.

We also find categorical response regression models very useful, and will conclude with a few remarks about them. The models we have so far found most convenient are two types that handle ordered categorical responses. If Y is a categorical variable which can take on values $1, 2, \ldots, c$ with probabilities $P_i(\mathbf{x})$ $i = 1, \ldots, c$, where $\sum P_i(\mathbf{x}) = 1$ and $\mathbf{x}$ is a vector of covariates, then the two types can be represented as follows:

(a) Following McCullagh (1980) and others, we take

$$\gamma_j(\mathbf{x}) = \Pr(Y \leq j \mid \mathbf{x})$$

to be of the form $F(\alpha_j + \mathbf{x}'\boldsymbol{\beta})$, where F is a continuous distribution function on $(-\infty, \infty)$ and $-\infty < \alpha_1 < \ldots < \alpha_c = \infty$. Two flexible models are the ones for which

$$F(u) = e^u/(1 + e^u) \text{ and } F(u) = 1 - \exp(-e^u),$$

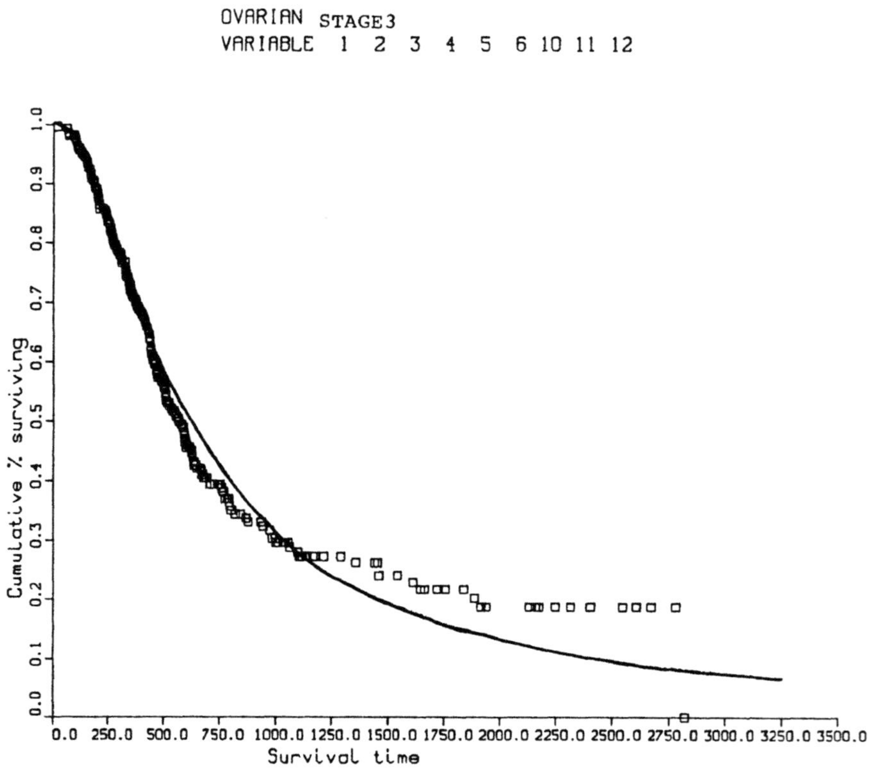

Figure 4. *Empirical (product-limit) and estimated survivor functions, stage III.*

respectively.

(b) In this case regression models are specified in a form that makes them convenient for handling grouped and possibly censored lifetime data (e.g., Lawless, 1982, section 7.3). These take

$$q_j(\mathbf{x}) = \Pr(Y = j \mid Y \geq j, \mathbf{x})$$

to be of the form $F(\alpha_j + \mathbf{x}'\beta)$, where F and the α_j's are as in (a).

Between them the models in (a) and (b) provide good flexibility for exploring categorical data. Regarding software, we note that the package ISMOD discussed in Section 4 also handles these methods, providing the same sorts of features described and illustrated above for the continuous survival time models.

ACKNOWLEDGMENTS

We would like to thank Drs. A. Dembo and A. Ciampi of the Ontario Cancer Institute for providing the ovarian cancer data, and the aforementioned and Steve McKinney for discussions about the data. During part of the work on this paper, K. Singhal was partially supported by a Medical Research Council grant to A. Ciampi.

REFERENCES

Andersen, P. K. (1982), "Testing goodness of fit of Cox's regression and life model." *Biometrics* **38**, 67–77.

Bennett, S. (1983a), "Log-logistic regression models for survival data". *Applied Statistics* **32**, 165–171.

Bennett, S. (1983b), "Analysis of survival data by the proportional odds model". *Statistics in Medicine* **2**, 273–277.

Cain, K. C., and N. T. Lange (1984), "Approximate case influence for the proportional hazards regression model with censored data". *Biometrics* **40**, 493–499.

Ciampi, A., R. Bush, M. Gospodarowicz, and J. E. Till (1981), "A classification of survival experience of non-Hodgkin's Lymphoma patients". *Cancer* **47**, 621–627.

Cook, R. D., and S. Weisberg (1983), "Diagnostics for heteroscedasticity in regression". *Biometrika* **70**, 1–10.

Cook, R. D., and S. Weisberg (1982), *Residuals and Influence in Regression*. New York: Chapman and Hall.

Cox, D. R. (1972), "Regression models and life tables" (with discussion). *Journal of the Royal Statistical Society*, Series B **34**, 187–202.

Cox, D. R., and E. J. Snell (1968), "A general definition of residuals". *Journal of the Royal Statistical Society*, Series B **30**, 826–838.

Cox, D. R., and D. Oakes (1984), *Analysis of Survival Data*. London: Chapman and Hall.

Crowley, J., and B. E. Storer (1983), "Comment on 'A reanalysis of the Stanford heart transplant data', by M. Aitkin, N. Laird and B. Francis". *Journal of the American Statistical Association* **78**, 277–281.

Dembo, A. J. (1984), "Postoperative abdominopelvic irradiation in patients with epithelial cancer of ovary". *Journal of Cancer Research and Clinical Oncology* **107**, 91–93.

Dembo, A. J., R. S. Bush, and T. C. Brown (1982), "Clinicopathological correlates in ovarian cancer". *Bulletin du Cancer, (Paris)*, **69**, 292–298.

Gore, S. M., S. J. Pocock, and G. R. Kerr (1984), "Regression models and nonproportional hazards in the analysis of breast cancer survival". *Applied Statistics* **33**, 176–195.

Hall, G. J., W. H. Rogers, and D. Pregibon (1982), "Outliers matter in survival analysis". Unpublished manuscript.

Kalbfleisch, J. D., and R. L. Prentice (1980), *The Statistical Analysis of Failure Time Data*. New York: Wiley and Sons.

Kay, R. (1977), "Proportion hazard regression models and the analysis of censored survival data". *Applied Statistics* **26**, 227–237.

Keys, A., C. Aravanis, *et al.* (1972), "Coronary heart disease: overweight and obesity as risk factors". *Annals of Internal Medicine* **77**, 15–27.

Lawless, J. F. (1982), *Statistical Models and Methods for Lifetime Data*. New York: Wiley and Sons.

Lawless, J. F., and K. Singhal (1978), "Efficient screening of non-normal regression models". *Biometrics* **34**, 318–327.

Lawless, J. F., and K. Singhal (1985), "ISMOD: An all-subsets regression program for generalized linear models: Parts I and II". Unpublished manuscript.

Louis, T. A. (1981), "Nonparametric analysis of an accelerated failure time model". *Biometrika* **68**, 318–390.

McCullagh, P. (1980), "Regression models for ordinal data". *Journal of the Royal Statistical Society*, Series B **42**, 109–127.

Mock, M. B., I. Ringqvist, L. D. Fisher *et al.* (1982), "Survival of medically treated patients in the Coronary Artery Bypass Study (CASS) registry". *Circulation* **66**, 562–568.

Moolgavkar, S. H., E. D. Lustbader, and D. Venzon (1984), "A geometric approach to non-linear regression diagnostics with application to matched case-control studies". *Applied Statistics* **12**, 816–826.

Singhal, K. (1985), *ISMOD User's Guide*. Department of Statistics and Actuarial Science, University of Waterloo.

Storer, B. E., and J. Crowley (1985), "A diagnostic for Cox regression and general conditional models". *Journal of the American Statistical Association* **80**, 139–147.

Tadikamalla, P. R. (1980), "A look at the Burr and related distributions". *International Statistical Review* **48**, 337–344.

Wagner, A. E., and W. Q. Meeker (1985), "A survey of statistical software for life data analysis". Paper presented at the American Statistical Association Meeting, August, 1985, Las Vegas, Nevada.

A. Ciampi, [1] C.-H. Chang, [2] S. Hogg [3] and S. McKinney [4]

RECURSIVE PARTITION: A VERSATILE METHOD FOR EXPLORATORY DATA ANALYSIS IN BIOSTATISTICS

ABSTRACT

In clinical and epidemiological studies, interactions among covariates are of primary importance. In the absence of clearly stated *a priori* hypotheses, it is usual to build predictive models by means of stepwise logistic or Cox regression, a practice that tends to emphasize main effects and overlook possible interactions.

An alternative approach is that of Recursive Partition (RP), a well known clustering method adapted by us to the case of response variables frequently encountered in Biostatistics (e.g., censored survival times and discrete response).

We will discuss the application of RP to the determination of prognostic strata, propensity strata, and treatment response strata.

In each case we look for a finite number of subgroups of a large patient population, homogeneous with respect to prognosis, probability of being assigned to a certain treatment, and response to treatment, respectively.

We will also tackle the general problem of evaluating the statistical soundness of exploratory analyses. The use of cross-validation, bootstrapping and model selection criteria, such as the Akaike Information Criterion, will be discussed.

[1] Montreal Children's Hospital Research Institute, 2300 Tupper Street, Montreal, Quebec H3H 1P3

[2] Department of Rheumatology, Wellesley Hospital, 160 Wellesley Street East, Toronto, Ontario M4Y 1J3

[3] Department of Preventive Medicine and Biostatistics, 12 Queen's Park Crescent West, Toronto, Ontario M5S 1A8

[4] Biostatistics Department, The Ontario Cancer Institute, Princess Margaret Hospital, 500 Sherbrooke Street, Toronto, Ontario M4X 1K9

I. B. MacNeill and G. J. Umphrey (eds.), Biostatistics, 23–50.

1. INTRODUCTION

In clinical and epidemiological studies it is often useful to summarize complex information by identifying and describing homogeneous strata, i.e., subpopulations of a given population which are defined by values of certain clinical variables and are homogeneous in a specified sense. We will discuss here three examples of particular importance, although many more situations could be cast in a similar framework.

a) Prognostic Stratification
One attempts to identify subpopulations of homogeneous prognosis.

b) Propensity Stratification
The aim is to find subpopulations within which the probability of being assigned to a certain treatment is homogeneous. This is a useful preliminary step before testing treatment effect (Rosenbaum and Rubin, 1984).

c) Treatment Response Stratification
Subpopulations are sought within which the treatment-response relationship is homogeneous, with the aim of separating cases for whom the treatment is effective, from those for whom it is useless or harmful (Gail and Simon, 1985).

The problem of finding an appropriate stratification is usually solved by *ad hoc* adaptations of regression techniques. For example, Byar (1982) used Cox and Weibull regressions to define a score $\mathbf{s} = \hat{\boldsymbol{\beta}} \cdot \mathbf{z}$, where $\hat{\boldsymbol{\beta}}$ is the estimated regression parameter vector and $\mathbf{z}$ is a covariate vector; the strata are then defined by splitting the score into appropriate intervals. A similar procedure could be applied to the propensity score in order to obtain propensity strata; in this case the score is constructed from a logistic regression model for the probability of receiving a certain treatment (Rosenbaum and Rubin, 1984).

These procedures have two main shortcomings: they lead to descriptions which are not always clinically interpretable, and they are based on regression models which, for practical reasons, are often built without proper consideration for interactions among clinical variables.

This work is devoted to an alternative approach which avoids the above shortcomings and has the additional advantage of providing a general and flexible framework for the problem of stratification. It is based on the idea of "growing trees", i.e., of successively dividing the population into subpopulations of increasing homogeneity until one obtains populations for which the degree of homogeneity cannot be increased by further splitting. An essential feature of the approach is that splits are induced by statements on certain "predictor" variables so that, by definition, the strata are described in direct clinical terms. An algorithm based on this approach is termed Recursive Partition (RP) (Hartigan, 1975, Ch. 18). An early example of RP is the AID algorithm and its generalizations (Sonquist *et al.*, 1973; Gillo and Shelly, 1974).

More recently, the monograph by Breiman *et al.* (1984) has introduced a number of ground-breaking new ideas and has described a computer program named CART (Classification and Regression Trees).

The present work develops the tree-growing approach for the situations prevalent in Biostatistics, where one often encounters non-standard, non-normal regression models.

2. DEFINITIONS AND NOTATION

We consider data in the form $(\mathbf{u}^{(i)}, \mathbf{z}^{(i)})$, $i = 1, \ldots, N$, where i labels individual and $\mathbf{u}$, $\mathbf{z}$ are (row) vectors of variables to be further described below.

We shall refer to the components of $\mathbf{u}$ as *criterion variables* and to those of $\mathbf{z}$ as *predictor variables*. *Predictors* contain background information used to define strata which are homogeneous according to a *criterion* based on the $\mathbf{u}$ variables; thus, for each homogeneous stratum one can define a unique *criterion quantity* independent of the $\mathbf{z}$ variables given the stratum.

For *prognostic stratification* the criterion variables are of the form:

$$(t^{(i)}, \delta^{(i)}), \qquad i = 1, \ldots, N,$$

where t is the observed survival time and δ is the censorship indicator, with $\delta = 0$ if t is censored and $\delta = 1$ if it is an observed failure time. (We recall that a survival time is said to be (right) censored if it is a lower bound to the actual failure time.) The criterion quantity is the survival curve. The $\mathbf{z}$ variables are known as prognostic factors.

For *propensity stratification* we have a single criterion variable u, a categorical variable indicating treatment group. The criterion quantity is the vector of probabilities of being assigned to each treatment group.

For *treatment response stratification,* the criterion variables are of the form

$$\mathbf{u}^{(i)} = (\mathbf{y}^{(i)}, \mathbf{x}^{(i)}), \qquad i = 1, \ldots, N,$$

where $\mathbf{y}$ is the response variable (generally of the form (t, δ) as above), and $\mathbf{x}$ is a vector variable defining treatment. The criterion quantity is the appropriate regression equation which represents the treatment-response relationship.

We will assume that *predictors* are categorical variables, i.e., $\mathbf{z} = (z_1, z_2, \ldots, z_k)$ and for each j, z_j has a finite number of levels $\ell_1, \ell_2, \ldots, \ell_{m_j}$. A categorical variable z_j is *ordered* if there is a natural order relationship among its levels, e.g., $\ell_1 \leq \ell_2 \leq \cdots \leq \ell_{m_j}$, and *unordered* otherwise.

Our aim is to build a *binary tree* with *nodes* representing subpopulations. In particular the *root* of the tree represents the whole population and the *terminal nodes* represent the homogeneous strata (see Figure 1a). Any node which is not terminal is split into two branches by a statement about $\mathbf{z}$; by convention, the node issuing from the left branch represents the subpopulation satisfying the statement and the one issuing from the right branch represents the subpopulation not satisfying it (Figure 1b).

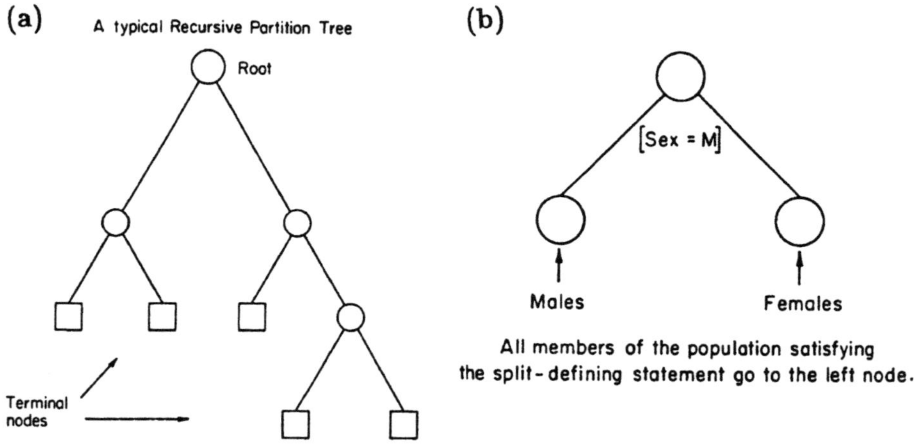

Figure 1.

There will be usually some restriction on the statements defining splits. We shall consider two classes of Split Defining Statements (SDS).

The class of simple SDS, SS

The class consists of statements of the form $z_j \in A_j$, where z_j is a component of $\mathbf{z}$ and A_j is a subset of the m_j levels of z_j. For z_j *unordered*, A_j can be any of the 2^{m_j-1} nontrivial subsets of $\{\ell_1, \ell_2, \ldots, \ell_{m_j}\}$. For z_j *ordered*, A_j can be any of the $m_j - 1$ subsets of the form $A_j = [\ell_1, \ell]$, $\ell = \ell_1, \ldots, \ell_{m_j-1}$.

The class of Boolean Combinations, SB

This class includes statements of the form '$z_{j_1} \in A_{j_1}$ and $z_{j_2} \in A_{j_2}$ and $\ldots$ and $z_{j_r} \in A_{j_r}$', $r = 1, \ldots, k$.

We shall use the following abbreviations:

$[z_j \in A_j]$ for $\{$all individuals $i \mid z_j^{(i)} \in A_j\}$,

$[z_j \leq \ell]$ for $[z_j \in [\ell_1, \ell]] = \{$all individuals $i \mid z_j^{(i)} \leq \ell\}$,

$[z_{j_1} \in A_{j_1},\ z_{j_2} \in A_{j_2}, \ldots, z_{j_r} \in A_{j_r}]$ for $\cap_{s=1}^{r} [z_{j_s} \in a_{j_s}]$,

etc.

3. DISSIMILARITY MEASURES AND HOMOGENEITY CRITERIA

In an RP tree each non-terminal node is split by a statement of the SDS class into two nodes which represent subpopulations as dissimilar as possible from the point of view of the criterion quantity. A terminal node represents a population which cannot be split any further and is therefore to be considered homogeneous. Thus, in order to specify an RP algorithm we need a *dissimilarity measure* between two disjoint populations and a homogeneity criterion to decide whether a node should be split further.

The likelihood ratio statistic (LRS) provides a natural measure of dissimilarity as follows. Let P_1, P_2 be disjoint populations and let P denote their union. We shall assume that the criterion quantity is represented by a parameter $\boldsymbol{\theta}$ which may take different values $\boldsymbol{\theta}_1$, $\boldsymbol{\theta}_2$ for P_1 and P_2 and that likelihood functions $L_1(\boldsymbol{\theta}_1)$, $L_2(\boldsymbol{\theta}_2)$ can be defined for P_1, P_2. We shall also assume that the likelihood function for P is of the form:

$$L(\boldsymbol{\theta}_1, \boldsymbol{\theta}_2) = L_1(\boldsymbol{\theta}_1) L_2(\boldsymbol{\theta}_2). \tag{1}$$

Consider now the hypothesis

$$H_o : \boldsymbol{\theta}_1 = \boldsymbol{\theta}_2 = \boldsymbol{\theta}$$

and the alternative

$$H : \boldsymbol{\theta}_1 \neq \boldsymbol{\theta}_2.$$

Then the LRS of H versus H_o is defined as:

$$\rho(H \mid H_o) = 2\ \log\{[L_1(\hat{\boldsymbol{\theta}}_1) L_2(\hat{\boldsymbol{\theta}}_2)]/[L_1(\hat{\boldsymbol{\theta}}) L_2(\hat{\boldsymbol{\theta}})]\}, \tag{2}$$

where $\hat{\boldsymbol{\theta}}_1$, $\hat{\boldsymbol{\theta}}_2$ are the maximum likelihood estimates (mle's) of $\boldsymbol{\theta}_1$ and $\boldsymbol{\theta}_2$ under H and $\hat{\boldsymbol{\theta}}$ is the mle of $\boldsymbol{\theta}$ under H_o. Clearly, the larger $\rho(H \mid H_o)$ is, the greater is the evidence in the data that P_1 and P_2 are dishomogeneous with respect to the criterion quantity. It therefore provides a reasonable and general measure of dissimilarity:

$$d(P_1, P_2) = \rho(H \mid H_o). \tag{3}$$

It is easy to generalize the definition of dissimilarity between two disjoint populations to an arbitrary number of disjoint populations $P_1, P_2, \ldots, P_n$. The LRS of the hypothesis H : $\boldsymbol{\theta}_i$ are not all the same, versus H_o : all $\boldsymbol{\theta}_i$ are equal, can be written

$$\rho(H \mid H_o) = 2 \log \left[\prod_{\ell=1}^{n} L_\ell(\hat{\boldsymbol{\theta}}_\ell) / \prod_{\ell=1}^{n} L_\ell(\hat{\boldsymbol{\theta}}) \right], \tag{4}$$

and provides a measure of *global dissimilarity* among the n populations. This expression is useful in evaluating tree structured predictors and will be taken up again in Section 5.

The LRS also provides a natural tool to define *homogeneity*. We recall that under regularity conditions $\rho(H \mid H_o)$ is asymptotically $\chi^2_{(k)}$, where $k = \dim(\boldsymbol{\theta})$ for the case of two populations [for n populations $k = (n-1)\dim(\boldsymbol{\theta})$]. Thus, it would seem reasonable to consider a population homogeneous if for every possible split $\{P_1, P_2\}$, $d(P_1, P_2)$ is *not* statistically significant at some preset significance level α. On the other hand, we also wish to avoid splits into subpopulations which are too small; if for no other reason, because the χ^2 approximation for the LRS would not be valid. These considerations lead to the following:

Definition

A split $\{P_1, P_2\}$ is *admissable* if for preset $\alpha \in [0, 1]$ and m an integer:

a) P_1 and P_2 have at least m individuals; (in the case of censored survival data we may require that in each population there be at least m *observed* failures).

b) $d(P_1, P_2)$ is significant at the α significance level.

A population is *homogeneous* if no split generated by the SDS class is admissable.

We shall now discuss the choice of dissimilarity measures for the three stratification problems studied in this work.

In the case of *prognostic stratification* we may assume that the time to failure is exponentially distributed, i.e., the survival function for P_ℓ, $\ell = 1, 2$, is:

$$S_\ell(t) = \exp\{-\lambda_\ell t\}$$

or, equivalently, the density function is:

$$f_\ell(t) = \lambda_\ell \exp\{-\lambda_\ell t\},$$

where λ_ℓ is the parameter of interest. Then, as is well known (Lawless, 1982), we have, in the hypothesis of random censoring:

$$\log L_\ell(\lambda_\ell) = \sum_{i \in P_\ell} \delta^{(i)} \log f_\ell(t^{(i)}) + (1 - \delta^{(i)}) \log S_\ell(t^{(i)}).$$

More realistic (and more complex) models (Weibull, gamma, etc.) can be used, but computations become very costly. Alternatively, one can define a LRS based on Cox pseudo-likelihood (Cox, 1972), as follows. Let us assume that the hazard function $\lambda(t) = f(t)/S(t)$ of one population is proportional to that of the other. Let γ be an indicator variable which takes value ℓ for an individual in P_ℓ, $\ell = 1, 2$. Equivalently the proportional hazards (PH) assumption can be written:

$$\lambda(t;\gamma) = e^{\Delta}\gamma\lambda_0(t),$$

where $\lambda_0(t)$ is an unspecified baseline hazard function. Then the LRS can be defined as:

$$\rho(H \mid H_o) = -2\ \log\left\{L^{\text{Cox}}(\hat{\Delta})/L^{\text{Cox}}(0)\right\}$$

(For a definition of L^{Cox} see Lawless, 1982, pp. 344–372).

It must be noticed that this approach rests on the PH assumption; however, this seems to be weaker than the assumption of exponential failure times.

In the case of ***propensity stratification*** the appropriate likelihood for P_1 and P_2 is the multinomial likelihood and the parameter of interest is $\mathbf{p} = (p_1, p_2, \ldots, p_s)$, the vector whose jth component is the probability of being assigned to the jth treatment.

We have:

$$L_\ell(\hat{\theta}) = \prod_{j=1}^{s} p_{j\ell}^{n_{j\ell}}, \qquad \ell = 1, 2, \tag{5}$$

where $n_{j\ell}$ is the number of individuals in P_ℓ assigned to the jth treatment.

In the case of ***treatment response stratification,*** we may assume that the treatment variable $\mathbf{x}$ influences response for P_ℓ by a PH model:

$$\lambda_\ell(t;\mathbf{x}) = e^{\boldsymbol{\beta}_\ell \cdot \mathbf{x}}\lambda_{o\ell}(t),$$

where $\boldsymbol{\beta}_\ell$ is the vector of regression coefficients for the ℓth population.

Then

$$L_\ell(\boldsymbol{\beta}_\ell) = L^{\text{Cox}}(\boldsymbol{\beta}_\ell), \qquad \ell = 1, 2, \tag{6}$$

and

$$\rho(H \mid H_o) = 2\ \log\left\{\left[L_1^{\text{Cox}}(\hat{\boldsymbol{\beta}}_1)L_2^{\text{Cox}}(\hat{\boldsymbol{\beta}}_2)\right] / \left[L_1^{\text{Cox}}(\hat{\boldsymbol{\beta}})L_2^{\text{Cox}}(\hat{\boldsymbol{\beta}})\right]\right\}.$$

4. RECURSIVE PARTITION AND AMALGAMATION ALGORITHMS

In order to define an RP algorithm one needs to specify three objects: (a) an SDS class S, (b) a dissimilarity measure $d(P_1, P_2)$ defined for any pair of disjoint subpopulations, (c) a criterion of homogeneity C (or C_α when we wish to emphasize its dependence on the chosen significance level α). Then an RP algorithm is defined as follows:

RP Algorithm

Choose S, d, C_α.

Denote by $\mathbf{N} = \{N_1, N_2, \ldots, N_r\}$ the current collection of nodes.

(i) To initialize, set $r = 1$ and let N_1 represent the whole population.

(ii) For every $N_j \in \mathbf{N}$ and every split $\{P_{1_j}, P_{2_j}\}$ defined by an element of S, compute $d(P_{1_j}, P_{2_j})$ and check admissability.

(iii) Using the results of (ii), check C_α for each node.

(iv) If all nodes of $\mathbf{N}$ pass the homogeneity criterion, declare the nodes of the current collection *terminal* and STOP. ELSE:

(v) Among all nodes which do not pass C_α, choose the node N_i^* corresponding to the split $\{P_{1_j}^*, P_{2_j}^*\}$ with largest dissimilarity and replace N_i^* by two nodes representing $P_{1_j}^*$ and $P_{2_j}^*$. Use the resulting collection of nodes as *current* and go to (ii).

In drawing the tree, we start from the whole population which we represent as the *root;* each node which is split is joined by branches to its two "sons".

The search for the largest dissimilarity in tree construction is the most expensive part of the algorithms and the cost increases with the size of S. If S is SB, the Boolean Combination SDS class, the cost is prohibitive. An approximate search in SB was proposed by Breiman *et al.* (1984, pp. 136–138) and we have used it here, with obvious adaptations.

From Breiman *et al.* (1984, pp. 140–150) we have also adapted the method for handling missing data, based on the notion of *surrogate variables,* and the definition of variable *importance* as an aid to interpretation.

While the terminal nodes of an RP tree are homogeneous by definition (in the sense specified by C_α), there is no guarantee that terminal nodes with a different "parent" node have truly different *criterion quantities.* For example, it is possible that nodes 4 and 5 of the tree in Figure 2 have indistinguishable survival curves. It is therefore useful to have an algorithm for amalgamating nodes which are not distinguishable from the point of view of the criterion quantity.

Let $P_1, P_2, \ldots, P_n$ be disjoint populations which are of "sufficient" size, since they may be thought of as arising from an RP algorithm. We need a

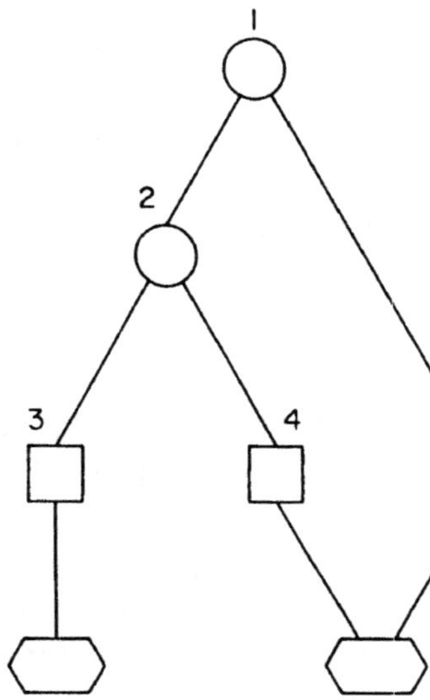

Figure 2. ***"Similar" terminal nodes 4 and 5 are joined by the Amalgamation Algorithm.***

joining criterion to decide whether any two populations are similar enough to be merged. Again, we can use the LRS to compare populations:

Definition

Let P_1, P_2 be two disjoint populations and let α be a preset SL. We shall say that P_1 and P_2 are *joinable* if $d(P_1, P_2)$ is *not* significant at the α level.

Formally, an Amalgamation Algorithm (AA) is as follows (Hartigan, 1975).

Amalgamation Algorithm

Choose d and a joining criterion J_α. Denote by $P = \{P_1, P_2, \ldots, P_n\}$ the current collection of populations.

(i) To initialize, choose the original collection $\{P_1, P_2, \ldots, P_n\}$ as *current*.

(ii) Calculate the matrix of $d(P_i, P_j)$, $i, j = 1, \ldots, r$, and check J_α for each pair.

(iii) If no pair meets J_α conclude that the current collection is a collection of distinct populations and STOP. ELSE:

(iv) Merge the pair $\{P_i^*, P_j^*\}$ with *smallest* dissimilarity and substitute for P_i^*, P_j^* their union. Declare *current* the collection thus obtained. Go back to (ii).

Notice that an AA is much simpler than an RP algorithm. After each merge, for example, only the i^*th row and the j^*th column of the new dissimilarity matrix must be computed.

If α is chosen so large that the joining criterion is always satisfied, an AA also leads to the construction of a rooted tree which is built "backwards", starting from the terminal nodes. There is no reason to expect that by applying an AA with a very large α to the terminal nodes of an RP tree we should obtain the original RP tree.

5. FINDING AN "HONEST" TREE: RESAMPLING PLANS AND THE AIC

Although the following discussion is also applicable to Amalgamation Algorithms, and indeed to all exploratory methods such as stepwise regression, we will return to RP and limit our discussion, for the sake of simplicity, to RP trees.

An RP tree is constructed according to a homogeneity criterion C_α — again for the sake of simplicity we will assume that the minimum size of admissable subpopulations is fixed once and for all—and the SL α is used as a "stopping rule", to decide when a node is terminal. This choice, however, is vulnerable to the criticism that we have ignored the problem of multiple comparisons. Since at each node many splits are examined and many SL's are computed, the SL α is at best nominal. Indeed, no matter what stopping rule is used, an RP tree would still have the tendency to overfit the data, that is, to capitalize on the peculiarities of the particular data from which it has been constructed. In order to avoid making unwarranted generalizations, it is advisable to have a separate collection of data on which to evaluate a tree constructed in an exploratory mode; however, such separate collections are rarely available. An alternative is to split the original sample into two halves, one for tree construction, the other for tree evaluation. By doing this, however, one sacrifices sample size, hence statistical power, which may be a serious drawback unless the data set is extremely large. A third possibility is that of generating a number of pseudo-samples from the original data set by one of several "resampling schemes" (Efron, 1982); these samples are then used for evaluating trees which have been generated on the original data set.

Let D denote a data set of size N. By "resampling plans" we mean a procedure to generate a number M of pseudo-samples from D: $D^{(1)}, D^{(2)}, \ldots, D^{(M)}$. The following are some of the most popular resampling schemes. In *jackknife* and *cross-validation* $D^{(i)}$, $i = 1, \ldots, N$, is obtained by dropping the ith individual from D; we then obtain N samples of size $N-1$. In *v-fold cross-validation* one constructs $M = N/V$ samples of size $N - N/V$ from D by dropping N/V groups of size V. The ***bootstrap*** builds $M = B$ samples (B arbitrary and large) of size N by sampling N elements with replacement from D, B times.

We propose to generate a nested family of trees from D and to use $D^{(i)}$, $i = 1, \ldots, M$, in order to select one that is less vulnerable to the criticism of overfitting—an "honest" tree. For each tree T_k and each pseudo-sample $D^{(i)}$, we use the global dissimilarity ρ (defined by equation (4), Section 4), of the terminal nodes to measure the ***adequacy*** $A_k^{(i)}$ of T_k for $D^{(i)}$. For the ***bootstrap*** and the ***jackknife*** we simply estimate the parameters of the likelihoods and calculate ρ on $D^{(i)}$. In ***cross-validation***, on the other hand, we estimate the unknown parameter on $D^{(i)}$ and calculate ρ for the data points in $D^{(i)} - D$. The algorithm described below is a generalization of one originally proposed by Stone (1974), who used ***cross-validation***, and of the approach of Breiman ***et al.*** (1984) to tree construction.

Tree Selection Algorithm (TSA)

(i) Choose a sequence of SL's $\alpha_1 < \alpha_2 < \cdots < \alpha_n$.

(ii) For each α_k build an RP tree, T_k, with C_α as homogeneity criterion. Clearly, $T_1 \leq T_2 \leq \cdots \leq T_n$. (In particular, if $\alpha_1 = 0$, T_1 is the trivial tree with only one node representing the whole population, and if $\alpha_n = 1$, T_n is a tree with terminal nodes such that any further split results in subpopulations which are too small to satisfy the homogeneity criterion.)

(iii) By an appropriate resampling scheme build M samples $D^{(i)}$, $i = 1, \ldots, M$, from D.

(iv) Let $A_k^{(i)}$ denote the adequacy of the tree T_k for $D^{(i)}$. Compute

$$\text{Ave}(A_k) = \frac{1}{M}\sum_{i=1}^{M} A_k^{(i)},$$

$$\text{SD}(A_k) = \left\{\frac{1}{M-1}\sum_{i=1}^{M}\left[\rho_k^{(i)} - \text{Ave}(A_k)\right]^2\right\}^{1/2}.$$

(v) Choose the tree corresponding to maximum $\text{Ave}(A_k)$. Or choose the simplest tree among those for which $\text{Ave}(A_k)$ falls within ℓ SD's from the "best" tree (ℓ fixed and usually $\ell = 1, 2, 3$).

The above TSA is computationally expensive because of the resampling necessary to calculate $\text{Ave}(A_k)$ and $\text{SD}(A_k)$. A great simplification is introduced by using the Akaike Information Criterion (AIC) instead of $\text{Ave}(A_k)$

and by selecting the tree with the smallest AIC. The AIC for T_k is:

$$\mathrm{AIC}_\alpha = -\rho_k + 2p_k,$$

where p_k is, in this case, the number of terminal nodes of T_k. This is justified by a result of Stone (1977) who showed that the AIC is approximately equivalent to $-\mathrm{Ave}(A_k)$ for (1-fold) *cross-validation*.

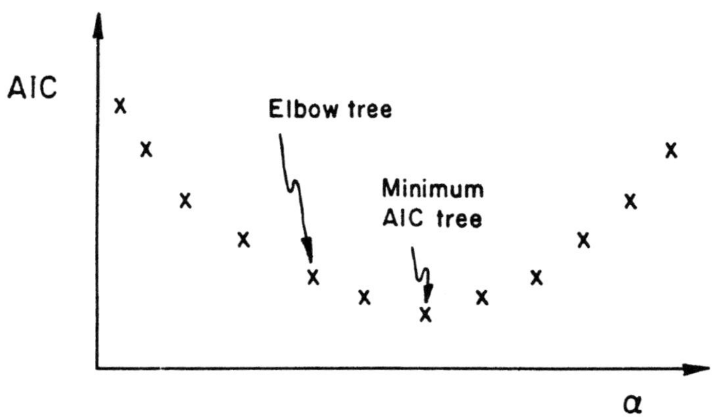

Figure 3. *Tree selection with the aid of Akaike's Information Criterion.*

It is useful to draw a profile of α_k versus AIC_α, which will have, in general, the shape of Figure 3. The shape of the AIC curve may suggest choosing a simpler tree than the one corresponding to the minimum AIC. If the curve has an elbow (Figure 3), one can argue that the *elbow tree* is a good choice. This is analogous to the situation in factor analysis, when one chooses the number of factors corresponding to the elbow of the SCREE plot.

6. EXAMPLES

a) Prognostic Stratification

There are several examples in the literature of prognostic strata defined by various prognostic variables for multiple myeloma (Bataille *et al.*, 1985). However, the variables used are not always available at every institute. Certain blood tests or other diagnostic tests are not performed uniformly by all cancer centers which treat myeloma patients. Hence, it is desirable to define prognostic stratification rules using the variables that are measured at a given institute (in this case the Princess Margaret Hospital). The variables conveniently available at PHM are shown in Table 1, and were used by the Recursive Partition and Amalgamation Algorithms to define the tree shown in Figure 4. For the Recursive Partition Algorithm, the class SB of SDS was used to derive admissable splits, along with an exponential model likelihood dissimilarity measure. The criterion of homogeneity C_α was: declare node N homogeneous if for any split $\{P_{1_j}, P_{2_j}\}$, *either* one of P_{1_j} and P_{2_j} has less than 25 individuals, *or* $d(P_{1_j}, P_{2_j})$ is *not* significant at the α level. The results presented here are for $\alpha = .05$.

Table 1. *Multiple Myeloma Variables Used by the Recursive Partition Algorithm for Prognostic Stratification*

1.	Haemoglobin level	normal, mildly depressed, severely depressed
2.	Calcium level	normal, mildly elevated, severely elevated
3.	Creatinine level	normal, mildly elevated, severely elevated
4.	Alkaline Phosphatase level	normal, mildly elevated, severely elevated
5.	White Blood Cell % Plasma	0%, > 0%
6.	Bone Marrow % Plasma	normal, trace, elevated
7.	Lytic Lesion count	0–3 lesions, 4 or more lesions
8.	Age	0–45, 45–60, over 60 years
9.	Symptom Duration	0–29, 30 or more months
10.	M-protein risk factor	Low risk, High risk

For the Amalgamation Algorithm, the same dissimilarity measure (exponential model) was used, in conjunction with the joining rule J_α with $\alpha = 0.05$.

The algorithms yield the three terminal clusters labelled as stages 1 to 3, shown in Figure 4. The Kaplan-Meier estimates of the survival curves for each stage appear in Figure 5. For the practicing clinician, such information is immediately useful. The strata-defining rules are phrased in clinically meaningful terms.

A total of at most three decisions are required to classify any patient, and the prognosis in terms of expected years of survival is quickly available

Table 2. *Multiple Myeloma Variables Used by the Stepwise Cox Proportional Hazards Regression Algorithm (BMDP2L)*

Z_1	=	I (Haemoglobin = mildly depressed)
Z_2	=	I (Haemoglobin = normal)
Z_3	=	I (Calcium = mildly elevated)
Z_4	=	I (Calcium = severely elevated)
Z_5	=	I (Creatinine = mildly elevated)
Z_6	=	I (Creatinine = severely elevated)
Z_7	=	I (Alk. Phos. = mildly elevated)
Z_8	=	I (Alk. Phos. = severely elevated)
Z_9	=	I (White Blood Cell % Plasma > 0%)
Z_{10}	=	I (Bone Marrow % Plasma = trace)
Z_{11}	=	I (Bone Marrow % Plasma = elevated)
Z_{12}	=	I (4 or more lytic lesions)
Z_{13}	=	I (Age 45–60 years)
Z_{14}	=	I (Age over 60 years)
Z_{15}	=	I (Symptoms lasted 30 or more months)
Z_{16}	=	I (M-protein risk factor = high risk)
$I(x)$	=	1 if the logical statement X is true.
$I(x)$	=	0 if the logical statement X is false.

from the estimated survival curves. Contrast this with the usual current statistical methodology for evaluating prognostic factors—analysis via the Cox model. The same prognostic factors as shown in Table 1 were recoded as 0–1 indicator variables as shown in Table 2, and used in a stepwise step-up Cox model algorithm (BMDP2L). The α-level to enter was set at $\alpha = 0.05$. The final variables selected appear in Table 3. Patients could be divided into prognostic strata by examining the distribution of the regression score $\mathbf{s} = \boldsymbol{\beta} \cdot \mathbf{z}$ of the 302 patients, and splitting this distribution into three or four disjoint subgroups. Such strata are, however, difficult to justify and explain to clinicians, as the strata-defining rules will not in general be clinically meaningful.

We can compare the recursive partition model to the stepwise Cox model, by defining two indicator variables for the three stages of Figure 4. The indicator variables are demonstrated in Table 4. Fitting a Cox model yields the results shown in Table 5. The RP model requires two parameters instead

Table 3. *Step-up Stepwise Fitted Cox Model ($\alpha = 0.05$ Level of Significance)*

Variable	Description	Parameter estimate	Chi-square	p-value
Z_2	HG = normal	$\hat{\beta}_2 = -0.6769$	23.70	$< 10^{-4}$
Z_4	CA = sev. elev.	$\hat{\beta}_4 = 0.5805$	7.25	0.0071
Z_{14}	Age > 60	$\hat{\beta}_{14} = 0.3327$	6.76	0.0093
Z_{15}	Symp. $\geq$ 30 months	$\hat{\beta}_{15} = -1.2100$	9.99	0.0016
Z_8	AP sev. elev.	$\hat{\beta}_8 = 1.0879$	6.87	0.0088
Z_{11}	BM % P elevated	$\hat{\beta}_{11} = 0.4111$	6.42	0.011
Z_{10}	BM % P trace	$\hat{\beta}_{10} = 0.3986$	4.33	0.038

No. of patients = 302 No. of parameters = 7
log likelihood = −1267.70 AIC = 2549.416

of seven, and is to be preferred in terms of Akaike's Information Criterion.

b) Propensity Stratification

When performing retrospective analyses on a group of variously treated patients, any identified difference in behaviour of one subgroup versus another may be due to the differing treatments used. If proportionately more males receive treatment A than B, while more females receive B than A, an analysis including sex may show sex to be an important factor when in fact, sex is unimportant but treatment A is superior to B. Conversely, when examining treatment A and B, a perceived benefit for A may really be reflecting the biological fact that males fare better than females for the given disease. Hence, identifying treatment imbalances among patient subgroups is a valuable aid in understanding effects observed in retrospect.

In a retrospective analysis of small cell lung cancer patients at Princess Margaret Hospital, various prognostic variables were identified. However, prophylactic cranial irradiation (PCI) treatment was used on some of the patients. Before making conclusions about the prognostic factors identified, the distribution of PCI among the patients needs to be clearly understood. Table 6 shows four of the most important prognostic factors, within which the distribution of PCI needs to be investigated. For most patients, a record

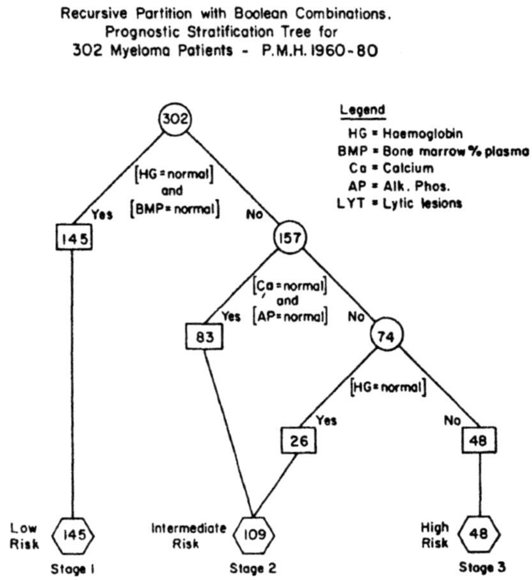

Figure 4.

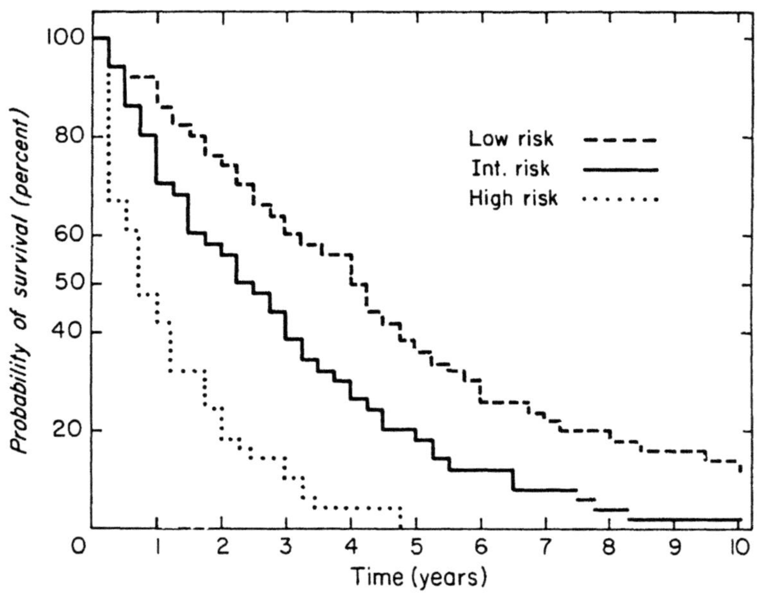

Figure 5.

Table 4. *Variables Defined by the Terminal Nodes of the Recursive Partition Tree of Figure 4*

$$Z_1 = \mathrm{I}\ (\text{patient} = \text{stage } 2)$$

$$Z_2 = \mathrm{I}\ (\text{patient} = \text{stage } 3)$$

Table 5. *Cox Model Defined by Recursive Partition and Amalgamation Algorithms*

(α = 0.05 Level of Significance)

Variable	Description	Parameter estimate	Chi-square	p-value	Risk (relative to Low risk)
Z_1	Stage 2	$\hat{\beta}_1 = 0.6187$	20.09	$< 10^{-4}$	1.8565
Z_2	Stage 3	$\hat{\beta}_2 = 1.6201$	63.11	$< 10^{-4}$	5.0537

No. of patients = 302 No. of parameters = 2
log likelihood = −1270.1835 AIC = 2544.367

of whether or not PCI was administered can be found in their medical charts, but for ten percent of the patients no reference to PCI can be found. Hence there are three treatment categories to investigate: (1) PCI administered, (2) PCI not administered, and (3) PCI assignment status unknown. The response variable PCI thus has 3 levels, suggesting the use of a multinomial model.

For the recursive partition algorithm, the class SB of SDS was used, with a multinomial likelihood dissimilarity measure. The criterion of homogeneity C_α, $\alpha = 0.05$ was used as previously described, however, in this case $d(P_{1_j}, P_{2_j})$ is distributed (asymptotically) as $\chi^2_{(2)}$ since there are three classes for the response variable.

The recursive partition algorithm yields the tree shown in Figure 6, and reveals three subgroups with widely differing PCI assignment rates. All of the "PCI assignment not stated" (N/S) patients are limited disease

Table 6. *Small Cell Lung Cancer Variables Used by the Recursive Partition Algorithm For Propensity Stratification*

Stratifying variables:	
Age	0–49, 50–59, 60–69, 70–99
Sex	Male, Female
Performance Status	1, 2, 3, 4.
Extent of Disease	Limited, Extensive
Response variable:	
Treatment includes Prophylactic Cranial Irradiation (PCI)?	Yes, No, Not stated

patients (disease localized in the lung), a good prognostic subgroup. Since the distribution of N/S patients is so systematic, some understanding of their probable PCI assignment may be possible through further discussions with the clinicians involved. Also, within this good prognostic subgroup, the PCI assignment rate is the lowest, as one would expect from a medical treatment decision point of view. Males with extensive disease (disease not localized to the lung alone), one of the prognostically worst subgroups, show a PCI assignment rate of 60%, whereas females with extensive disease show a rate of 53%.

Hence, there is some confounding of PCI with sex for patients with extensive disease, and an overall confounding of PCI with extent of disease. These facts must be borne in mind when investigating either the benefits of PCI, or the prognostic value of sex and extent of disease.

c) Treatment-Response Stratification

Often in clinical trials, the magnitude of the difference in effect of two treatments will be larger within some subgroups of the patient population than the magnitude within other subgroups. Identification of such subgroups is important in deciding where the treatments should be used, especially if one treatment is more toxic or costly and doing relatively little better than

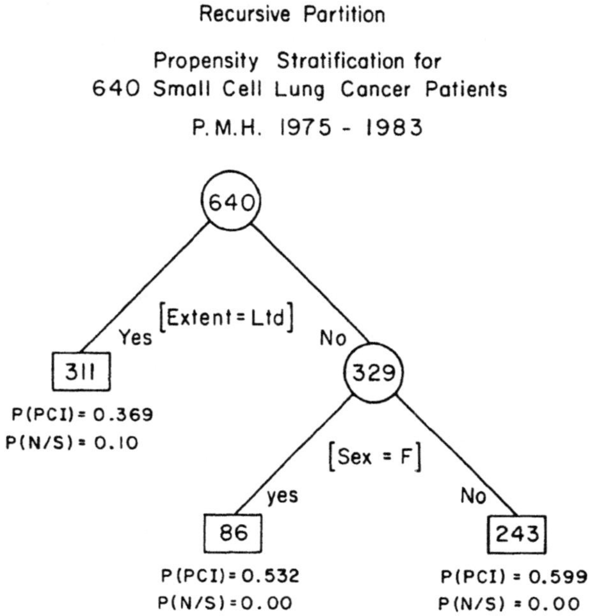

Figure 6.

another for a portion of the patients.

From 1971 to 1979 a series of clinical trials for ovarian cancer patients were conducted at Princess Margaret Hospital comparing pelvic irradiation and abdominopelvic irradiation. The patients were all FIGO state IB, II and asymptomatic stage III; with a completed bilateral salpingo oophorectomy hysterectomy. The patients received either radiation to the pelvic area (pelvic irradiation), or radiation to the pelvic area plus to the abdominal lymph node area (abdominopelvic irradiation). Overall, abdominopelvic irradiation was shown to be superior to pelvic irradiation, especially for delaying recurrence (Dembo and Bush, 1982). But is the observed effect true for the entire patient population investigated?

Five of the most important prognostic variables are shown in Table 7a; these are the predictor variables $\mathbf{z}$. The treatment variable x, a 0–1 indicator variable for treatment, is shown in Table 7b. This variable and the censored survival time constitute the criterion variables $\mathbf{u}$.

For the Recursive Partition Algorithm, the class SB of SDS was used, along with a Cox likelihood dissimilarity measure to test for differences of regression equations (see equation (6), Section 3). The indicator variable x, ($x = 0$ or 1 according as the patient received pelvic or abdominopelvic

Table 7. *Ovarian Cancer Variables Used by the Recursive Partition Algorithm for Treatment-Response Stratification*

a) Criterion variables, b) Treatment variable

a)

1	Age	< 40, ≥ 40.
2	Histology	undifferentiated, serous, endometroid, mucinous, clear cell.
3	Grade	well differentiated, moderately diff., poorly diff.
4	Stage	IB, II, III.
5	Residuum	none, small amount.

b)

$X = \text{I}$ (treatment = Abdominopelvic irradiation).

irradiation), was used as the regression treatment covariate. The criterion of homogeneity C_α was as in the previous example.

The algorithm yields the tree shown in Figure 7. The relative risk estimates (R.R.) from the Cox model for each stratum are also shown (R.R. $= \exp(\hat{\beta}_j)$, $j = 1, 2$). The splitting rule identifies a favourable subgroup of patients (no residual disease after BSOH, age ≥ 40, and Stage = IB or II). Within this favourable subgroup, the addition of abdominal lymph node irradiation reduces the rate of disease recurrence by a factor of $1/0.39 \simeq 2.5$. The accompanying Kaplan-Meier estimates of the relapse-free curves for the two treatment groups are shown in Figure 8. The parameter estimate $\hat{\beta}_1$ is significantly different from zero ($p = 0.0027$). Within the complementary subgroup, the parameter estimate $\hat{\beta}_2$ is not significantly different from zero ($p = 0.068$), indicating little difference between the two treatment methods (Figure 9). Hence, there appears to be a subgroup of the selected patient population that benefits from the extra abdominal irradiation, while the complementary patient population shows no benefit.

Identification of such subgroups is an important aspect within the framework of a long-term series of clinical trials, so that treatments can eventually be matched to that subgroup of patients where the effects are greatest.

Recursive Partition with Boolean Combinations
Subgroup Regression Tree
276 Ovarian Cancer Patients – P.M.H. 1971-79

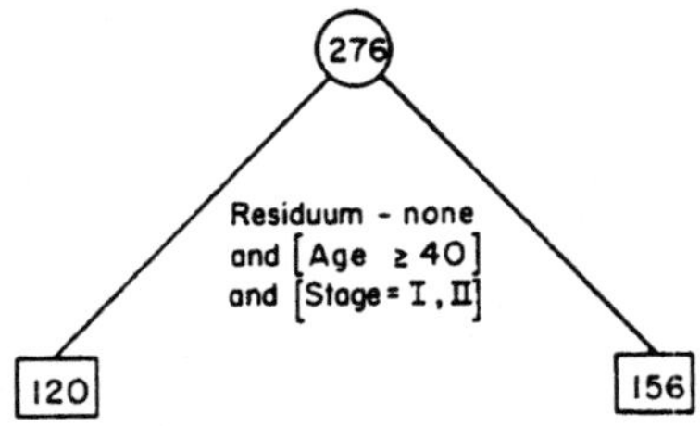

Relative Risk Estimates

	RR (No)	RR (No)
Pelvic XRT	1.00(42)	1.00 (40)
Abdominopelvic XRT	0.39 (78)	1.54 (116)

Figure 7.

d) Minimum AIC plots

Returning to the myeloma data example, we can vary the level of α of the C_α criterion and calculate the AIC for the resulting trees. The AIC's found by varying α from 0.01 to 0.90 are shown in Figure 10. The AIC plot suggests that the trees corresponding to $\alpha = 0.15$ (the "elbow" tree) or $\alpha = 0.20$ are in some sense to be preferred. The elbow tree is shown in Figure 11. The prognostic stratification rule now yields four patient subgroups. The Kaplan-Meier survival curve estimates for each subgroup are shown in Figure 12. Table 8 shows the three indicator variables defined by the elbow tree, and Table 9 lists the results of the Cox model derived from these indicator variables. Notice that the AIC for this elbow tree is the smallest AIC of the three Cox models fitted to the myeloma data (Tables 3, 5 and 9). The likelihood for this elbow tree is also the smallest of the three model likelihoods. The AIC plot as a guide to model selection is much more informative than merely setting $\alpha = 0.05$, as is traditional in classical statistics.

7. DISCUSSION

We have outlined a general framework for defining strata from biostatistical data.

The main novelty of this formulation is its generality and flexibility. One

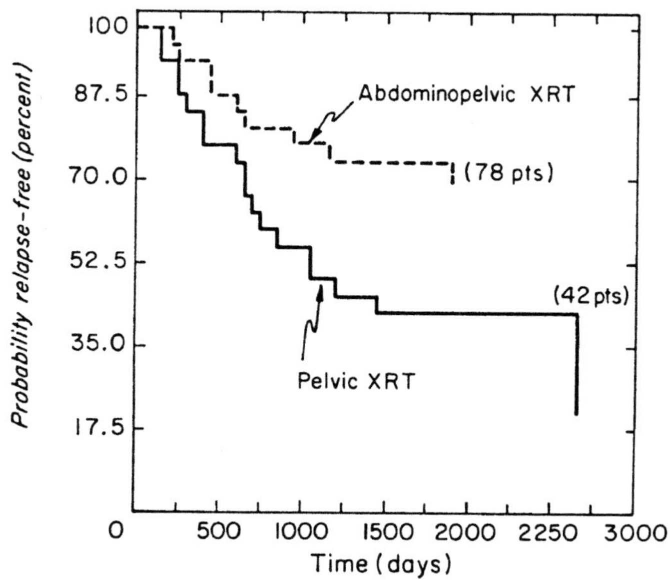

Figure 8.

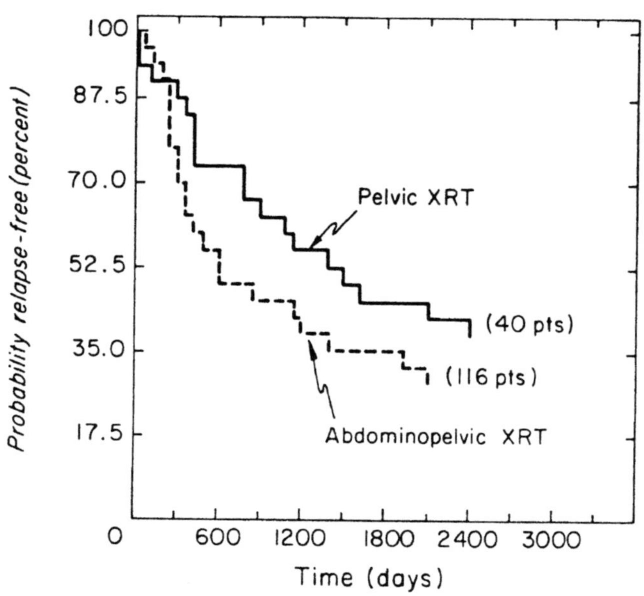

Figure 9.

Table 8. *Variables Defined by the Terminal Nodes Of the Recursive Partition Tree of Figure 11*

$$Z_1 = \text{I (patient = stage 2)}$$
$$Z_2 = \text{I (patient = stage 3)}$$
$$Z_3 = \text{I (patient = stage 4)}$$

Table 9.

Variable	Description	Parameter estimate	Chi-square	p-value	Risk (relative to Stage 1)
Z_1	Stage 2	$\hat{\beta}_1 = 0.6067$	17.06	$< 10^{-4}$	1.83
Z_2	Stage 3	$\hat{\beta}_2 = 1.1527$	20.18	$< 10^{-4}$	3.16
Z_3	Stage 4	$\hat{\beta}_3 = 1.7895$	69.44	$< 10^{-4}$	5.98

No. of patients = 302 No. of parameters = 3
log likelihood = −1265.7825 AIC = 2537.56

needs to specify the ***language*** in which strata are to be defined, the ***objective*** of the stratification, and the ***homogeneity criterion*** to decide whether a population is homogeneous in the sense specified by the objective of the stratification. In our formulation, the language is formed from the SDS class, the objective is embodied in the definition of dissimilarity, and the homogeneity criterion is formulated on the basis of the statistical properties of the dissimilarity measure (large sample approximation). Once the choices are made, RP, and possibly amalgamation, provide algorithms for obtaining strata.

The general framework permits one to generalize the "node impurity measures" used by Breiman *et al.* (1984) to the situation in which the objective of the stratification is defined on the basis of censored survival data (e.g., prognostic stratification), of categorical response (e.g., propen-

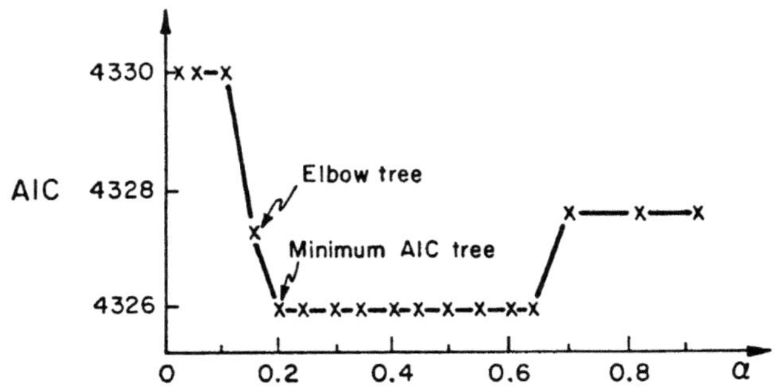

α	0.01 - 0.10	0.15	0.20 - 0.65	0.70 - 0.90
AIC	4330.15	4327.65	4326.25	4327.91

Figure 10. *Choice of tree with the aid of Akaike's Information Criterion.*

sity stratification), or of a relationship between variables (e.g., treatment response stratification). In particular, we have discussed a class of dissimilarity measures based on likelihood ratio statistics which covers many of the situations encountered in biostatistics. Within this framework, most of the ideas developed by Breiman *et al.* (1984) can be implemented with minor modifications. Indeed, our work strongly relies on appropriate adaptations of some of these ideas (missing data, variable importance, Boolean combinations, etc.).

We have also attempted to formulate the problem of model selection and to suggest an approach slightly more general than that of Breiman *et al.* (1984) via the idea of resampling schemes. This formulation and the use of likelihood based measures of model significance allows us to use the AIC instead of the more expensive cross-validation or bootstrap calculations for the selection of "honest" trees.

The examples discussed in this paper illustrate the flexibility of the framework and its usefulness in the clinical context. They show that RP produces results of clear interpretability and are therefore attractive to physicians. From a statistical point of view, and if we adopt the AIC as a criterion for model comparison, the examples show that RP based models compare well or favourably with those obtained by stepwise regression.

The emphasis of this paper has been on the general formulation. Clearly many points need clarification and further investigation. For example, the choice of dissimilarity measures other than the likelihood ratio could be more desirable in certain applications. We have implemented, in the case of prog-

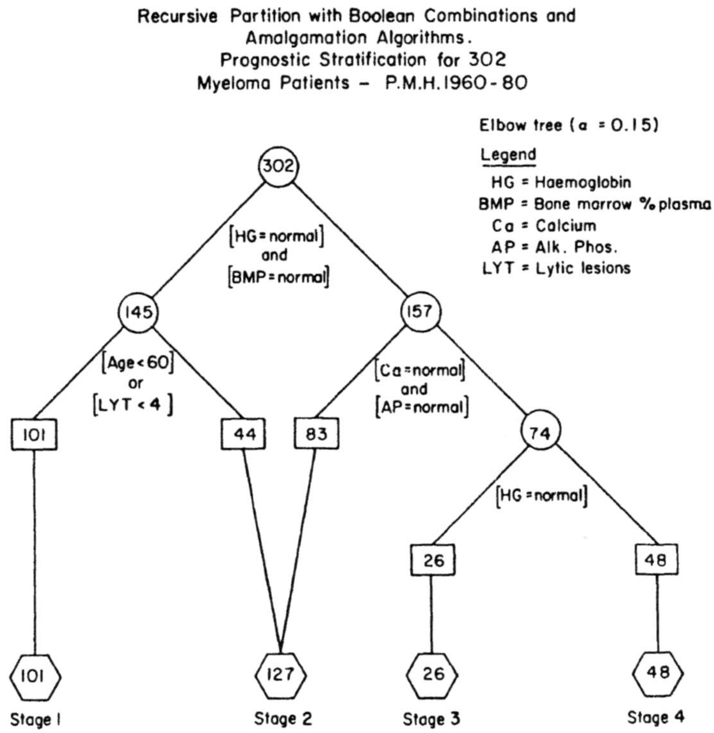

Figure 11.

nostic stratification, a variety of alternatives to the exponential model and the Cox model likelihood ratio statistics. The results, however, do not seem dramatically different from those obtained with the choices discussed in this work; moreover, there seems to be no substitute for the AIC criterion in "honest" tree selection. The AIC criterion is equivalent to a cross-validation criterion only under large sample approximation (Stone, 1977). However, the validity of the approximation has not been discussed here and will be the object of a separate study. Also, the comparative advantages and disadvantages of cross-validation, bootstrap and jackknife in model selection have not been studied in this work but merit serious attention.

Comparisons between regression models and RP trees also need a more systematic approach than the one outlined here. In fact it would be highly desirable to integrate RP and regression. It is by no means clear that one should perform uniformly better than the other. On the contrary, the two approaches appear to be complementary.

There is a fundamental criticism that can be raised against RP methods,

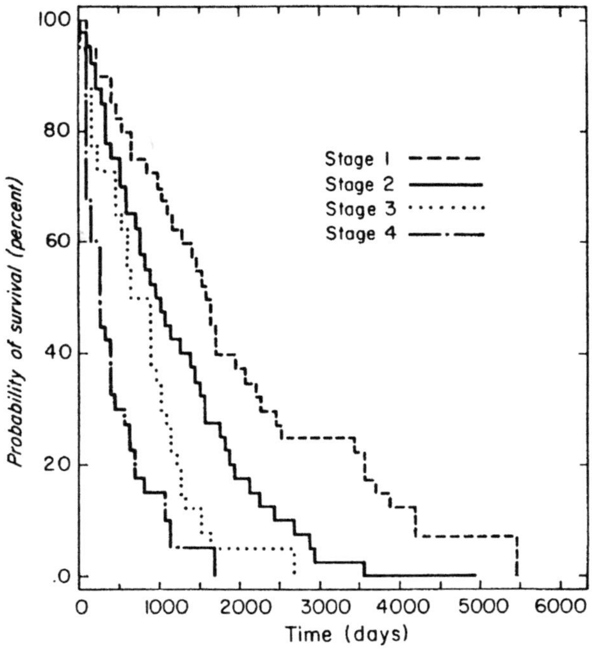

Figure 12.

or rather, against their careless use: RP, just like stepwise regression and-factor analysis, remains principally an *exploratory* method and should not be used to make extraordinary claims. There is no substitute for independently collected data if one wishes to evaluate the merit of a data generated hypothesis. Similarly, there is no substitute for controlled clinical studies if one needs a reasonable degree of certainty about a medical theory.

On the other hand, methods like RP, especially if strengthened by appropriate model selection procedures such as those outlined in Section 5, are a useful tool in obtaining as much information as possible from existing data. They have the advantage over more traditional stepwise regression techniques of producing results in clinical terms, thus facilitating the complex task of hypothesis formulation.

The problem remains that exploratory methods are difficult to evaluate from a statistical point of view. There seems to be no alternative to the use of intense, complex and costly simulations in order to explore the limits of validity of a method like RP. The task is thankless, but necessary and important.

ACKNOWLEDGMENTS

The authors appreciate the excellent secretarial assistance received from Ms. Debbie Vaz. The invaluable collaboration of members of the Princess Margaret Hospital Departments of Medical and Radiation Oncology, Drs. Alon Dembo, Ronald Feld, Leonard Kaizer, and Uri Sagman in particular, is gratefully acknowledged.

This research was supported in part by grants from the Medical Research Council, and the National Cancer Institute (Canada).

REFERENCES

Bataille, R. G., G. M. Durie, J. Grenier, and J. Sany (1985), "Prognostic factors and staging in multiple myeloma: a re-appraisal". *Journal of Clinical Oncology* (in press).

Breiman, L., J. H. Friedman, R. A. Olshen, and C. J. Stone (1984), *Classification and Regression Trees*. Belmont, CA: Wadsworth.

Byar, D. P. (1982), "Analysis of survival data: Cox and Weibull models with covariates". In *Statistics in Medical Research*, ed. V. Mike and K. E. Stanley, pp. 365–401. New York: Wiley and Sons.

Cox, D. R. (1972), "Regression models and life tables" (with discussion). *Journal of the Royal Statistical Society*, Series B **34**, 187–208.

Dembo, A. J., and R. S. Bush (1982), "Choice of postoperative therapy based on prognostic factors". *International Journal of Oncology Biology Physics* **8**, 893–897.

Efron, B. (1982), *The Jackknife, the Bootstrap, and other Resampling Plans*. Philadelphia: Society for Industrial and Applied Mathematics.

Gail, M., and R. Simon (1985), "Testing for qualitative interactions between treatment effects and patient subsets". *Biometrics* **41**, 361–372.

Gillo, M. W., and M. W. Shelly (1974), "Predictive modeling of multivariable and multivariate data". *Journal of the American Statistical Association* **69**, 646–653.

Hartigan, J. A. (1975), *Clustering Algorithms*. New York: Wiley and Sons.

Lawless, J. (1982), *Statistical Models and Methods for Lifetime Data*. New York: Wiley and Sons.

Rosenbaum, P. R., and D. B. Rubin (1984), "Reducing bias in observational studies using subclassification on the propensity score". *Journal of the American Statistical Association* **79**, 516–524.

Sonquist, J. A., E. L. Baker, and J. N. Morgan (1973), *Searching for Structure: An Approach to Analysis of Substantial Bodies of Micro-Data and Documentation for a Computer Program* (Revised Edition). Ann Arbor: Institute for Social Research, University of Michigan.

Stone, M. (1974), "Cross-validatory choice and assessment of statistical predictions" (with discussion). *Journal of the Royal Statistical Society*, Series B **36**,

111–147.

Stone, M. (1977), "An asymptotic equivalence of choice of model by cross-validation and Akaike's criterion". *Journal of the Royal Statistical Society, Series B* **39**, 44–47.

J. W. Chapman, [1] J. Etezadi-Amoli, [2] P. J. Selby, [3]
N. F. Boyd, [4] and D. Dalley [5]

STATISTICAL RAMIFICATIONS OF LINEAR ANALOGUE SCALES IN ASSESSING THE QUALITY OF LIFE OF CANCER PATIENTS

ABSTRACT

In evaluating cancer treatments, it is important to know both about the quality and duration of the patient's survival. This study was initiated to investigate the properties of scales that could be used to assess quality of life. 115 breast cancer patients from the Princess Margaret Hospital in Toronto were asked to evaluate their quality of life using linear analogue scales. Thirty-one variables covered the areas of general health and major problems associated with breast cancer derived from clinical experience and the opinions of patients with the disease. Patients found the linear analogues simple to use, and the broad areas considered by each scale made the mental effort much less than what would be involved with extensive evaluative inventories.

Our general health items were compared with the Sickness Impact Profile from which they were derived to assess the validity of the method. Our patient population produced reliable test-retest results that were stable for both factor analysis and multivariate regressions. Also, we can report the results of some sample size investigations for both reliability and stability of data collected in this and other studies.

[1] Statistical Consultant for Cancer Research, 11 Dayman Court, Kitchener, Ontario N2M 3A1

[2] Decision Aid Consulting, 950 Yonge Street, Toronto M4W 2J4

[3] Royal Marden Hospital, Sutton, England 5MZ 5PT

[4] Ontario Cancer Institute, Princess Margaret Hospital, Toronto, Ontario M4X 1K9

[5] St. Vincent's Hospital, Darlinghurst, New South Wales, Australia 2010

I. B. MacNeill and G. J. Umphrey (eds.), Biostatistics, 51–60.

1. INTRODUCTION

In evaluating cancer treatments, it is important to know both about the quality and duration of the patient's survival. A lot of work has been directed at survival analyses, but relatively little has been done to assess the patient's quality of life. Our data were to be collected in a clinical setting so the approach needed to be simple and quick; large health inventories would be unwieldy for general use.

Priestman and Baum (1976) looked at the quality of life in advanced breast cancer patients with Linear Analogue Self-Assessment (LASA) scales. They used a 10 centimeter line with the ends of the line labelled with descriptive extremes of a symptom. The patients marked the position on the line corresponding to their own feeling about the symptom. The results from all ten indices were then summed to give an overall value. There was a high correlation (0.87) between scores obtained in the presence of a doctor and those 24 hours later, and significant changes in LASA scores were observed to correspond with clinical impressions.

We used thirty-one scales to cover twenty-eight of the patient's general health and specific disease-related items and one scale as a global assessment of overall lifestyle. These issues have been discussed elsewhere (Selby *et al.*, 1984) and attention will be focused here on statistical points.

2. DATA

The study population was 115 breast cancer cases in clinics at the Princess Margaret, a cancer treatment hospital in Toronto. The cases had recurrent disease or were receiving adjuvant therapy. We wanted patients to assess the effects of disease or treatment on their lives so they were instructed to consider their lives relative to what would be normal *for them.*

e.g. Physical Activity

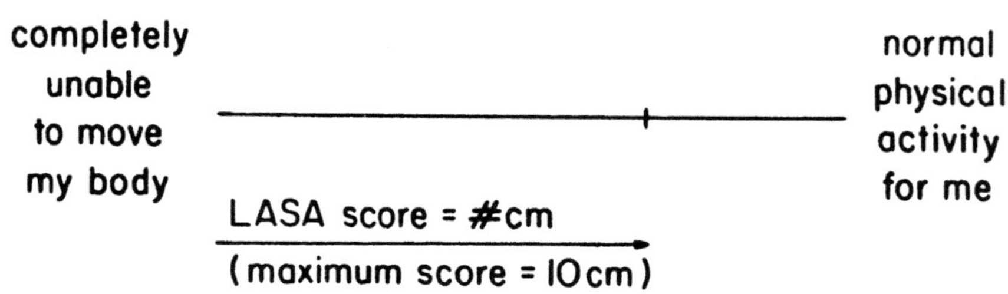

The areas covered are listed in Table 1.

Table 1. *Areas Covered by Linear Analogue Scales*

General Health	*Disease Related*
Work	Dysuria
Writing	Pain
Concentration	Bowels
Recreation	Sore Mouth
Housework	Breathing
Social Life	Fatigue
Physical Activity	Nausea
Mobility	Vomiting
Self Care	Attractiveness
Family Relations	Hair Loss
Eating	Appearance
Sleep	Information (about disease)
Speech	
Anger	
Depression	
Anxiety	
Global	
Uniscale — global assessment of overall quality of life	

Our general health categories were derived from the Sickness Impact Profile (SIP) of Bergner *et al.* (1981) so a group of 52 patients completed both forms to permit comparison of the two approaches. SIP involves a long list of questions for which the patient checks the presence or absence of. The responses to questions are then weighted and grouped to give the patient's status for particular areas of life; the overall index is derived directly from the scores for the subcategories. The same group of patients had a physician independently assess their status with linear analogue scales and the Karnofsky performance scale.

Our disease-related items were obtained from clinical experience and the opinions of patients with the disease. Three areas were given two scales: sleeping less or more, eating less or more, and constipation/diarrhoea. These were then combined to make three continuous variables for sleeping, eating, and bowels.

Medical information was also available on the patient's age, presence of other syndrome, whether she had had a mastectomy, site of disease, and times from diagnosis to LASA, recurrence to LASA, and diagnosis to recurrence. These medical variables were considered because they might affect responses about quality of life.

3. METHODOLOGY

The properties of these scales were investigated from a pragmatic viewpoint. The scales appeared to be acceptable for patients in that they responded positively to taking part in the study 89% of the time, and the form usually took only about 5 minutes to complete. It should be noted that the methods and results reported here form a summary of several related investigations on different subsets of data so that not all the patients were used for each.

Reliability was examined in two different ways. Ninety-six patients completed LASA in outpatient clinics and then 9–12 hours later completed it again at home. It was not anticipated that a patient's clinical status would change in this time frame so product moment correlation coefficients were calculated to examine the relationship between the two sets of scores. As well, intraclass correlation coefficients were used for LASA's completed in the clinic, at home 9–12 hours later, and a week after their first LASA. The product moment and intraclass correlation coefficients were repeated without the group of people who had fairly normal lives, with scores ≥ 9.7.

We looked at the validity of the results with 52 patients. Product moment correlation coefficients were used to compare the patient's LASA scores with those linear analogues given by the physician (LAPA) as well as to compare areas in common with SIP and the Karnofsky performance status. Meanwhile, the LASA scores for metastic and adjuvant chemotherapy patients were compared to see if changes in quality of life associated with treatment or the progression of disease could be inferred from changes in LASA scores.

Multivariate analyses were used in two contexts. Factor analysis was employed to examine the relationships of the twenty-eight specific LASA categories. Varimax and promax rotations were used to define five factors, with each variable within a factor having loading greater than 0.3. The patient's sum of the variable values for each factor was used to calculate product moment correlation coefficients for the two LASA's completed within 12 hours.

There is a medical interest in the relationship between the global measure of quality of life, UNI, and individual variables. Priestman and Baum (1976) gave uniform weights to their items while Bergner *et al.* (1981) used the opinions of professionals, and factor analysis defines the association of variables by quantity of impairment. We felt that none of these approaches adequately handled the situation since in the first instance a woman might not consider restrictions in some areas of life as important as in others, in the second she might have different views from a professional on what was important to her, and lastly even within a factor analysis grouping she might consider some variables of greater or lesser relevance. Our UNI scores were derived from a separate questions about overall lifestyle so a simplistic approach was chosen of performing a regression with UNI as the dependent and the other LASAs as the independent variables. This amounts to an *ad hoc* way of incorporating a patient's view of item relevance.

The original data were very non-normal. Priestman and Baum (1976) had used an arcsine transformation on their data. After trying several transformations, we decided to use log{arccos(variable/10)} which incorporates the aesthetic flipping of the unaffected LASA scores of 10 to 0. Forward stepwise linear regressions were used on the LASA and LAPA scores. The residuals were looked at with P-P plots as well as for any evidence of non-linear relationships.

Sample size considerations can be deduced from comparison with similar patients in other studies where various sample sizes were used to answer different medical questions. Product moment correlation coefficients were computed for samples of size 25, 60, and 96 to measure the repeatability of the LASA scores, where the 96 patients are from the present data set. In a social science context, absolute values of say ≥ 0.70 are usually used as measures of reliabilty, but we have considered a statistical criterion here of significance levels for testing whether a correlation coefficient is greater than 0. The stability of regressions can be compared for samples of 60 and 75 caes, where the 75 cases are from the present study.

4. RESULTS

Table 2 shows the distribution of correlation coefficients for the comparison of data from outpatient clinics and values 9–12 hours later. All the correlations were significantly greater than 0 at the 1% level. The intraclass correlations yielded similar results except for nausea and vomiting which were very close to zero; it should be noted that these were two of the three variables with product-moment correlations less than 0.60. Restricting the cases considered to those LASA values < 9.7 did not affect the results greatly.

The LASA and LAPA scores are compared in Table 3 while the LASA scores are compared with SIP and Karnofsky in Table 4. There is no real evidence of disagreement between the LASA scores and the other indicators of patient status. Substantial differences were also noticed between LASA

Table 2. *Reliability of Linear Analogue Scales*

Ranges for Correlation Coefficients	*# Variables*
[0.80,1.00]	11
[0.70,0.80)	13
[0.60,0.70)	5
[0.00,0.60)	3
	32

Table 3. *Comparison of Patient and Physician Scores*

Ranges for Correlation Coefficients	*# Variables*
[0.80,1.00]	5
[0.70,0.80)	13
[0.60,0.70)	6
[0.00,0.60)	8
	32

scores for metastatic and adjuvant chemotherapy patients.

The five factors defined by factor analysis are given in Table 5. In broad terms the factors may be described as covering the areas of activity level and mental alertness, physical status, emotional, digestive, and physical attributes. Table 6 contains the correlations for these factors between repeated measure of the LASA 9–12 hours apart. All the factors except for the digestive one, composed of nausea, vomiting, and eating, had good repeatability with correlations over 0.80.

Stable regressions were obtained with around 75 cases; the addition or removal of about 10 cases had only a minor affect on the results. The final model could be considered as including physical activity, anxiety, social life, appearance, depression, housework, speech, and diagnosis to recurrence time, and accounted for 70–84% of the variation. The LAPA scores were more precise than the LASA scores with stable regressions with about 45 cases, but only physical activity was in common with the LASA model (80% of the variation was accounted for by the LAPA model). There was no

Table 4. *Comparison of Linear Analogue Scales With Other Measures of Patient Status*

With Sickness Impact Profile:

Ranges for Correlation Coefficients	*# Variables*
[0.80,1.00]	1
[0.70,0.80)	3
[0.60,0.70)	2
[0.00,0.60)	1
	7

Common areas covered: overall score, mobility, housework, work, recreation, eating less, concentration vs. alertness behaviour.

With KARNOFSKY PERFORMANCE STATUS:
Correlation Coefficient with LASA UNI = 0.62

Table 5. *Factors Defined by Factor Analysis*[1]

Factor 1: Mobility, Physical Activity, Recreation, Fatigue, Social Life, Housework, Writing, Concentration

Factor 2: Pain, Physical Activity, Bowel Habit, Breathing

Factor 3: Depression, Anger, Anxiety, Appearance, Concentration, Family Relations

Factor 4: Nausea, Vomiting, Eating

Factor 5: Attractiveness, Family Relationships, Hair Loss

[1]All variables had factor loadings greater than 0.3.

evidence of substantial departures from normality in the P-P plots after the regression or in the other residual plots that the linear model is inadequate. One concern arose from the regressions; when the same individuals were

Table 6. *Comparison of Factors for Repeated Linear Analogues 9–12 Hours Apart*

Factor	*Correlation Coefficient*
1	0.86
2	0.86
3	0.81
4	0.42
5	0.88

involved in several repeated LASAs, the intervariable correlations increased markedly with the number of times that the LASA was repeated.

Table 7 contains the results of sample size observations. With 96 cases, all the correlation coefficients were significantly greater than 0 at the 1% level while with around 75 cases there appeared to be quite stable regression results.

Table 7. *Sample Size Observations*

RELIABILITY:	
# Cases	*Observation*
96[1]	all the correlation coefficients were significantly greater than 0 at the 1% level
60	5 variables had correlation coefficients *not* significantly greater than 0 at the 5% level
25	9 variables had correlation coefficients *not* significantly greater than 0 at the 5% level
REGRESSIONS:	
# Cases	*Observation*
75[1]	in 4 similar regressions 6 of 9 'final' subset variables are the same, 2 other variables are included in 3 of the final models
60	only 4 of 9 or 10 'final' subset variables are the same in various regressions

[1]Cases are subsets of the 115 being looked at here; all others are from other studies.

5. DISCUSSION

Linear analogue scales are simple to use and accepted readily by patients in a clinic setting. It appears that they produce quite repeatable results in directions that are supported by other measure of patient status.

The scales also yielded data sets with fairly stable multivariate models. It is quite reasonable that the digestive factor from factor analysis might have much lower repeatability than the others. It is also reasonable that the physician scores would be more precise given that they lack the interperson variability of the LASA scores and were completed by a healthy individual. The physician though would not likely have known the true impact of disease or treatment on a patient's whole lifestyle; this could explain the overlap of the observable physical activity in the patient and physician regression models. It seems important from a medical point of view that a patient's opinions of relevance of scale items be considered rather than the alternatives of uniform weighting, weighting by nondiseased experts or the weights by statistical association of impairment. Some caution would be advisable if many repeated measures are planned to follow treatment or disease progression as there was evidence that the dimensions may be collapsing. Further investigation is required to ascertain better the nature of this.

The linear analogues could be used to evaluate the patient needs on a group or individual basis with a view to providing assistance, especially with very toxic regimens. They could also be used to follow patient status and needs with time as disease progresses or with treatment as suggested by Priestman and Baum (1). However, caution would be urged, at least at present, regarding this latter use of the LASA scales discussed here, pending further clarification of the effects of repeated measures.

ACKNOWLEDGMENTS

The authors wish to acknowledge and thank Heather Sutherland for data collection. Joanne Campbell and Catherine Selby acted as the Research Officers for this study, and their skill and contribution was fundamental to its success.

This study was supported in part by a grant from the National Cancer Institute of Canada.

REFERENCES

Bergner, M., R. A. Bobbit, W. B. Carter, and B. S. Gilson (1981), "The sickness impact profile: development and final revision of a health status measure".

Medical Care **19**, 787–805.

Priestman, T. J., and M. Baum (1976), "Evaluation of quality of life in patients receiving treatment for advanced breast cancer". *Lancet* **1**, 899–901.

Selby, P. J., J. W. Chapman, J. Etezadi-Amoli, D. Dalley, and N. F. Boyd (1984), "The development of a method for assessing the quality of life of cancer patients". *British Journal of Cancer* **50**, 13–22.

D. Krewski [1 2], R. T. Smythe [3] and D. Colin[1]

TESTS FOR TREND IN BINOMIAL PROPORTIONS WITH HISTORICAL CONTROLS: A PROPOSED TWO-STAGE PROCEDURE

ABSTRACT

Historical control information may be used to strengthen inferences made concerning the carcinogenic potential of substances subjected to long term bioassay. Previously proposed tests which incorporate historical control information assume at least tacitly that the concurrent and historical control data arise from the same distribution. In this paper, we introduce a two-stage procedure which utilizes historical data only if the response rate in the concurrent control group falls within a suitable tolerance interval defined in terms of the historical control series. The asymptotic distribution theory required to use the procedure is presented and applied to an actual set of bioassay data for which historical control information is available.

1. INTRODUCTION

Long term bioassays with small rodents continue to be widely used to assess the carcinogenic potential of chemical substances (Bickis and Krewski, 1985a). These studies generally involve the administration of different levels of the test substance to groups of animals throughout the major portion of their lifespan, along with a similar group of unexposed animals comprising the concurrent controls. At the end of the study, the data are examined for the presence of a dose related increase in the occurrence of tumours among

[1] Health Protection Branch, Health and Welfare Canada, Ottawa, Ontario K1A 0L2

[2] Department of Mathematics and Statistics, Carleton University, Ottawa, Ontario K1S 5B6

[3] Department of Statistics, George Washington University, Washington, D.C. 20052

I. B. MacNeill and G. J. Umphrey (eds.), Biostatistics, 61–69.

the exposed animals relative to controls.

The use of historical control data in conjunction with the concurrent control data as a means of strengthening statistical tests was first proposed in the bioassay context by Tarone (1982). Since that time, related procedures have been proposed by Dempster *et al.* (1983), Hoel (1983), Krewski *et al.* (1985) and Smythe *et al.* (1986). A potential limitation shared by all of these procedures is their presumption that the concurrent and historical control data arise from the same distribution. Because of this, we propose to consider two-stage tests in which the historical control data is employed only if it appears to be consistent with the concurrent control results.

In Section 2, the Cochran-Armitage test for increasing trend in binomial proportions commonly applied in the absence of historical control data is reviewed. The incorporation of historical control information into this procedure in the spirit of Tarone (1982) and Krewski *et al.* (1985) is discussed in Section 3. The proposed two-stage procedure is then developed in Section 4. This procedure involves the construction of a suitable tolerance interval based on the historical control series, with the historical data used only if the response rate in the concurrent control group falls within this interval. (See Hoel and Yanagawa, 1986, for a related approach in the finite-sample case.) The application of the proposed procedure is illustrated in Section 5 using an actual set of bioassay data for which historical control information is also available.

2. TESTS FOR TREND IN BINOMIAL PROPORTIONS

Consider an experiment with $k+1$ dose levels $0 = d_0 < d_1 < \ldots < d_k$ in which x_i of the n_i animals at dose d_i respond ($i = 0, 1, \ldots, k$). We assume that x_i follows a binomial distribution $B(n_i, p_i)$ where the response probability $p_i = P(d_i)$, with $P(d)$ satisfying the logistic dose response model

$$P(d) = [1 + \exp\{-(a + bd)\}]^{-1} \tag{2.1}$$

($-\infty < a, b < +\infty$) for $d \geq 0$. The likelihood function for the unknown parameters a and b given the data $\mathbf{x} = (x_0, x_1, \ldots, x_k)$ is

$$L(a, b \mid \mathbf{x}) = \prod_{i=0}^{k} \binom{n_i}{x_i} p_i^{x_i} (1 - p_i)^{n_i - x_i}. \tag{2.2}$$

Treating a as a nuisance parameter, the score statistic for testing the null hypothesis $H_0 : b = 0$ versus the one-sided alternative $H_1 : b > 0$ is given by

$$\begin{aligned} T_{CA} &= \left.\frac{\partial \log L}{\partial b}\right|_{a=\hat{a},b=0} \\ &= \sum x_i d_i - \hat{p} \sum n_i d_i, \end{aligned} \tag{2.3}$$

where $\hat{p} = x/n$ is a consistent estimator of $p = p_0$ with $x = \sum x_i$ and $n = \sum n_i$, and $\hat{a} = -\log((1-\hat{p})/\hat{p})$ is the maximum likelihood estimator of a under $H_0 : b = 0$ (Tarone and Gart, 1980). The variance of this statistic is

$$\begin{aligned} n^{-1}V(T_{CA}) &= n^{-1}p(1-p)\{\sum n_i d_i^2 - (\sum n_i d_i)^2\} \\ &\sim p(1-p)\sigma_d^2. \end{aligned} \tag{2.4}$$

Here, $\sigma_d^2 = \sum \lambda_i (d_i - \bar{d})^2$ with $\bar{d} = \sum \lambda_i d_i$ and $\sim$ denotes asymptotic equivalence as $n \to \infty$ with $n_i/n \to \lambda_i > 0$.

The standardized test statistic $S_{CA} = T_{CA}/[V(T_{CA})]^{1/2}$ is commonly called the Cochran-Armitage test statistic after Cochran (1954) and Armitage (1955). Because the x_i are independent random variables, it follows from standard central limit arguments that

$$\lim_{n\to\infty} \Pr\{S_{CA} \le x\} = \Phi(x), \tag{2.5}$$

where Φ denotes the standard normal cumulative distribution function.

3. INCORPORATION OF HISTORICAL CONTROLS

To incorporate historical controls, we suppose that $p = [1 + \exp(-a)]^{-1}$ varies from experiment to experiment according to a beta distribution with density

$$f(p \mid \alpha, \beta) = \frac{\Gamma(\alpha+\beta)}{\Gamma(\alpha)\Gamma(\beta)} p^{\alpha-1}(1-p)^{\beta-1}, \tag{3.1}$$

for $0 < p < 1$ $(\alpha, \beta > 0)$. Then $a = -\log((1-p)/p))$ has density

$$g(a \mid \alpha, \beta) = \frac{\Gamma(\alpha+\beta)}{\Gamma(\alpha)\Gamma(\beta)} \left(\frac{1}{1+e^{-a}}\right)^{\alpha} \left(\frac{e^{-a}}{1+e^{-a}}\right)^{\beta}. \tag{3.2}$$

The marginal likelihood for b is then given by

$$L^*(b \mid \mathbf{x}; \alpha, \beta) \propto \int_{-\infty}^{\infty} g(a \mid \alpha, \beta) L(a, b \mid \mathbf{x}) da. \tag{3.3}$$

The score statistic for testing $H_0 : b = 0$ is now given by

$$\begin{aligned} T &= \frac{\partial \log L^*}{\partial b}\Big|_{b=0} \\ &= \sum x_i d_i - \tilde{p} \sum n_i d_i, \end{aligned} \tag{3.4}$$

where $\tilde{p} = (x + \alpha)/(n + \alpha + \beta)$ (Krewski *et al.*, 1985). Under H_0, $E(T) = 0$ and

$$\begin{aligned} n^{-1}V(T) &= n^{-1}\frac{\alpha\beta}{(\alpha+\beta)(\alpha+\beta+1)}\left\{\sum n_i d_i^2 - \frac{1}{(n+\alpha+\beta)}(\sum n_i d_i)^2\right\} \\ &\sim \frac{\alpha\beta}{(\alpha+\beta)(\alpha+\beta+1)}\sigma_d^2. \end{aligned} \tag{3.5}$$

The limiting null distribution of $S = T/[V(T)]^{1/2}$ is given by

$$\lim_{n\to\infty} \Pr\{S \leq x \mid \alpha, \beta\} = E_p \Phi(x/r_p), \tag{3.6}$$

where $r_p^2 = p(1-p)(\alpha\beta)^{-1}(\alpha+\beta)(\alpha+\beta+1)$. The asymptotic distribution of S is thus a mixture of normal distributions with mean zero and variance $E(r_p^2) = 1$. For applications, the critical value of S for testing $H_0 : b = 0$ is well approximated by that based on the standard normal distribution (see Krewski *et al.*, 1985, for further details).

The test for trend using historical control data based on the statistic S may be expected to perform well when the historical data and the concurrent controls arise from the same distribution. Because of systematic differences between experiments, however, it is possible that the concurrent control response probability may follow a beta distribution with parameters $(\alpha^*, \beta^*) \neq (\alpha, \beta)$. In particular, if $\alpha^*/(\alpha^* + \beta^*) > \alpha/(\alpha+\beta)$, the value of T in (3.4) will tend to be inflated thereby leading to an excessive Type I error rate.

4. A TWO-STAGE TEST WITH HISTORICAL CONTROLS

In order to protect against inflation of the Type I error rate in cases where the historical and concurrent controls do not follow the same distribution, we propose the use of a two-stage procedure in which the historical control data is used only if it appears to be consistent with the concurrent control data. Specifically, let $I_{n_0} = [k_1(n_0), k_2(n_0)]$ be a $100\gamma\%$ tolerance interval for x_0 satisfying

$$\sum_{x_0=k_1(n_0)}^{k_2(n_0)} \Pr(x_0) \geq \gamma \tag{4.1}$$

$(0 < \gamma < 1)$, where the marginal distribution of x_0 is specified by

$$\Pr(x_0) = \binom{n_0}{x_0} \frac{\Gamma(\alpha+\beta)}{\Gamma(\alpha)\Gamma(\beta)} \frac{\Gamma(x_0+\alpha)\Gamma(n_0-x_0+\beta)}{\Gamma(n_0+\alpha+\beta)} \tag{4.2}$$

$(x_0 = 0, 1, \ldots, n_0)$. The test statistic T^* is then selected as

$$T^* = \begin{cases} T & \text{if } x_0 \in I_{n_0} \\ T_{CA} & \text{otherwise} . \end{cases} \tag{4.3}$$

The mean and variance of x_0 based on (4.2) are given by

$$E(x_0) = n_0\theta \tag{4.4}$$

and

$$V(x_0) = n_0\{\theta(1-\theta) + (n_0 - 1)\Lambda\}, \tag{4.5}$$

where $\theta = \alpha/(\alpha+\beta)$ and $\Lambda = \alpha\beta(\alpha+\beta)^{-2}(\alpha+\beta+1)^{-1}$. The asymptotic distribution of $z = [x_0 - E(x_0)]/\{V(x_0)\}^{1/2}$, is given by

$$\lim_{n\to\infty} \Pr\{z \le y\} = \begin{cases} 1 & \text{if } z > (1-\theta)\Lambda^{-1/2}, \\ F(\theta + \Lambda^{1/2}y) & \text{if } -\theta\Lambda^{-1/2} \le z \le (1-\theta)\Lambda^{-1/2}, \\ 0 & \text{if } z < -\theta\Lambda^{-1/2}, \end{cases} \tag{4.6}$$

where F is the cumulative distribution of the beta density in (3.1). The limiting density of z is thus

$$\phi(z) = \Lambda^{1/2} f(\theta + \Lambda^{1/2} z). \tag{4.7}$$

In large samples, condition (4.1) may now be expressed as

$$\int_{c_1}^{c_2} \phi(z)dz = \gamma, \tag{4.8}$$

where $c_1 < 0$ and $c_2 > 0$ may be uniquely defined by setting

$$\int_{-\theta\Lambda^{-1/2}}^{c_1} \phi(z)dz = \int_{c_2}^{(1-\theta)\Lambda^{-1/2}} \phi(z)dz = (1-\gamma)/2. \tag{4.9}$$

The interval $I_{n_0} = [k_1(n_0), k_2(n_0)]$ is then approximated by $k_i \sim n_0(\theta + c_i\Lambda^{1/2})$.

To obtain an approximate critical region for the two-stage procedure, we note that under the null hypothesis,

$$\begin{aligned} \Pr\{\text{reject} H_0\} = {} & \Pr\{\text{reject} H_0 \mid T^* = T\}\Pr\{T^* = T\} \\ & + \Pr\{\text{reject} H_0 \mid T^* = T_{CA}\}\Pr\{T^* = T_{CA}\}. \end{aligned} \tag{4.10}$$

Under the assumption that $(\alpha^*, \beta^*) = (\alpha, \beta)$, $\Pr\{T^* = T\} = \gamma$ and $\Pr\{T^* = T_{CA}\} = 1 - \gamma$. Thus, in order to obtain an overall Type I error rate of δ $(0 < \delta < 1)$, the Type I error rates given $T^* = T$ and given $T^* = T_{CA}$ may both be set equal to δ.

Using arguments similar to those given by Krewski *et al.* (1985), the first term on the right hand side of (4.10) may be approximated by

$$\Pr\{S > s_0 \mid k_1(n_0) \leq x_0 \leq k_2(n_0)\} \sim \gamma^{-1} \int_{\theta+c_1\Lambda^{1/2}}^{\theta+c_2\Lambda^{1/2}} \Phi\left(\frac{-s_0}{r_p}\right) g(p)dp \tag{4.11}$$

where r_p^2 is defined in (3.6). The critical value s_0 for S conditional on $x_0 \in I_{n_0}$ is thus obtained by setting

$$\gamma^{-1} \int_{\theta+c_1\Lambda^{1/2}}^{\theta+c_2\Lambda^{1/2}} \Phi(-s_0/r_p)g(p)dp = \delta. \tag{4.12}$$

Similarly, the critical values for S_{CA} conditional on $x_0 \notin I_{n_0}$ are determined by setting

$$\Pr\{S_{CA} > s_1 \mid x_0 < k_1(n_0)\} \sim \Phi(-s_1) \tag{4.13}$$

and

$$\Pr\{S_{CA} > s_2 \mid x_0 > k_2(n_0)\} \sim \Phi(-s_2) \tag{4.14}$$

equal to δ

5. AN EXAMPLE

As an example of the application of the proposed procedure, consider the data given in Table 1 on the occurrence of alveolar-bronchiolar adenomas in mice taken from the data base examined by Bickis and Krewski (1985b). As shown in Table 2, the Cochran-Armitage test applied without the use of historical controls does not lead to a significant result at the nominal 5% level of significance. Application of the single-stage test with historical controls, on the other hand, yields a p-value of 0.017. Because the response rate $6/54 = 0.11$ in the concurrent control exceeds the mean response rat $\alpha/(\alpha+\beta) = 0.08$ in the historical controls, however, it is not clear which of these two tests is more appropriate.

The results of the two-stage test shown in Table 2 indicate that with $\gamma = 0.75$ or 0.90, the historical control data is used, with the p-values conditional on $x_0 \in I$ being only trivially larger than the unconditional p-value of 0.017. With $\gamma = 0.60$, however, $x_0/n_0 = 0.11 > 0.105 = k_2/n_0$ and the Cochran-Armitage statistic is used.

Table 1. *Occurrence of Alveolar-Bronchiolar Adenomas in Mice Following 102 Weeks on Test.*

(no. with tumours/no. at risk)

Historical Controls[a]

0/20	0/12	1/19	2/25	1/10	8/49
0/20	0/12	1/19	4/47	2/20	3/18
0/19	0/10	1/17	2/22	3/20	4/20
0/17	1/20	1/15	2/20	3/20	

Current Experiment

Dose:	0	1	2
x_i/n_i:	6/54	5/54	10/54

[a] $\alpha = 14.2$, $\beta = 160.0$

6. CONCLUSIONS

The use of historical control information can appreciably increase the sensitivity of tests for increasing trend in tumour occurrence with increasing dose. When the underlying response rate in the concurrent control group is substantially greater than the mean response rate in the historical controls, however, the Type I error rate of this procedure may be in excess of the nominal level. Because of this, a two-stage procedure is proposed in which the historical control data is used only if the concurrent control response rate falls within a suitable tolerance interval defined in terms of the historical control distribution.

Before this two-stage procedure can be recommended for routine use in practice, further study of its properties is required. In particular, the degree to which the Type I error rate of the single-stage test with historical controls may be inflated is currently under investigation. Also under study is the selection of a suitable tolerance level γ for use in defining the tolerance interval to be used in the first stage of the two-stage procedure.

Table 2. *Tests for Trend With and Without Historical Controls*

Historical Controls	Test Procedure	Test Statistic	p-value
No	Cochran-Armitage	$S_{CA} = 1.15$	0.126
Yes	One-Stage	$S = 2.13$	0.017
Yes	Two-Stage $\gamma = 0.90$; $\frac{x_0}{n_0} \in I = [0.047, 0.123]$	$S^* = S = 2.13$	0.018
	$\gamma = 0.75$; $\frac{x_0}{n_0} \in I = [0.054, 0.111]$	$S^* = S = 2.13$	0.019
	$\gamma = 0.60$; $\frac{x_0}{n_0} \notin I = [0.059, 0.1052]$	$S^* = S_{CA} = 1.15$	0.126

ACKNOWLEDGMENTS

This work was supported in part by grant no. A8664 from the Natural Sciences and Engineering Research Council to D. Krewski and grant no. DMS–8503774 from the National Science Foundation to R. T. Smythe.

REFERENCES

Armitage, P. (1955), "Tests for linear trend in proportions and frequencies". *Biometrics* **11**, 375–386.

Bickis, M., and D. Krewski (1985a), "Statistical design and analysis of the long-term carcinogenicity bioassay". In *Toxicological Risk Assessment I, Biological and Statistical Criteria*, ed. D. Clayson, D. Krewski, and I. Munro, pp. 125–147. Boca Raton, Florida: CRC Press.

Bickis, M., and D. Krewski (1985b), "Statistical issues in the analysis of the long term carcinogenicity bioassay in small rodents: an empirical evaluation of statistical decision rules". Technical Report No. 65, Laboratory for Research in Statistics and Probability, Carleton University.

Cochran, W. G. (1954), "Some methods of strengthening the common χ^2 tests".

Biometrics **10**, 417–451.

Dempster, A. P., M. D. Selwyn, and B. J. Weeks (1983), "Combining historical and randomized controls for assessing trends in proportions". *Journal of the American Statistical Association* **78**, 221–227.

Hoel, D. G. (1983), "Conditional two-sample tests with historical controls". In *Contributions to Statistics: Essays in Honor of Norman L. Johnson,* ed. P. K. Sen, pp. 229–236. Amsterdam: North-Holland.

Hoel, D. G., and I. Yanagawa (1986), "Incorporating historical controls in testing for a trend in proportions". Submitted.

Krewski, D., R. Smythe, and R. Burnett (1985), "The use of historical control information in testing for trend in quantal response carcinogenicity data". In *Proceedings of the Symposium on Long-Term Animal Carcinogenicity Studies: A Statistical Perspective*, pp. 56–62. Washington, D.C.: American Statistical Association.

Smythe, R., D. Krewski and D. Murdoch (1986), "The use of historical control information in modelling dose response relationship in carcinogenesis. *Statistics and Probability Letters*, **4**, 87–93.

Tarone, R. E. (1982), "The use of historical control information in testing for a trend in proportions". *Biometrics* **38**, 215–220.

Tarone, R. E., and J. J. Gart (1980), "On the robustness of combined tests for trend in proportions. *Journal of the American Statistical Association* **75**, 110–116.

S. D. Walter [1]

POINT ESTIMATION OF THE ODDS RATIO IN SPARSE 2×2 CONTINGENCY TABLES

ABSTRACT

The performance of a number of alternative point estimators of a single odds ratio and its logarithm is evaluated in defined sets of 2×2 contingency tables with small cell frequencies. The estimators considered are: four versions of the unconditional maximum likelihood estimators, modified to be defined when zero cells occur in the table; the conditional maximum likelihood estimator, similarly modified; and two pseudo-count or "shrinkage" estimators, proposed by Fienberg and Holland (1970).

The sets of contingency tables are generated by considering all of the possible outcome frequencies in the table which are consistent with a single fixed margin; the probability of each table in the set is derived as a product of binomial probabilities associated with the realisations in each row of the table, corresponding to the table having arisen through the drawing of two independent binomial samples with known probabilities of "success" in each. The mean, variance, bias, mean squared error and average absolute error of each of the estimators are obtained and compared. Contrary to previous simulation work in $2 \times 2 \times k$ tables, it is found that the unconditional MLE is usually preferred to the conditional MLE in single 2×2 tables. It is possible that this finding is due in part to the modifications used to make the estimators be defined when zero cell frequencies occur.

1. INTRODUCTION

We will consider the estimation of the odds ratio in a single 2×2 con-

[1] Department of Clinical Epidemiology and Biostatistics, McMaster University, Hamilton, Ontario L8N 3Z5

I. B. MacNeill and G. J. Umphrey (eds.), Biostatistics, 71–102.

tingency table, denoted as follows:

	# of Successes	*# of Failures*	*Sample Size*
Sample 1	a	b	n_1
Sample 2	c	d	n_2

We will suppose that such a table has been generated by drawing two independent binomial samples of size n_1 and n_2, with probabilities of "success" on each sample of p_1 and p_2 respectively.

The odds ratio in such a table is defined as $p_1(1-p_2)/[p_2(1-p_1)]$, and will be denoted by ρ. We will be concerned with the estimation of ρ, and also of its logarithm, $\beta = \ln(\rho)$. Previous literature suggests that there may be some advantage of working with β rather than ρ, because it tends to have a more symmetric and approximately normal sampling distribution. ρ has a range of 0 to ∞, with a null value of 1, whereas β has a range of $-\infty$ to ∞, with a null value of zero.

The odds ratio has been widely used as a measure of association in contingency tables, enjoying a number of advantages of estimability and interpretability not possessed by other indices. Indeed, the odds ratio is the fundamental parameter underlying the analysis of multidimensional contingency tables by log-linear models. The odds ratio is used frequently in epidemiology and other medical research, perhaps because it may be estimated in each of three major study designs, the prospective cohort study, the retrospective case-control study, and the cross-sectional survey. It has also been pointed out that the odds ratio forms a useful approximation to the relative risk in retrospective studies when the incidence of the disease in question is rare. The coefficients estimated in a logistic regression involving continuous and/or discrete independent variables can also be interpreted as log odds ratios.

Other useful properties of the odds ratio include:

(i) its value does not depend on which way round the variables are assigned to the rows and columns of the table; this makes it attractive when the causal directionality of the association is not well defined, and

(ii) an exchange of the rows or columns of the table results simply in an inversion of the odds ratio; this is because the odds ratio has a similar functional dependence on both the row and column margins.

The basic concepts of statistical variability in contingency table data are now widely appreciated by investigators in epidemiology and other branches of medical research. This is reflected by the almost uniform requirement of biomedical journals to report statistical findings of this kind with suitable tests of significance, and (to a lesser extent) with confidence intervals for estimated parameters, such as the odds ratio or relative risk. As methodologic support for these calculations, there now exists a considerable statistical literature concerned with testing and estimation of odds ratios in a variety of situations, especially when an odds ratio is to be evaluated over a set of 2×2 tables, with each table containing the data from within a stratum defined by confounding or nuisance variables. A brief review of various methods for confidence interval construction is given by Fleiss (1979).

Despite this, relatively little is known about the statistical properties of the various point estimators of the odds ratio which have been suggested. This is unfortunate, because indeed it is often the point estimate itself which is quoted as the single statistic which best summarises the data. For example, one might hear that the odds ratio (or relative risk) between smokers and non-smokers of contracting lung cancer is about 10, or that the relative risk of heart disease in a particular study is 1.5 in men as compared to women. Such statements are often expressed in the lay press, and to some extent by scientists, but they are frequently made without the benefit of associated confidence intervals or p-values. Indeed, often great stress is made of numerical comparisons between the relative risks reported in different studies, again in purely numerical rather than statistical terms.

For these reasons, it seems important to be able to provide an unbiased and precise point estimate of the odds ratio, even though such an estimate *should* be accompanied by appropriate tests and confidence intervals for thorough scientific reporting. It is the purpose of this paper to compare a number of alternative estimators, especially when the sample sizes are small, because it is in small samples where estimation problems are most acute.

2. REVIEW OF AVAILABLE ESTIMATORS

A simple estimator of ρ is given by the empirical cross-product ratio in the table: i.e.,

$$\hat{\rho}_{ML} = ad/(bc), \tag{1}$$

where ML denotes maximum likelihood. The corresponding estimate of the log odds ratio is defined by $\hat{\beta}_{ML} = \ln(\hat{\rho}_{ML})$. It should be noted that the maximum likelihood estimate (MLE) of p is a/n, and so by the invariance principle the MLE of the logit $p_1/(1-p_1)$ is a/b. Similarly the MLE of $p_2/(1-p_2)$ is c/d. Hence $\hat{\rho}_{ML}$ is also the MLE of ρ; in particular, we will refer to this as the unconditional MLE (the UMLE), because the only quantities regarded as fixed are the sample sizes n_1 and n_2. [It should be noted that $\hat{\rho}_{ML}$ is equal to the Mantel-Haenszel estimate of the odds ratio for single 2×2 tables.] In contrast, the conditional MLE (to be discussed below), also fixes the other marginal totals, $(a+c)$ and $(b+d)$.

If b or c are zero in a particular sample, $\hat{\rho}_{ML}$ is undefined. Similarly if *any* of the cell frequencies are zero, $\hat{\beta}_{ML}$ is undefined. To avoid this difficulty, it has been suggested that one add a positive constant ε to each cell, to create modified estimators of the form

$$\hat{\rho}_\varepsilon = (a+\varepsilon)(d+\varepsilon)/[(b+\varepsilon)(c+\varepsilon)] \tag{2}$$

with, correspondingly, $\hat{\beta}_\varepsilon = \ln(\hat{\rho}_\varepsilon)$. Haldane (1956), commenting on a paper by Woolf (1955), advocated using $\varepsilon = 1/2$, based on Taylor series expansion of $\ln(\hat{\rho})$ which shows that the first order bias term can be eliminated with this choice. It is interesting to note that this order of bias elimination can be achieved by using a value of ε which does not depend on p_1 or p_2. It has also been suggested that a similar modification be made to the expression for the standard errors of $\hat{\rho}_\varepsilon$ and $\hat{\beta}_\varepsilon$, so that these too would be defined even if a zero cell frequency occurred (Gart and Zweifel, 1967).

Hitchcock (1962) considered other values of ε, in the context of bioassay analysis using a logistic model, when there are more than two dose levels of the experimental stimulus. In particular, she found that using $\varepsilon = 1/2$ sometimes created more bias in the estimate of the logistic regression coefficient (or equivalently in the log odds ratio) than if the correction was not used at all. As a compromise, she suggested using $\varepsilon = 1/4$ when there are three dose levels. This suggestion was taken up by Hauck *et al.* (1982), who used this modification to define one of the estimators which they considered in sets of 2×2 tables.

In the context of testing the significance of the association within a 2×2 table, there has been considerable debate about whether it is necessary to condition on both margins of the table, as is done in the usual chi-square test. The row margins, n_1 and n_2, are essentially fixed by virtue of the binomial sampling design. The reason usually advanced for also conditioning on the second margin (the column totals) is that the margins do not contain information on the association, for example on the *difference* between the two outcome proportions p_1 and p_2; hence conditioning is a useful way of eliminating the effect of the nuisance parameter, which in this situation could be taken as the overall probability of success in the combined samples. In fact it has been shown that the margins *do* contain information on the association within the table. However, the amount of information thus conveyed is sufficiently small that conditioning on the second margin is generally considered to be appropriate. This issue has been considered at length by Yates (1984), and by discussants of his paper. Space does not permit a full discussion of all the issues here, but suffice it to say that the majority opinion appears to favour the conditional inference approach for significance testing in this context.

Returning to estimation, the MLE of ρ which *does* condition on the second margin of the table is defined by setting the observed value of one of the cell frequencies (say the "a" cell) equal to its expectation under the non-central hypergeometric distribution, whose non-centrality parameter is ρ; i.e.

$$a = E(a \mid n_1, n_2, a+c, b+d, \rho) \tag{3}$$

(see Breslow and Day, 1980; equation 4.5). Computation of this CMLE will in general involve a polynomial of high order in ρ, whose numerical solution would require an iterative approach. We should note that the CMLE is undefined if a zero frequency occurs in any cell of the table.

For completeness, we will mention some other odds ratio estimators which have been suggested, but which will not be considered further here, for a variety of reasons. The first, given in terms of β, is due to Birch (1964), and is defined by

$$\hat{\beta}_B = \frac{(ad - bc)(n_1 + n_2)(n_1 + n_2 - 1)}{n_1 n_2 (a+c)(b+d)}.$$

The corresponding estimator of ρ would be obtained simply by exponentiation. Birch's estimator was not considered in this paper because it is known to be inconsistent. (Hauck *et al.*, 1982)

Next, Tukey (cited by Anscombe, 1956) and Berkson (1953) have proposed estimators for a single logit, and which are "corrected" so that they are defined when zeros occur. It would be possible to extend their ideas to cover the situation of a logit difference (or equivalently a log odds ratio), but this was not pursued, largely on the basis of a rather poor showing in a comparison of the single logit estimators with Haldane's approach, by Gart and Zweifel (1967).

Finally, Schaefer (1983) has recently suggested a small sample bias correction for the estimated coefficients in a multiple logistic regression analysis. His corrected estimates are given by

$$\hat{\beta}_S = \hat{\beta}_{ML} + \frac{1}{2}(\mathbf{X}'\mathbf{V}\mathbf{X})^{-1}\mathbf{X}'\mathbf{V}\left\{(1 - 2\pi_k)x_k'(\mathbf{X}'\mathbf{V}\mathbf{X})^{-1}x_k\right\},$$

where $\mathbf{X}$ is the $(N \times p)$ design matrix, $\mathbf{V} = \text{diag}\{\pi_i(1-\pi_i)\}$, $\{a_k\}$ denotes a vector with elements a_k, and π_i is the response probability for the ith individual, being the usual multiple logistic function of the independent variables.

For the situation with exactly one dichotomous independent regression variable, the problem reduces to the estimation of a single log odds ratio; it may be shown (Walter, 1985) that the general form of Schaefer's correction produces a corrected odds ratio estimate equal to

$$\hat{\beta} = \ln\left(\frac{ad}{bc}\right) + \frac{1}{2}\left[\frac{b-a}{ab} + \frac{c-d}{cd}\right].$$

This estimator was also not taken further in this paper, because it is undefined when a zero cell occurs.

3. EXTENSIONS TO SETS OF 2 × 2 TABLES

Several of these estimation methods have been extended to cover the case of a $2 \times 2 \times k$ contingency table. Each 2×2 table within the set of k such tables is defined by a level of a covariate, which would usually be a confounding or nuisance variable. For example, in a study of the association of alcohol consumption and the occurrence of a certain cancer, the k strata might be formed by 5-year age groups of the study subjects. The objective is to compute an overall odds ratio for the combined data; the within-strata odds ratios are often regarded as constant, although tests for heterogeneity are possible (Mantel *et al.*, 1977). A number of comparisons have been made of the estimators for sets of 2×2 contingency tables. These have mostly been by computer simulation, using Monte Carlo methods. Although the present paper is restricted to a consideration of single 2×2 tables, the findings from investigations of multiple tables are obviously relevant.

Early discussion of a number of MLE's of the odds ratio is given by Gart, for both the unconditional sampling space (Gart, 1962) and for the conditional approach (Gart, 1970). Specific numerical examples are provided, but there is no systematic evaluation of the estimators over a variety of tables.

A more extensive investigation using Monte Carlo methods was carried out by McKinlay (1975), who had the objective of comparing matched and unmatched sampling strategies. The estimators considered by her were: an unstratified cross-product estimate (given by applying equation 1 to the marginal 2×2 table obtained by collapsing across the k strata), Woolf's estimate modified with 1/2 added to both the stratum-specific estimates and to their standard errors, the Mantel-Haenszel estimate (1959), Birch's estimate, and an estimate based on the pair-matching of the data. It was found that the pair-matched estimate was one of the least precise, and demonstrated considerable bias. Woolf's estimator was precise, but showed high bias, especially when there were a large number of strata so that the individual 2×2 tables were sparsely populated.

A second set of simulation results was described by McKinlay (1978). Here various sets of 2×2 tables were defined, including some where interaction was present, i.e., where the odds ratio varied across strata. The same estimators as above were used, plus a modification of Birch's estimate due to Goodman (1969), and an asymptotic conditional maximum likelihood estimate. In this case, however, the 1/2 correction to cell frequencies was not used, the stated reason being that "the overall frequencies were never zero"; although not explicitly stated, this implies a slightly different use of the Haldane correction, which would be applied only as necessary when zero cells occur, rather than all the time. The difference in the sampling distributions which is implied between correcting (i) as necessary, versus (ii) all the time, will tend to be less important in large samples (because of the relatively small effect of the ε correction when large cell frequencies are involved), but this may not be so in small samples. This suggests that the properties of the two estimators, defined with these alternative corrections, should be compared. Incidentally, the strategy of correcting for zero cells only as necessary is probably used often in practice; it is likely that many data analyses proceed using the uncorrected estimators until the problem of a zero actually occurs, at which time one of the "corrected" methods is hurriedly brought into play! It is of course difficult to document that this has occurred in actual study reports, because only the final version of the analysis is given.

Before examining her numerical situation results, McKinlay suggested that "the standard error of the [conditional] estimate of the odds ratio should be smaller than its unconditional counterpart". However, her results showed that the CMLE did *not* perform well in terms of bias or precision. The Mantel-Haenszel estimate (which is actually the first approximation in the iterative computation of the CMLE) did equally well, and this estimate was recommended by McKinlay overall, especially with large samples and when there are more than two strata. The Goodman and Birch estimates performed variably well. Woolf's estimate was the most precise, but was unstable in its mean value, especially when there were many strata; McKinlay recommended avoiding the Woolf method in this situation.

Breslow (1981) examined a slightly different situation, again by simulation. He supposed that the number of 2×2 tables increases as the sample size increases; this might occur, for instance, with pair-matched data, where the pairs are regarded as being uniquely matched, such as when sibship matching is used. The estimators used here were the Woolf, the UMLE and CMLE, and the Mantel-Haenszel estimate. The Woolf estimate was adapted for zeros in a way that was slightly different again from the Haldane and McKinlay methods, in that 1/2 was added only *as necessary,* and *only* to those cells with zero entries. It was found that this strategy reduced the bias in the

Woolf estimate for larger sample sizes, but aggravated its instability in small samples; overall Breslow did not recommend using the Woolf method. The best estimator, in his opinion, was the CMLE, but it was recognised that considerably more computation is required for this than the MH or UMLE methods. The MH estimate was always preferred to the UMLE; however the bias in the UNLE was somewhat "predictable", and so the method might be judiciously avoided for situations where one is alert to the possibility of bias and where such bias might be expected based on the simulation results.

Finally Hauck *et al.* (1982) reported simulation results for sets of 2×2 tables, using the UMLE and MH estimate with ε corrections of 1/4 and 1/2, and two new "pseudo-count" estimators. In the last two, data-dependent corrections are made to the cell frequencies before computing the odds ratio with one of the usual methods; in the context of multiple tables, these estimators correspond to shrinking towards tables of constant odds ratio, or to tables of independence. The CMLE was not included.

Yet another strategy was adopted by Hauck *et al.* to deal with zero cells which actually occurred; tables with any zero frequency cell were *dropped* altogether from the calculations for the UMLE when a correction was used. (For the UMLE with no ε correction, tables with zeros do not contribute to the overall likelihood.) The statistical properties of these corrected estimators were therefore calculated only over the (possibly) restricted set of tables where the estimator could be computed.

A rather complicated set of recommendations emerged from Hauck's study. The pseudo-count methods had small variances when the odds ratio was one, but elsewhere had much larger bias than the other estimators. Among the rest, the recommended choice of estimator depended on whether reducing bias or mean squared error is more important, and also on whether the odds ratio or its logarithm is the primary parameter to be estimated. The 1/2 and 1/4 corrected UMLE's showed underestimation of both the odds ratio and its logarithm, especially when the association in the table was strong. The UMLE, modified by the addition of 1/4 to each cell was suggested for the estimation of the log odds ratio, partly because this estimator showed less bias than the 1/2 corrected version. The Mantel-Haenszel method modified with $\varepsilon = 1/4$ and the UMLE method with $\varepsilon = 1/2$ were proposed for the odds ratio itself; the latter showed lower variance than the 1/4 corrected estimate, but it also had large bias when the odds ratio was large.

4. REPORTING OF ZERO CELLS IN DATA

As noted above, a number of different strategies have been used by different authors to deal with zero-celled tables when they arise. These may be summarised as follows:

Haldane	Add 1/2 always to cells of 2×2 table
Hitchcock	Add 1/4 always to cells of $2 \times k$ table for logistic regression, especially with $k = 3$.
McKinlay	Add 1/2 as necessary, i.e., to all cells of tables containing a zero frequency.
Breslow	Add 1/2 only to zero cells as they occur.
Hauck	Discard tables containing zeros from the analysis, when using ε corrections in $2 \times 2 \times k$ tables.

Hauck's approach is simple and pragmatic for *sets* of 2×2 tables to be combined, and will not lose very much of the information content in the data if only occasional zeros occur in a small proportion of the tables. However, it does not seem useful to extend this approach to situations where there is only *one* 2×2 table. For instance, suppose the following table occurred in a study to compare the frequency of adverse side effects in persons given one of two alternative vaccines:

	Side Effects	
	Yes	No
Vaccine A	10	990
Vaccine B	0	1000

Simply because a zero happened to occur in one of the cells would not be compelling reason to discard the data altogether. Such data could reasonably be reported using Fisher's exact test for the significance of the differences in the rate of side effects in the two groups, and exact one-sided confidence intervals for the odds ratio.

The occurrence of a zero may imply a qualitative difference in the results, as compared to a table where the cell frequencies are all non-zero. Consider the alternative table:

	Side effects	
	Yes	No
Vaccine A	10	990
Vaccine B	1	999

In this second table, the *existence* of a patient with a side effect in group B shows that side effects are possible in this group, whereas the zero in the first table is much more problematic. This would be an important distinction in tables arising from diagnostic testing for the presence of biochemical abnormalities or disease; certain laboratory tests are assumed by their nature to be incapable of giving a "false positive" result. For instance, a test for the presence of a viral antibodies might be assumed, for pathognomonic reasons, to give positive result *only* when infection has taken place; the demonstration of a positive result for some other reason would be a crucial finding.

For the first table above, one could compute and quote one of the odds ratio estimates which is defined even if a zero is present, for example the modified Woolf estimate with $\varepsilon = 1/2$. However, it would be most unwise for an investigator to report this figure without also reporting the fact that there were *no* side effects at all in the group with vaccine B; one might wish to compute the one-sided confidence interval for the rate of side effects in this group specifically, using the methods described by Hanley and Lippman-Hand (1983).

In the first table, the odds ratio estimate using Haldane's 1/2 correction is 21.2; in the second table, using the same method, the estimate is 7.06. This indicates the instability of the results in such a small sample, and the investigator would do well to report the whole table.

What is of concern in the present paper is not so much how a table with a zero might be reported, but how to incorporate the tables with zero cells into the entire distribution of possible outcome tables. When the probability of a zero occurring is not small, it seems inappropriate to simply discard the tables with zeros and work within the restricted and conditional distribution of the other tables. Therefore we will consider the behaviour of a number of estimators over the *entire* set of tables, including those with zeros. Recognising the likely strategy of many data analysts in practice, some of the estimators to be considered here will respond to zeros only as they arise; others will be computed in the same way for all tables, whether or not they contain zeros.

The issue of how to deal with zero cells has also been discussed in the context of constructing smoothed test statistics of hypotheses in log-linear models of larger contingency tables. The strategy of adding a positive constant only to zero counts was proposed by Grizzle *et al.* (1969), but Cox (1970) and Goodman (1970) have suggested always adding such a constant. In a Monte Carlo simulation, Bhapkar and Hosmane (1984) recommended the latter in the construction of an adjusted Wald statistic with optimal properties. Koopman (1984) has also recently proposed a 1/2 correction to deal with zeros in the estimation of the relative risk (p_1/p_2). Finally, Jewell (1984) has compared the small sample properties of the MLE, jackknife and other estimators of the odds ratio from matched data.

5. ESTIMATORS TO BE EVALUATED

With the rationale above in mind, the following estimators were chosen for comparison:

(i) $\hat{\rho}_1$, Woolf's estimator, defined as in equation (2) with $\varepsilon = 1/2$, following Haldane's modification of the uncorrected UMLE;

(ii) $\hat{\rho}_2$, a variant of $\hat{\rho}_1$, with the 1/2 correction added only when zeros occur in the table; specifically

$$\hat{\rho}_2 = \begin{cases} \hat{\rho}_1 & \text{if } abcd = 0 \\ \hat{\rho}_{ML} & \text{otherwise.} \end{cases}$$

(iii) $\hat{\rho}_3$, Hitchcock's estimator, defined as in equation (2), with $\varepsilon = 1/4$;

(iv) $\hat{\rho}_4$, a modified Hitchcock estimator with a correction of 1/4 used only when zero cells occur, and defined similarly to $\hat{\rho}_2$;

(v) $\hat{\rho}_5$, the CMLE defined in equation (3). As noted earlier, this estimator is undefined if a zero cell occurs. Therefore the following *ad hoc* modifications will be adopted, which is defined in all tables. First, if two zeros occur in the same column ($a + c = 0$ or $b + d = 0$) we define $\hat{\rho}_5 = 1.0$. Second, if $a = 0$ or $d = 0$ only (i.e., there is a single zero cell in the main diagonal), we define $\hat{\rho}_5$ as the solution to

$$a + 0.5 = E(a \mid n_1, n_2, a + c, b + d, \rho). \tag{4}$$

Finally, if $b = 0$ or $c = 0$, we define $\hat{\rho}_5$ as the solution to

$$a - 0.5 = E(a \mid n_1, n_2, a + c, b + d, \rho). \tag{5}$$

Lastly the two pseudocount estimators considered by Hauck *et al.* were also evaluated. These are defined below for the 2×2 table situation.

(vi) The first pseudocount estimator is $\hat{\rho}_6 = a'd'/(b'c')$, where a', b', c' and d' are modified cell frequencies, defined as follows. For the each cell, we replace its entry (Y, say) by

$$[Y + f/4]/[1 + f/4],$$

where $f = [N^2 - \| X \|^2]/[\| X \|^2 - N^2/4]$. Here $\| X \|^2 = a^2 + b^2 + c^2 + d^2$ unless $a = b = c = d$, in which case $\| X \|^2 = N^2/4$ and $f = N^2 - \| X \|^2$. In $2 \times 2 \times k$ tables, the generalisation of this estimator corresponds to shrinking the odds ratio estimates towards a constant value across the sub-tables. It should be noted that the marginal totals are *not* preserved using this estimator.

(vii) The second pseudo-count estimator is denoted by $\hat{\rho}_7$. In order to compute it, we first calculate the modified frequency a', defined as

$$a' = \left[a + f' n_1 (a + c)/N^2\right] / \left[1 + f'/N\right],$$

where

$$f' = \begin{cases} [N^2 - \| X \|^2]/[4(a-c)^2] & \text{if } a \neq c, \\ (N^2 - \| X \|^2)/2 & \text{if } a = c. \end{cases}$$

Then the other modified frequencies are computed as $b' = n_1 - a'$, $c' = a + c - a'$, and $d' = n_2 - c'$; thus the original margins are preserved. $\hat{\rho}_7$ is then computed as $a'd'/(b'c')$. In $2 \times 2 \times k$ tables, this estimator shrinks the unadjusted empirical estimate back towards the null value of one, corresponding to independence.

Both pseudo-count estimators ($\hat{\rho}_6$ and $\hat{\rho}_7$) are defined if there is exactly one zero cell or a zero diagonal. If there is a zero column, $\hat{\rho}_6$ is defined, with the value 1 if $n_1 = n_2$. The calculation of the pseudocounts for $\hat{\rho}_7$ with a zero column leaves the table unchanged. As for $\hat{\rho}_5$, we again assigned the value 1 for $\hat{\rho}_7$ if this occurred.

6. EXAMPLES

In the first example, shown in Table 1, the cell frequencies are small, and we see considerable variation between the seven estimators. $\hat{\rho}_2$, which in this case amounts to the uncorrected UMLE, is the largest, and $\hat{\rho}_6$ is the smallest. The difference between the 1/2 correction, the 1/4 correction, and no correction is quite marked.

In the second example, shown in Table 2, the total sample size has increased from 10 to 100. The variation between estimates is now much smaller, especially between the various corrected UMLE's ($\hat{\rho}_1$ to $\hat{\rho}_4$). $\hat{\rho}_2$ is again the largest and $\hat{\rho}_6$ the smallest.

The final example (see Table 3) is taken from a paper published in the New England Journal of Medicine (Whitley *et al.*, 1977), concerning the association of survival with brain biopsy status in a series of herpes simplex virus patients. Although the sample size is moderate in total, the presence of a single small cell (the "1") leads again to some disparity in the estimates.

Table 1. *Example of Variation Between Estimates in a Small Sample*

	# of Successes	# of Failures	Sample Size
Sample 1	3	2	5
Sample 2	1	4	5

$\hat{\rho}_1 = 4.200$; $\hat{\rho}_2 = 6.000$; $\hat{\rho}_3 = 4.911$; $\hat{\rho}_4 = 6.000$;
$\hat{\rho}_5 = 4.918$; $\hat{\rho}_6 = 1.970$; $\hat{\rho}_7 = 3.315$.

Table 2. *Example of Variation Between Estimates in a Moderate Sample*

	# of Successes	# of Failures	Sample Size
Sample 1	35	15	50
Sample 2	25	25	50

$\hat{\rho}_1 = 2.290$; $\hat{\rho}_2 = 2.333$; $\hat{\rho}_3 = 2.312$; $\hat{\rho}_4 = 2.333$;
$\hat{\rho}_5 = 2.313$; $\hat{\rho}_6 = 1.829$; $\hat{\rho}_7 = 2.040$.

Depending on the estimate chosen, we would conclude from a doubling to almost a quadrupling in the odds of death for biopsy positive patients as compared to biopsy negative patients.

7. NUMERICAL EVALUATION OF ESTIMATORS

The numerical comparison of the estimates was made over a series of 2×2 contingency tables, where the sample sizes n_1 and n_2 were fixed, and the outcome probabilities p_1 and p_2 were taken as known. Sets of tables with a range of equal binomial sample sizes were used, with $n_1 = n_2 = n = 5, 10, 25$ and 50. A number of other enumerations were also made with unequal binomial sample sizes, but the results were not materially different.

Table 3. *Mortality in Herpes Simplex Patients Treated with Adenine Arabinoside, by Brain Biopsy Status.*

	Dead	Alive	Total
Biopsy Positive	5	13	18
Biopsy Negative	1	10	11

$\hat{\rho}_1 = 2.852$; $\hat{\rho}_2 = 3.846$; $\hat{\rho}_3 = 3.429$; $\hat{\rho}_4 = 3.846$;
$\hat{\rho}_5 = 3.692$; $\hat{\rho}_6 = 2.012$; $\hat{\rho}_7 = 2.668$.

Second, the outcome probabilities were defined as 0.1, 0.3, 0.5, and 0.7, with p_1 being taken as the greater, so that tables with "positive" associations were included. These values yield odds ratios ranging from 1 to 21, thus covering most practical situations. Note that the results for the log odds ratio in tables with higher values of p_1 and p_2 may be obtained by symmetry.

For each estimator, the mean, variance, bias, mean squared error (MSE), and average absolute error (AAE) were computed over the set of all possible tables with the given margins n_1 and n_2. The value for each table-specific estimate was weighted by the (unconditional) probability of the table's occurrence. As noted earlier, this evaluation was restricted to estimators which are defined for all tables, so tables with zero cells or columns were included and weighted appropriately. It should be emphasised that this represents a complete numerical evaluation of the sampling distribution of each estimator, rather than a simulation experiment.

All calculations were made using a FORTRAN program, with the MicroSoft compiler (version 3.2), running under MS-DOS 2.10 and using double precision throughout. The program was executed on an IBM-XT microcomputer, fitted with an 8087 math co-processor.

8. RESULTS

Table 4 shows the results for all the sample sizes and p-values considered. We will discuss some typical results, first for $n_1 = n_2 = 50$. Among the modified MLE's for ρ ($\hat{\rho}_1$ to $\hat{\rho}_5$), the Woolf estimator is uniformly best on bias, MSE and AEE; also the log equivalent, $\hat{\beta}_1$ is the best of this group. In particular, the CMLE's appear to be inferior to the Woolf estimator everywhere, although some of the differences in performance are small numerically.

For the odds ratio itself, the pseudo-count estimators $\hat{\rho}_6$ and $\hat{\rho}_7$ are superior to $\hat{\rho}_1$ for a good many problems. However, $\hat{\rho}_7$ tends to have very high MSE on occasion, particularly when the contingency table shows a strong association. (See, for example, the configuration $p_1 = 0.7$, $p_2 = 0.1$.)

We will now turn to the results for $n_1 = n_2 = 10$, considering first the estimation of the log odds ratio. As before, $\hat{\beta}_1$ is almost always the least biased estimate; $\hat{\beta}_2$ did perform slightly better for a few problems where p_2 was small. Of the first 5 estimators, $\hat{\beta}_1$ also performed well on MSE and AAE, although the CMLE $\hat{\beta}_5$ did slightly better in a number of problems with low values of p_2. However, $\hat{\beta}_6$ had the lowest MSE and AAE overall in most problems with moderate or high p_2.

For the odds ratio, $\hat{\rho}_6$ performed well overall, and either it or $\hat{\rho}_5$ were the least biased for all problems. The Hitchcock estimators $\hat{\rho}_3$ and $\hat{\rho}_4$ did poorly, with high bias and/or MSE.

Comparing the Woolf and CML estimators, the latter [$\hat{\rho}_5$] was preferred in all respects when p_2 was small; otherwise the results were mixed. $\hat{\rho}_7$ again showed very high MSE for some problems, and occasional high bias, especially when the odds ratio was large.

Some but not all of these trends were repeated in the results for the smallest samples considered, with $n_1 = n_2 = 5$. Comparing the unconditional and conditional MLE's $\hat{\beta}_1$ and $\hat{\beta}_5$, the Woolf estimate was generally better on bias, and it also performed quite well on MSE and AAE, although there was no clear "winner" in this respect.

Indeed for some problems, the ranking of several of the estimators on the basis of the MSE criterion was the reverse of the ranking for bias. Consider, for instance, the results when $p_1 = 0.5$ and $p_2 = 0.3$, for the estimators $\hat{\beta}_1$, $\hat{\beta}_5$, $\hat{\beta}_6$ and $\hat{\beta}_7$. On the criterion of least bias, the preference would be for estimator 7, followed by 1, 5 and 6. However, on the basis of MSE, the rank order would be 6, 5, 1, and 7, exactly the reverse!

As in earlier results, the Hitchcock estimators $\hat{\beta}_3$ and $\hat{\beta}_4$ appeared to be inferior in most cases to the Woolf estimators $\hat{\beta}_1$ and $\hat{\beta}_2$. For the odds ratio, $\hat{\rho}_3$ and $\hat{\rho}_4$ again performed poorly and are not recommended.

The CMLE $\hat{\rho}_5$ did well, with the lowest MSE everywhere; also it often had the lowest AAE, with $\hat{\rho}_6$ sometimes being better. The UMLE $\hat{\rho}_1$ was less biased than $\hat{\rho}_5$ for a number of problems with $p_2 = 0.1$. [This was the reverse of the finding in larger samples, where the UML approach was better with larger values of p_2.]

Table 4. *Results of Complete Enumeration of Alternative Odds Ratio Estimators in a Series of Fourfold Tables (Next 12 Pages)*

Definitions of Estimators Used:

1 – Unconditional MLE (add 1/2 always) – [Haldane-Woolf]
2 – Unconditional MLE (add 1/2 if zeros)
3 – Unconditional MLE (add 1/4 always) – [Hitchcock]
4 – Unconditional MLE (add 1/4 if zeros)
5 – Conditional MLE
6 – Pseudocount # 1 (shrinkage to constant)
7 – Pseudocount # 2 (shrinkage to independence)

Abbreviations:

EST – Estimator
n_1, n_2 – sample sizes
p_1, p_2 – binomial outcome probabilities
OR – odds ratio
L-OR – log odds ratio
VAR – variance
MSE – mean squared error
AAE – average absolute error

Table 4 (continued)

Total Sample Size = 100, $n_1 = n_2 = 50$

		STATISTICS FOR ODDS RATIO					STATISTICS FOR LOG ODDS RATIO				
	EST	MEAN	VAR	BIAS	MSE	AAE	MEAN	VAR	BIAS	MSE	AAE
$p_1 = .10$	1	1.295	1.655	.295	1.742	.655	.000	.490	.000	.490	.532
	2	1.379	2.241	.379	2.384	.765	.000	.614	.000	.614	.594
$p_2 = .10$	3	1.389	4.292	.389	4.443	.762	.000	.581	.000	.581	.567
	4	1.437	4.652	.437	4.843	.824	.000	.650	.000	.650	.601
OR = 1.0000	5	1.359	1.912	.359	2.040	.742	.000	.592	.000	.592	.586
	6	1.241	1.042	.241	1.100	.585	.000	.418	.000	.418	.494
L-OR = .0000	7	1.106	.479	.106	.490	.289	.000	.184	.000	.184	.250
$p_1 = .30$	1	4.712	18.375	.855	19.105	2.148	1.352	.341	.002	.341	.452
	2	5.336	26.286	1.479	28.473	2.640	1.437	.407	.087	.415	.492
$p_2 = .10$	3	5.215	52.582	1.358	54.426	2.585	1.396	.389	.046	.391	.473
	4	5.563	57.452	1.706	60.364	2.868	1.440	.426	.090	.434	.495
OR = 3.8571	5	5.223	23.397	1.366	25.264	2.554	1.423	.398	.073	.404	.487
	6	3.888	6.284	.031	6.284	1.631	1.215	.264	−.135	.282	.423
L-OR = 1.3499	7	4.194	20.678	.337	20.791	2.146	1.199	.394	−.150	.417	.513
$p_1 = .50$	1	10.921	96.785	1.921	100.477	4.900	2.199	.325	.002	.325	.441
	2	12.597	141.558	3.597	154.494	6.140	2.304	.388	.107	.400	.480
$p_2 = .10$	3	12.203	283.735	3.203	293.995	5.961	2.253	.372	.056	.375	.462
	4	13.134	311.326	4.134	328.417	6.677	2.308	.407	.111	.419	.484
OR = 9.0000	5	12.206	124.703	3.206	134.984	5.871	2.280	.379	.082	.386	.474
	6	8.251	28.580	−.749	29.140	3.759	1.967	.264	−.231	.318	.457
L-OR = 2.1972	7	11.223	223.255	2.223	228.197	5.551	2.171	.384	−.026	.385	.477

Table 4 (continued)

Total Sample Size = 100, $n_1 = n_2 = 50$

		STATISTICS FOR ODDS RATIO					STATISTICS FOR LOG ODDS RATIO				
	EST	MEAN	VAR	BIAS	MSE	AAE	MEAN	VAR	BIAS	MSE	AAE
$p_1 = .70$	1	25.731	568.697	4.731	591.077	11.827	3.047	.341	.002	.341	.453
	2	30.302	859.464	9.302	945.995	15.127	3.172	.406	.127	.423	.494
$p_2 = .10$	3	29.060	1679.503	8.060	1744.462	14.512	3.111	.389	.066	.394	.474
	4	31.596	1859.030	10.596	1971.296	16.420	3.175	.425	.131	.442	.498
OR = 21.0000	5	28.745	712.883	7.745	772.865	14.053	3.129	.391	.085	.399	.484
	6	20.279	240.827	−.721	241.347	9.790	2.832	.321	−.212	.366	.488
L-OR = 3.0445	7	29.060	2505.782	8.060	2570.751	15.066	3.082	.412	.037	.413	.486
$p_1 = .30$	1	1.101	.262	.101	.273	.370	.000	.192	.000	.192	.346
	2	1.108	.285	.108	.297	.383	.000	.204	.000	.204	.357
$p_2 = .30$	3	1.104	.274	.104	.285	.376	.000	.198	.000	.198	.351
	4	1.106	.286	.108	.297	.383	.000	.204	.000	.204	.357
OR = 1.0000	5	1.106	.278	.106	.289	.379	.000	.200	.000	.200	.353
	6	1.071	.173	.071	.178	.302	.000	.136	.000	.136	.288
L-OR = .0000	7	1.054	.142	.054	.145	.216	.000	.103	.000	.103	.204
$p_1 = .50$	1	2.552	1.340	.219	1.388	.834	.847	.176	.000	.176	.334
	2	2.617	1.507	.284	1.588	.876	.868	.186	.020	.186	.342
$p_2 = .30$	3	2.584	1.420	.251	1.483	.854	.857	.181	.010	.181	.338
	4	2.617	1.508	.284	1.589	.876	.868	.186	.020	.186	.342
OR = 2.3333	5	2.588	1.435	.255	1.500	.857	.859	.181	.011	.182	.338
	6	2.062	.809	−.271	.882	.747	.646	.147	−.201	.187	.358
L-OR = .8473	7	2.316	1.302	−.017	1.302	.846	.738	.195	−.109	.207	.372

Table 4 (continued)

Total Sample Size $= 100, n_1 = n_2 = 50$

	EST	STATISTICS FOR ODDS RATIO MEAN	VAR	BIAS	MSE	AAE	STATISTICS FOR LOG ODDS RATIO MEAN	VAR	BIAS	MSE	AAE
$p_1 = .70$	1	6.014	8.580	.569	8.904	2.059	1.695	.192	.000	.192	.348
	2	6.302	10.258	.858	10.994	2.220	1.735	.204	.041	.206	.359
$p_2 = .30$	3	6.154	9.373	.710	9.877	2.134	1.715	.198	.020	.198	.353
	4	6.302	10.265	.858	11.001	2.220	1.735	.204	.041	.206	.359
OR = 5.4444	5	6.158	9.413	.713	9.922	2.137	1.715	.198	.021	.199	.353
	6	4.803	6.265	−.642	6.677	2.011	1.456	.220	−.238	.276	.425
L-OR = 1.6946	7	5.818	9.079	.373	9.218	2.110	1.650	.216	−.045	.218	.371
$p_1 = .50$	1	1.083	.205	.083	.212	.335	.000	.160	.000	.160	.317
	2	1.087	.216	.087	.224	.343	.000	.167	.000	.167	.324
$p_2 = .50$	3	1.085	.211	.085	.218	.339	.000	.163	.000	.163	.321
	4	1.087	.216	.087	.224	.343	.000	.167	.000	.167	.324
OR = 1.0000	5	1.085	.211	.085	.218	.339	.000	.163	.000	.163	.321
	6	1.025	.057	.025	.058	.143	.000	.048	.000	.048	.139
L-OR = .0000	7	1.047	.116	.047	.118	.203	.000	.089	.000	.089	.193
$p_1 = .70$	1	2.552	1.340	.219	1.388	.834	.847	.176	.000	.176	.334
	2	2.617	1.507	.284	1.588	.876	.868	.186	.020	.186	.342
$p_2 = .50$	3	2.584	1.420	.251	1.483	.854	.857	.181	.010	.181	.338
	4	2.617	1.508	.284	1.589	.876	.868	.186	.020	.186	.342
OR = 2.3333	5	2.588	1.435	.255	1.500	.857	.859	.181	.011	.182	.338
	6	2.062	.809	−.271	.882	.747	.646	.147	−.201	.187	.358
L-OR = .8473	7	2.316	1.302	−.017	1.302	.846	.738	.195	−.109	.207	.372

Table 4 (continued)

Total Sample Size = 50, $n_1 = n_2 = 25$

		STATISTICS FOR ODDS RATIO					STATISTICS FOR LOG ODDS RATIO				
	EST	MEAN	VAR	BIAS	MSE	AAE	MEAN	VAR	BIAS	MSE	AAE
$p_1 = .10$	1	1.608	4.178	.608	4.548	1.041	.000	.930	.000	.930	.729
	2	1.739	4.827	.739	5.373	1.212	.000	1.132	.000	1.132	.830
$p_2 = .10$	3	2.073	14.401	1.073	15.553	1.535	.000	1.363	.000	1.363	.852
	4	2.152	14.752	1.152	16.078	1.635	.000	1.482	.000	1.482	.909
OR = 1.0000	5	1.565	2.971	.565	3.290	1.006	.000	.913	.000	.913	.741
	6	1.419	1.966	.419	2.141	.820	.000	.697	.000	.697	.636
L-OR = .0000	7	1.240	1.543	.240	1.600	.482	.000	.382	.000	.382	.369
$p_1 = .30$	1	5.575	41.389	1.718	44.341	3.372	1.344	.660	−.006	.660	.631
	2	6.539	47.909	2.681	55.099	4.136	1.484	.759	.134	.777	.699
$p_2 = .10$	3	7.610	159.752	3.753	173.839	5.303	1.455	.891	.105	.902	.705
	4	8.172	162.337	4.315	180.957	5.770	1.531	.942	.181	.975	.747
OR = 3.8571	5	6.028	34.141	2.171	38.854	3.666	1.442	.697	.092	.705	.672
	6	3.795	8.859	−.062	8.862	2.060	1.108	.427	−.242	.486	.565
L-OR = 1.3499	7	5.093	64.846	1.236	66.373	3.563	1.143	.754	−.207	.796	.723
$p_1 = .50$	1	12.865	218.472	3.865	233.412	7.710	2.192	.625	−.005	.625	.617
	2	15.529	256.824	6.529	299.456	9.647	2.372	.702	.175	.732	.678
$p_2 = .10$	3	17.898	876.978	8.898	956.156	12.376	2.324	.849	.127	.865	.688
	4	19.448	891.302	10.448	1000.455	13.565	2.421	.886	.224	.936	.727
OR = 9.0000	5	14.150	181.312	5.150	207.833	8.465	2.313	.648	.116	.662	.652
	6	7.735	45.279	−1.265	46.879	4.857	1.802	.447	−.395	.603	.643
L-OR = 2.1972	7	15.621	829.799	6.621	873.634	11.060	2.163	.889	−.034	.890	.721

Table 4 (continued)

Total Sample Size = 50, $n_1 = n_2 = 25$

		STATISTICS FOR ODDS RATIO					STATISTICS FOR LOG ODDS RATIO				
	EST	MEAN	VAR	BIAS	MSE	AAE	MEAN	VAR	BIAS	MSE	AAE
$p_1 = .70$	1	30.813	1523.563	9.813	1619.849	19.007	3.040	.660	−.004	.660	.636
	2	38.725	1947.049	17.725	2261.216	24.765	3.260	.740	.215	.786	.701
$p_2 = .10$	3	44.098	7202.496	23.098	7735.997	31.142	3.194	.891	.149	.913	.707
	4	48.720	7408.240	27.720	8176.649	34.760	3.311	.928	.266	.999	.752
OR = 21.0000	5	33.580	1115.260	12.580	1273.521	20.395	3.165	.667	.121	.681	.663
	6	20.238	652.836	−.762	653.418	13.656	2.663	.602	−.381	.747	.708
L-OR = 3.0445	7	48.769	28897.242	27.769	29668.349	36.812	3.149	1.020	.104	1.031	.749
$p_1 = .30$	1	1.219	.847	.219	.896	.563	.000	.390	.000	.390	.488
	2	1.257	1.124	.257	1.189	.615	.000	.447	.000	.447	.520
$p_2 = .30$	3	1.239	1.213	.239	1.270	.590	.000	.418	.000	.418	.503
	4	1.260	1.375	.260	1.442	.618	.000	.448	.000	.448	.520
OR = 1.0000	5	1.245	1.017	.245	1.077	.598	.000	.428	.000	.428	.509
	6	1.122	.377	.122	.392	.392	.000	.226	.000	.226	.357
L-OR = .0000	7	1.126	.592	.126	.608	.342	.000	.222	.000	.222	.297
$p_1 = .50$	1	2.814	4.451	.480	4.681	1.277	.848	.356	.001	.356	.471
	2	3.015	6.333	.681	6.797	1.443	.891	.398	.044	.400	.497
$p_2 = .30$	3	2.915	6.623	.582	6.961	1.360	.869	.376	.022	.377	.483
	4	3.022	7.732	.689	8.206	1.450	.891	.399	.044	.401	.497
OR = 2.3333	5	2.928	5.451	.594	5.804	1.371	.872	.380	.025	.381	.486
	6	2.039	1.844	−.294	1.931	.991	.575	.240	−.272	.314	.473
L-OR = .8473	7	2.488	5.695	.154	5.718	1.257	.707	.355	−.141	.375	.508

Table 4 (continued)

Total Sample Size = 50, $n_1 = n_2 = 25$

		STATISTICS FOR ODDS RATIO					STATISTICS FOR LOG ODDS RATIO				
	EST	MEAN	VAR	BIAS	MSE	AAE	MEAN	VAR	BIAS	MSE	AAE
$p_1 = .70$	1	6.738	34.238	1.294	35.912	3.230	1.696	.390	.002	.390	.493
	2	7.622	57.202	2.177	61.942	3.894	1.783	.446	.088	.454	.525
$p_2 = .30$	3	7.182	63.933	1.737	66.951	3.559	1.738	.418	.044	.420	.507
	4	7.659	78.067	2.214	82.970	3.931	1.783	.448	.088	.456	.525
OR = 5.4444	5	7.180	42.898	1.736	45.911	3.555	1.740	.420	.046	.422	.509
	6	4.711	18.170	−.733	18.708	2.836	1.320	.417	−.375	.557	.615
L-OR = 1.6946	7	6.580	116.776	1.136	118.066	3.502	1.616	.476	−.078	.482	.552
$p_1 = .50$	1	1.175	.540	.175	.571	.499	.000	.321	.000	.321	.446
	2	1.192	.623	.192	.659	.526	.000	.350	.000	.350	.466
$p_2 = .50$	3	1.183	.579	.183	.613	.512	.000	.335	.000	.335	.456
	4	1.192	.623	.192	.660	.526	.000	.350	.000	.350	.466
OR = 1.0000	5	1.183	.581	.183	.615	.513	.000	.335	.000	.335	.456
	6	1.053	.148	.053	.150	.212	.000	.099	.000	.099	.199
L-OR = .0000	7	1.103	.332	.103	.343	.310	.000	.188	.000	.188	.279
$p_1 = .70$	1	2.814	4.451	.480	4.681	1.277	.848	.356	.001	.356	.471
	2	3.015	6.333	.681	6.797	1.443	.891	.398	.044	.400	.497
$p_2 = .50$	3	2.915	6.623	.582	6.961	1.360	.869	.376	.022	.377	.483
	4	3.022	7.732	.689	8.206	1.450	.891	.399	.044	.401	.497
OR = 2.3333	5	2.928	5.451	.594	5.804	1.371	.872	.380	.025	.381	.486
	6	2.039	1.844	−.294	1.931	.991	.575	.240	−.272	.314	.473
L-OR = .8473	7	2.488	5.695	.154	5.718	1.257	.707	.355	−.141	.375	.508

Table 4 (continued)

Total Sample Size = 20, $n_1 = n_2 = 10$

		STATISTICS FOR ODDS RATIO					STATISTICS FOR LOG ODDS RATIO				
	EST	MEAN	VAR	BIAS	MSE	AAE	MEAN	VAR	BIAS	MSE	AAE
$p_1 = .10$	1	1.808	4.786	.808	5.439	1.273	.000	1.254	.000	1.254	.858
	2	1.859	4.960	.859	5.697	1.342	.000	1.337	.000	1.337	.903
$p_2 = .10$	3	2.680	17.363	1.680	20.185	2.198	.000	2.197	.000	2.197	1.127
	4	2.710	17.425	1.710	20.350	2.239	.000	2.247	.000	2.247	1.153
OR = 1.0000	5	1.330	1.516	.330	1.626	.598	.000	.555	.000	.555	.454
	6	1.443	1.628	.443	1.824	.852	.000	.769	.000	.769	.675
L-OR = .0000	7	1.463	3.759	.463	3.974	.790	.000	.694	.000	.694	.546
$p_1 = .30$	1	5.641	40.883	1.784	44.065	3.996	1.212	1.124	−.138	1.143	.868
	2	6.143	42.105	2.285	47.328	4.187	1.309	1.189	−.041	1.190	.868
$p_2 = .10$	3	9.371	174.528	5.513	204.926	7.554	1.454	1.740	.104	1.751	1.074
	4	9.670	173.425	5.813	207.212	7.672	1.509	1.759	.159	1.785	1.075
OR = 3.8571	5	4.418	17.422	.561	17.737	2.841	1.090	.873	−.260	.941	.772
	6	3.221	8.311	−.636	8.715	2.186	.872	.617	−.478	.846	.746
L-OR = 1.3499	7	6.202	228.781	2.345	234.280	5.182	1.069	1.263	−.281	1.342	.980
$p_1 = .50$	1	13.075	287.898	4.075	304.503	9.205	2.062	1.035	−.136	1.053	.838
	2	14.813	301.860	5.813	335.652	9.791	2.234	1.004	.037	1.006	.800
$p_2 = .10$	3	23.206	2001.254	14.206	2203.059	18.753	2.372	1.564	.175	1.595	1.039
	4	24.253	1993.632	15.253	2226.292	19.129	2.468	1.510	.270	1.583	1.019
OR = 9.0000	5	10.540	92.628	1.540	94.998	6.323	2.003	.774	−.194	.812	.705
	6	6.786	133.169	−2.214	138.072	5.934	1.495	.728	−.702	1.221	.912
L-OR = 2.1972	7	23.687	16571.110	14.687	16786.830	20.578	2.095	1.799	−.103	1.810	1.105

Table 4 (continued)

Total Sample Size = 20, $n_1 = n_2 = 10$

		STATISTICS FOR ODDS RATIO					STATISTICS FOR LOG ODDS RATIO				
	EST	MEAN	VAR	BIAS	MSE	AAE	MEAN	VAR	BIAS	MSE	AAE
$p_1 = .70$	1	33.286	2634.130	12.286	2785.086	24.417	2.911	1.124	−.133	1.142	.864
	2	38.562	2628.644	17.562	2937.083	26.272	3.150	1.025	.105	1.036	.809
$p_2 = .10$	3	70.850	31067.792	49.850	33552.828	60.370	3.290	1.740	.246	1.800	1.096
	4	74.108	30856.380	53.108	33676.842	61.593	3.421	1.628	.377	1.770	1.066
OR = 21.0000	5	24.021	440.694	3.021	449.823	14.298	2.847	.722	−.197	.761	.693
	6	20.996	2193.697	−.004	2193.697	19.702	2.317	1.169	−.727	1.698	1.078
L-OR = 3.0445	7	134.407	*********	113.407	*********	125.574	3.242	2.430	.197	2.469	1.255
$p_1 = .30$	1	1.679	6.395	.679	6.856	1.125	.000	.994	.000	.994	.759
	2	1.877	8.358	.877	9.126	1.360	.000	1.255	.000	1.255	.866
$p_2 = .30$	3	2.045	21.240	1.045	22.331	1.511	.000	1.283	.000	1.283	.841
	4	2.161	22.441	1.161	23.789	1.647	.000	1.432	.000	1.432	.900
OR = 1.0000	5	1.677	4.595	.677	5.053	1.189	.000	1.042	.000	1.042	.794
	6	1.268	1.479	.268	1.551	.582	.000	.446	.000	.446	.468
L-OR = .0000	7	1.593	21.840	.593	22.192	.903	.000	.730	.000	.730	.525
$p_1 = .50$	1	3.892	43.008	1.558	45.436	2.632	.850	.905	.002	.905	.741
	2	4.756	60.660	2.423	66.528	3.434	.958	1.131	.111	1.144	.834
$p_2 = .30$	3	5.064	220.464	2.730	227.919	3.776	.918	1.107	.071	1.112	.803
	4	5.573	231.690	3.239	242.182	4.251	.977	1.233	.130	1.250	.854
OR = 2.3333	5	4.021	26.660	1.687	29.507	2.730	.896	.970	.049	.973	.779
	6	2.319	16.465	−.015	16.465	1.534	.521	.467	−.327	.574	.637
L-OR = .8473	7	4.526	1421.567	2.193	1426.374	3.560	.728	.913	−.119	.928	.759

Table 4 (continued)

Total Sample Size = 20, $n_1 = n_2 = 10$

		STATISTICS FOR ODDS RATIO					STATISTICS FOR LOG ODDS RATIO				
	EST	MEAN	VAR	BIAS	MSE	AAE	MEAN	VAR	BIAS	MSE	AAE
$p_1 = .70$	1	9.907	376.101	4.463	396.018	7.109	1.699	.994	.005	.994	.781
	2	12.856	464.403	7.411	519.331	9.618	1.910	1.204	.215	1.250	.878
$p_2 = .30$	3	15.460	3257.898	10.016	3358.216	12.415	1.836	1.283	.142	1.303	.856
	4	17.240	3306.262	11.795	3445.389	14.001	1.951	1.396	.256	1.461	.919
OR = 5.4444	5	9.678	133.475	4.233	151.395	6.616	1.777	.982	.083	.988	.793
	6	5.815	241.782	.371	241.920	4.978	1.167	.836	−.528	1.115	.901
L-OR = 1.6946	7	20.531	36640.200	15.086	36867.793	18.384	1.635	1.506	−.059	1.509	.955
$p_1 = .50$	1	1.529	4.794	.529	5.073	.958	.000	.816	.000	.816	.697
	2	1.714	7.910	.714	8.420	1.172	.000	1.034	.000	1.034	.782
$p_2 = .50$	3	1.659	14.570	.659	15.003	1.102	.000	.931	.000	.931	.739
	4	1.765	16.586	.765	17.172	1.223	.000	1.051	.000	1.051	.784
OR = 1.0000	5	1.600	4.716	.600	5.076	1.044	.000	.916	.000	.916	.739
	6	1.170	1.374	.170	1.403	.402	.000	.279	.000	.279	.330
L-OR = .0000	7	1.439	56.382	.439	56.574	.736	.000	.573	.000	.573	.477
$p_1 = .70$	1	3.892	43.008	1.558	45.436	2.632	.850	.905	.002	.905	.741
	2	4.756	60.660	2.423	66.528	3.434	.958	1.131	.111	1.144	.834
$p_2 = .50$	3	5.064	220.464	2.730	227.919	3.776	.918	1.107	.071	1.112	.803
	4	5.573	231.690	3.239	242.182	4.251	.977	1.233	.130	1.250	.854
OR = 2.3333	5	4.021	26.660	1.687	29.507	2.730	.896	.970	.049	.973	.779
	6	2.319	16.465	−.015	16.465	1.534	.521	.467	−.327	.574	.637
L-OR = .8473	7	4.526	1421.567	2.193	1426.374	3.560	.728	.913	−.119	.928	.759

Table 4 (continued)

Total Sample Size $= 10, n_1 = n_2 = 5$

		STATISTICS FOR ODDS RATIO					STATISTICS FOR LOG ODDS RATIO				
	EST	MEAN	VAR	BIAS	MSE	AAE	MEAN	VAR	BIAS	MSE	AAE
$p_1 = .10$	1	1.729	4.439	.729	4.971	1.126	.000	1.139	.000	1.139	.755
	2	1.745	4.517	.745	5.073	1.147	.000	1.165	.000	1.165	.767
$p_2 = .10$	3	2.569	17.502	1.569	19.963	2.016	.000	2.074	.000	2.074	1.022
	4	2.579	17.539	1.579	20.031	2.029	.000	2.089	.000	2.089	1.029
OR = 1.0000	5	1.131	.605	.131	.623	.234	.000	.222	.000	.222	.178
	6	1.322	1.218	.322	1.322	.652	.000	.571	.000	.571	.532
L-OR = .0000	7	1.656	25.225	.656	25.655	.999	.000	.897	.000	.897	.617
$p_1 = .30$	1	4.953	50.564	1.096	51.766	3.655	.977	1.395	−.373	1.534	.958
	2	5.153	51.269	1.296	52.950	3.749	1.014	1.453	−.336	1.566	.964
$p_2 = .10$	3	9.078	407.943	5.220	435.196	7.669	1.247	2.308	−.103	2.318	1.268
	4	9.200	407.545	5.343	436.094	7.728	1.269	2.334	−.081	2.340	1.271
OR = 3.8571	5	2.597	8.061	−1.260	9.649	2.466	.572	.699	−.778	1.304	.970
	6	2.872	27.203	−.986	28.175	2.324	.659	.713	−.691	1.190	.855
L-OR = 1.3499	7	9.065	4208.134	5.208	4235.256	8.401	.966	1.687	−.384	1.835	1.140
$p_1 = .50$	1	11.595	311.143	2.595	317.876	8.850	1.798	1.377	−.400	1.537	.963
	2	12.218	307.646	3.218	318.003	8.945	1.890	1.364	−.307	1.458	.929
$p_2 = .10$	3	25.927	3663.949	16.927	3950.480	22.090	2.222	2.169	.025	2.169	1.200
	4	26.314	3651.531	17.314	3951.317	22.158	2.275	2.129	.077	2.135	1.181
OR = 9.0000	5	5.698	31.922	−3.302	42.828	5.537	1.327	.894	−.871	1.652	1.036
	6	7.046	267.553	−1.954	271.372	7.927	1.257	1.013	−.941	1.898	1.178
L-OR = 2.1972	7	48.916	50846.686	39.916	52440.011	46.366	1.984	2.675	−.213	2.720	1.307

Table 4 (continued)

Total Sample Size = 10, $n_1 = n_2 = 5$

		STATISTICS FOR ODDS RATIO					STATISTICS FOR LOG ODDS RATIO				
	EST	MEAN	VAR	BIAS	MSE	AAE	MEAN	VAR	BIAS	MSE	AAE
$p_1 = .70$	1	26.295	1116.638	5.295	1144.676	20.981	2.619	1.395	−.426	1.576	1.016
	2	27.382	1081.448	6.382	1122.181	19.894	2.745	1.227	−.300	1.317	.890
$p_2 = .10$	3	72.221	15666.118	51.221	18289.659	62.770	3.198	2.308	.153	2.331	1.250
	4	72.902	15583.286	51.902	18277.099	62.088	3.269	2.150	.224	2.200	1.179
OR = 21.0000	5	11.036	78.396	−9.964	177.675	12.585	2.085	.722	−.960	1.644	1.057
	6	19.375	1177.313	−1.625	1179.954	21.474	2.022	1.626	−1.023	2.672	1.371
L-OR = 3.0445	7	200.114	*********	179.114	*********	192.734	3.227	3.929	.182	3.963	1.599
$p_1 = .30$	1	2.262	19.314	1.262	20.908	1.763	.000	1.650	.000	1.650	.979
	2	2.490	21.383	1.490	23.602	2.026	.000	1.943	.000	1.943	1.092
$p_2 = .30$	3	3.531	134.056	2.531	140.460	3.069	.000	2.542	.000	2.542	1.196
	4	3.671	135.171	2.671	142.302	3.228	.000	2.715	.000	2.715	1.261
OR = 1.0000	5	1.679	4.466	.679	4.928	1.089	.000	1.049	.000	1.049	.741
	6	1.495	8.938	.495	9.183	.844	.000	.715	.000	.715	.578
L-OR = .0000	7	3.402	1217.708	2.402	1223.476	2.810	.000	1.687	.000	1.687	.837
$p_1 = .50$	1	5.296	116.919	2.963	125.695	4.258	.821	1.632	−.026	1.633	1.004
	2	6.067	121.406	3.733	135.343	4.943	.922	1.950	.075	1.955	1.118
$p_2 = .30$	3	10.084	1194.010	7.751	1254.084	8.985	.975	2.403	.128	2.419	1.185
	4	10.561	1193.234	8.228	1260.928	9.442	1.032	2.580	.185	2.614	1.262
OR = 2.3333	5	3.581	17.005	1.248	18.562	2.533	.724	1.228	−.123	1.243	.892
	6	3.184	85.135	.851	85.859	2.541	.513	.830	−.334	.942	.795
L-OR = .8473	7	16.343	14975.188	14.010	15171.464	15.569	.824	2.118	−.024	2.119	1.103

Table 4 (continued)

Total Sample Size = 10, $n_1 = n_2 = 5$

		STATISTICS FOR ODDS RATIO					STATISTICS FOR LOG ODDS RATIO				
	EST	MEAN	VAR	BIAS	MSE	AAE	MEAN	VAR	BIAS	MSE	AAE
$p_1 = .70$	1	12.010	443.990	6.566	487.099	9.837	1.642	1.650	−.052	1.653	1.043
	2	13.393	436.008	7.948	499.181	10.518	1.821	1.736	.126	1.752	1.057
$p_2 = .30$	3	28.089	5450.756	22.645	5963.541	25.506	1.950	2.542	.256	2.607	1.257
	4	28.951	5420.337	23.506	5972.889	26.043	2.051	2.556	.356	2.683	1.284
OR = 5.4444	5	6.674	41.392	1.230	42.904	4.343	1.454	1.052	−.241	1.111	.818
	6	7.866	400.021	2.421	405.883	7.426	1.115	1.292	−.579	1.627	1.090
L-OR = 1.6946	7	64.639	76843.791	59.194	80347.744	63.121	1.797	3.220	.103	3.231	1.432
$p_1 = .50$	1	2.335	30.272	1.335	32.055	1.833	.000	1.615	.000	1.615	.958
	2	2.738	34.644	1.738	37.664	2.283	.000	2.093	.000	2.093	1.126
$p_2 = .50$	3	3.620	255.948	2.620	262.814	3.146	.000	2.264	.000	2.264	1.110
	4	3.869	258.447	2.869	266.678	3.420	.000	2.546	.000	2.546	1.204
OR = 1.0000	5	1.950	7.167	.950	8.069	1.446	.000	1.403	.000	1.403	.926
	6	1.542	17.890	.542	18.184	.845	.000	.660	.000	.660	.502
L-OR = .0000	7	4.551	2848.040	3.551	2860.653	3.942	.000	1.707	.000	1.707	.802
$p_1 = .70$	1	5.296	116.919	2.963	125.695	4.258	.821	1.632	−.026	1.633	1.004
	2	6.067	121.406	3.733	135.343	4.943	.922	1.950	.075	1.955	1.118
$p_2 = .50$	3	10.084	1194.010	7.751	1254.084	8.985	.975	2.403	.128	2.419	1.185
	4	10.561	1193.234	8.228	1260.928	9.442	1.032	2.580	.185	2.614	1.262
OR = 2.3333	5	3.581	17.005	1.248	18.562	2.533	.724	1.228	−.123	1.243	.892
	6	3.184	85.135	.851	85.859	2.541	.513	.830	−.334	.942	.795
L-OR = .8473	7	16.343	14975.188	14.010	15171.464	15.569	.824	2.118	−.024	2.119	1.103

9. CONCLUSIONS

A complicated pattern of results has emerged, and there does not appear to be a simple answer to the question of which estimator to use in all problems. The optimal choice will depend on whether the odds ratio or its logarithm is of greater interest, and whether the estimator is to be judged on bias, precision, or mean squared error. The choice will also depend on the sample sizes involved, and the underlying binomial outcome probabilities; the latter suggests that prior estimates of p_1 and p_2 would be very useful in making the choice, if such estimates are available.

Some broad recommendations may be made, although it is clear that some of them cannot be made with great conviction, as the difference in performance between the various estimators is sometimes small, and their relative ranking may change rapidly as the problem specifications are altered.

1. Contrary to some suggestions in the literature, the conditional ML appoach does not consistently outperform the unconditional ML method. In fact the CMLE $\hat{\beta}_5$ was usually *inferior* to the Woolf modification of the UMLE, i.e., $\hat{\beta}_1$. The main exception to this was for small n (say $n \leq 10$) with small p_2, where the CMLE had lower MSE and AAE. This intriguing finding is at variance with Hauck's (1984) recent findings in simulations over sets of $2 \times 2 \times k$ tables, with $k = 5$ or 10.

2. Among the set of modified UMLE's for $\beta[\hat{\beta}_1$ to $\hat{\beta}_4]$ it appeared that $\hat{\beta}_1$ was best. Thus 1/2 should be added to the four contingency cells in all cases, and this will give a better estimator than with the strategy of adding 1/2 only as necessary when zeros arise, or of adding 1/4 in either way.

3. For the estimators of the odds ratio itself, the results are less clear cut. However, the UMLE $\hat{\rho}_1$ was much better than the CMLE $\hat{\rho}_5$ for large n, whereas the CMLE tended to be better (especially with respect to MSE and AAE) for smaller samples. Given that the CMLE requires considerably more computation time, it cannot be recommended for routine use. However in some problems, the gain in precision from using the CMLE can be substantial, in comparison to the UMLE. The difficulty here is in obtaining prior estimates of p_1 and p_2 so that those particular problems, where a gain in precision might be expected, may be accurately identified.

4. $\hat{\beta}_6$ and $\hat{\rho}_6$ appear to be promising estimators worthy of further consideration, frequently having the lowest MSE and AAE values; $\hat{\beta}_6$ was also competitive with respect to bias in a number of problems. However, we should bear in mind here Hauck's finding that both pseudo-count estimators generally had large bias when applied to sets of 2×2 tables.

5. Several estimators may reasonably be ruled out of consideration. The Hitchcock estimators $\hat{\rho}_3$ and $\hat{\rho}_4$ are nearly always inferior to the Woolf estimator $\hat{\rho}_1$, which outperforms $\hat{\rho}_2$ for odds ratio estimation, as well as for

log odds ratio estimation. $\hat{\rho}_7$ should be ruled out because of its very unstable MSE, especially in small samples, and $\hat{\beta}_7$ also performed indifferently.

In interpreting the numerical results, we should recall that the influence of tables with zero cells becomes greater as their probability of occurrence increases, i.e., in small samples. It is quite possible that the performance of some of the estimators has been substantially affected by the method selected to deal with zero cells, for example the solution of a substitute equation (equations (4) or (5)) for the CMLE, or the assignment of the value 1 for $\hat{\rho}_5$ and $\hat{\rho}_7$ when a zero column occurred. Although these adaptations are *ad hoc* and lack any theoretical foundation, it is evident that *some* kind of adaptation is necessary for the estimator to be defined in all problems which might arise. Further research is needed on strategies for dealing with zero cells, and this may clarify the optimal choice of estimation technique, especially in very small samples where prior estimates of the underlying binomial probabilities are not available.

ACKNOWLEDGMENTS

This research was supported in part by a National Health Scientist award, from the National Health Research and Development Program. The author would also like to acknowledge the helpful comments made on this work by Dr. C. W. Dunnett of McMaster University.

REFERENCES

Anscombe, F. J. (1956), "On estimating binomial response relations". *Biometrika* **43**, 461–464.

Berkson, J. (1953), "A statistically precise and relatively simple method of estimating the bio-assay with quantal response, based on the logistic function". *Journal of the American Statistical Association* **48**, 565–599.

Bhapkar, V. P., and B. S. Hosmane (1984), "Smoothing of Wald statistics for log-linear hypotheses in categorical data". University of Kentucky Department of Statistics Technical Report No. 221, 1–27.

Birch, M. W. (1964), "The detection of partial association, I: the 2×2 case". *Journal of the Royal Statistical Society*, Series B **26**, 313–324.

Breslow, N. (1981), "Odds ratio estimators when the data are sparse". *Biometrika* **68**, 73–84.

Breslow, N. E., and N. E. Day (1980), *Statistical Methods in Cancer Research, Volume 1—The Analysis of Case-Control Studies*. Lyon, France: International Agency for Research on Cancer.

Cox, D. R. (1970). *The Analysis of Binary Data*. London: Methuen.

Fienberg, S. E., and P. W. Holland (1970), "Methods for eliminating zero counts in contingency tables". In *Random Counts in Models and Structures*, ed. G. P. Patil. University Park: Pennsylvania State University Press.

Fleiss, J. L. (1979), "Confidence intervals for the odds ratio in case-control studies: the state of the art". *Journal of Chronic Diseases* **32**, 69–77.

Gart, J. J. (1962), "On the combination of relative risks". *Biometrics* **18**, 601–610.

Gart, J. J. (1970), "Point and interval estimation of the common odds ratio in the combination of 2 × 2 tables with fixed marginals". *Biometrika* **57**, 471–475.

Gart, J. J., and J. R. Zweifel (1967), "On the bias of various estimators of the logit and its variance, with application to quantal bioassay". *Biometrika* **54**, 181–187.

Goodman, L. A. (1969), "On partitioning χ^2 and detecting partial association in three-way contingency tables". *Journal of Royal Statistical Society*, Series B **31**, 486–498.

Goodman, L. A. (1970), "The multivariate analysis of qualitative data: interactions among multiple classifications". *Journal of the American Statistical Association* **65**, 226–256.

Grizzle, J. E., C. F. Starmer, and G. G. Koch (1969), "Analysis of categorical data by linear models". *Biometrics* **25**, 489–504.

Haldane, J. B. S. (1956), "The estimation and significance of the logarithm of a ratio of frequencies". *Annals of Human Genetics* **20**, 309–311.

Halperin, M., J. H. Ware, D. P. Byar, N. Mantel, C. C. Brown, J. Koziol, M. Gail, and S. B. Green (1977), "Testing for interaction in an $I \times J \times K$ contingency table". *Biometrika* **64**, 271–275.

Hanley, J. A., and A. Lippman-Hand (1983), "If nothing goes wrong, is everything all right? Interpreting zero numerators". *Journal of the American Medical Association* **249**, 1743–1745.

Hauck, W. W., S. Anderson, and F. J. Leahy III (1982), "Finite-sample properties of some new estimators of a common odds ratio from multiple 2 × 2 tables". *Journal of the American Statistical Association* **77**, 145–152.

Hauck, W. W. (1984), "A comparative study of conditional maximum likelihood estimation of a common odds ratio". *Biometrics* **40**, 1117–1123.

Hitchcock, S. E. (1962), "A note on the estimation of the parameters of the logistic function, using the minimum logit χ^2 method". *Biometrika* **49**, 250–252.

Jewell, N. P. (1984), "Small-sample bias of point estimators of the odds ratio from matched sets". *Biometrics* **40**, 421–435.

Koopman, P. A. R. (1984), "Confidence intervals for the ratio of two binomial proportions". *Biometrics* **40**, 513–517.

Mantel, N., C. Brown, and D. P. Byar (1977), "Tests for homogeneity of effect in an epidemiologic investigation". *American Journal of Epidemiology* **106**, 125–129.

Mantel, N., and W. Haenszel (1959), "Statistical aspects of the analysis of data from retrospective studies of disease". *Journal of the National Cancer Institute* **22**, 719–748.

McKinlay, S. M. (1975), "The effect of bias on estimators of relative risk for pair-matched and stratified samples". *Journal of the American Statistical Associa-

tion **70**, 859–864.

McKinlay, S. M. (1978), "The effect of nonzero second-order interaction on combined estimators of the odds ratio". *Biometrika* **65**, 191–202.

Schaefer, R. L. (1983), "Bias correction in maximum likelihood logistic regression". *Statistics in Medicine* **2**, 71–78.

Walter, S. D. (1985), "Small sample estimation of log odds ratios from logistic regression and fourfold tables". *Statistics in Medicine* **4**, 437–444.

Whitley, R. J., S. J. Soong, R. Dolin, G. J. Galasso, L. T. Ch'ien, C. A. Alford and the Collaborative Study Group (1977), "Adenine arabinoside therapy of biopsy-proved herpes simplex encephalitis". National Institute of Allergy and Infectious Diseases Collaborative Antiviral Study, *New England Journal of Medicine* **297**, 289–294.

Woolf, B. (1955), "On estimating the relation between blood group and disease". *Annals of Human Genetics* **19**, 251–253.

Yates, F. (1984), "Tests of significance for 2 × 2 contingency tables". *Journal of Royal Statistical Society*, Series A **147**, 426–463.

George A. Wells [1] and Allan Donner [2]

DEVELOPMENT OF FORMULAS FOR THE BIAS AND MEAN SQUARE ERROR OF THE LOGIT ESTIMATOR

ABSTRACT

Formulas for calculating the bias and mean square error of the logit estimator are developed. These formulations involve asymptotic series which should provide adequate approximations provided the parameters n and P of the binomial model are such that nP is not small. Although the asymptotic expansions are algebraically complicated, they have been mathematically expressed in such a way that general coefficients can be isolated and calculated.

1. INTRODUCTION

The logit λ is defined as the logarithm of the ratio of the probability of the occurrence of an event to the probability of its nonoccurrence. That is $\lambda = \ln[P/(1-P)]$ where X has a binomial distribution with parameters n and P. An estimator of λ is $\hat{\lambda} = \ln[X/(n-X)]$. There is a positive probability that $n - X$ can be zero and, thus, the moments of the distribution of $\hat{\lambda}$ are infinite. Provided $0 < P < 1$ and the sample size is large enough, the observed frequency will be positive and the logit estimate can be calculated. When the observed frequency is zero, suggested remedies may involve techniques anywhere from smoothing the observed frequencies to the more general practice of adjustment by adding a simple constant, commonly $\frac{1}{2}$, to each observed frequency. Gart and Zweifel (1967) have examined the small

[1] Bureau of Communicable Disease Epidemiology, Laboratory Centre for Disease Control, Tunney's Pasture, Ottawa, Ontario K1A 0L2

[2] Department of Epidemiology and Biostatistics, The University of Western Ontario, London, Ontario N6A 5C1

I. B. MacNeill and G. J. Umphrey (eds.), Biostatistics, 103–124.

sample bias of several of these logit estimators. In particular, the estimator

$$\hat{\lambda}_{HA} = \ln \frac{X + \frac{1}{2}}{n - X + \frac{1}{2}}$$

is an unbiased estimator of λ to order n^{-1} (that is, except for terms of $O(n^{-2})$). Haldane (1955) was the first to suggest this estimator. Commenting on Woolf's estimator (1955) of log odds, he used logarithmic expansions of the bias to arrive at this estimator (Anscombe (1956) independently suggested the same estimator in his examination of the parameters of the logistic function using the Berkson's minimum logit χ^2 method). Unpublished results of Goodman, given in a paper by Gart *et al.* (1985), uses this approach. As noted by Naylor (1967), Haldane's paper is "hard to follow and may be incomplete". A more systematic and complete development of this approach is required.

For a binomial distribution with parameters n and P, methods and formulas for calculating the bias and mean square error of the logit estimator are developed. These formulations involve asymptotic series in powers of n^{-1} which for moderately large n and for values of P not too extreme, provide adequate approximations to these quantities. Although the asymptotic expansions are algebraically complicated, they have been mathematically expressed in such a way that general coefficients can be isolated and calculated. These coefficients can then be implemented for any choice of the parameter n and P. Computer programs are available to calculate these various coefficients up to and including terms of any desired order of n^{-1}.

2. METHOD

Consider a discrete random variable X that has a binomial distribution with parameters n and P. Let $Q = 1 - P$. The logit is $\lambda = \ln[P/(1-P)]$ and an estimator of it is given by

$$\hat{\lambda}(t) = \ln\left(\frac{X+t}{n-X+t}\right),$$

where t is some constant to be established.

Formulations of the bias and mean square error of the logit estimator are developed using an approach that is a generalization of the one considered by Haldane (1955), involving the logarithmic expansion of the logit estimator. Two results that are required in this development are given in the following two lemmas. The first result is a closed form expression for the central moments of the binomial distribution. Using the principle of mathematical induction, the following lemma is established.

Lemma 1. Consider a random variable X distributed as a binomial with parameters n and P. The central moments $\mu_r = E(\gamma^r)$, where $\gamma = X - nP$, can be expressed as:

$$\mu_r = (-1)^r \sum_{h=0}^{[r/2]} n^h \sum_{\ell=1}^{r-2h+1} P^{r-h+1-\ell} Q^{h-1+\ell} (-1)^{\ell-1} C_{h,r-2h+1,\ell} \tag{1}$$

$$= -\sum_{h=0}^{[r/2]} n^h (PQ)^h \sum_{\ell=1}^{\left[\frac{r-2h+2}{2}\right]} (-1)^\ell C_{h,r-2h+1,\ell} (PQ)^{\ell-1}$$
$$(Q^{r-2h-2\ell+2} + (-1)^r P^{r-2h-2\ell+2}) \tag{2}$$

where the constants

$$C_{h,m,\ell} = (2h+m-2)C_{h-1,m,\ell} + (h+m-\ell)C_{h,m-1,\ell-1} + (h+\ell-1)C_{h,m-1,\ell}$$

subject to

$$\begin{cases} C_{h,m,0} = C_{h,m,m+1} = 0 & \text{for } h,\ m = 0,1,2,\dots \\ C_{0,m,\ell} = 0 & \text{for } m,\ \ell = 0,1,2,\dots \\ & \text{(except } m = \ell = 1) \\ C_{0,1,1} = 1 & \end{cases}$$

A second fundamental result that will be required involves certain basic, but convenient, relationships involving the parameter P. These relationships are given in the following lemma.

Lemma 2. If P and Q are such that $0 \le P \le 1$ and $Q = 1 - P$, then for any positive integer i,

$$Q^i - (-1)^\kappa P^i = (Q-P)^{1-\kappa} \sum_{j=1}^{\left[\frac{i+1+\kappa}{2}\right]} d_{ij\kappa} (-1)^{j-1} (PQ)^{j-1} \tag{3}$$

where kappa is a constant that can be set to 0 or 1 and

$$d_{ij\kappa} = \binom{i-j}{i-2j+1} + \binom{i-j}{i-2j+2} 2\kappa.$$

Some common functions used in the various infinite series are:

$$\kappa_i = \begin{cases} 0 & \text{if } i \text{ is an even integer or zero} \\ 1 & \text{otherwise,} \end{cases}$$

$$[i/2] = \text{largest integer } \leq i/2,$$

$$\delta_a = \begin{cases} 0 & \text{if } a < 0 \\ 1 & \text{if } a \geq 0, \end{cases}$$

$$\delta_{ij} = \begin{cases} 1 & \text{if } i \neq j/2 \\ 0 & \text{if } i = j/2, \end{cases}$$

$$\underline{i} = \begin{cases} 1 & \text{if } i = 0 \\ i & \text{if } i > 0. \end{cases}$$

Also the combination formula $\binom{a}{b}$ is defined to be zero if a is negative, b is negative, or $b > a$.

For the estimator $\hat{\lambda}(t)$ of the logit λ, both the bias and mean square error of this estimator can be expressed as an asymptotic series of powers of n^{-1}. More specifically, Corollaries 1 and 2 establish, respectively, that the bias can be expressed as

$$b(t) = (Q - P) \sum_{w=1}^{\infty} \sum_{s=0}^{w-1} \sum_{g=2s-w+1}^{w} A_{w,s,g} t^{w-g} (PQ)^s (nPQ)^{-w} \tag{4}$$

and the mean square error can be written as

$$m(t) = \sum_{w=1}^{\infty} \sum_{s=0}^{w-1} \sum_{g=2s-w}^{w} F_{w,s,g} t^{w-g} (PQ)^s (nPQ)^{-w}. \tag{5}$$

In these expressions, $A_{w,s,g}$ and $F_{w,s,g}$ are constants with respect to n, P and t.

To derive these expressions, the logarithmic functions

$$\ln\left(1 + \frac{t+\gamma}{nP}\right) \quad \text{and} \quad \left(1 + \frac{t-\gamma}{nQ}\right)$$

must be considered. For both of these functions, the function and its derivatives of all orders exist at $(t+\gamma)/(nP) = 0$ and $(t-\gamma)/(nQ) = 0$ respectively. A Maclaurin series can be considered to expand each function as a power series in its argument. This expansion of a function represents that function for only those values that are in the interval of convergence of the series. For such functions this interval is $(-1, 1]$.

The actual domain of possible values of $(t+\gamma)/(nP)$ and $(t-\gamma)/(nQ)$, under the assumption that t is positive, each has a lower limit greater than -1 but the upper limits may be greater than 1. In regards to the latter, if $0 < t < nP$ and $0 < t < nQ$, then both upper limits approach values less than or equal to 1 only for large n and moderate P. Thus, except for these special cases, there is a failure to include the entire range of values in the interval of convergence.

Still a valid asymptotic series of powers of n^{-1} can be derived for $b(t)$ and $m(t)$. This series may not converge, but has the asymptotic property that it remains bounded as w increases. As noted by Hotelling (1953), when confronted with a similar problem in the study of the correlation coefficient, "Though presumably divergent, it will nevertheless provide through its first terms a good approximation to the true moment if the sample is large enough".

In Theorem 1 the logarithmic functions are expanded and asymptotic series of powers of n^{-1} are derived for the bias and mean square error.

Theorem 1. Let X be a random variable distributed as a binomial with parameters n and P. Consider the estimator

$$\hat{\lambda}(t) = \ln\left(\frac{X+t}{n-X+t}\right),$$

where t is constant, of $\ln(P/Q)$. The bias and mean square error of this estimator can be expressed as an asymptotic series of powers of n^{-1}. That is, respectively,

$$b(t) = \sum_{w=1}^{\infty}\sum_{h=0}^{w}\sum_{j=2h}^{h+w}\sum_{\ell=1}^{j-2h+1}\Bigg[\binom{w+h}{j}(-1)^{w+h+j+\ell}(w+h)^{-1}C_{h,j-2h+1,\ell}$$
$$\left(Q^{w+h}+(-1)^{j+1}P^{w+h}\right)P^{j-2h+1-\ell-w}Q^{\ell-1-w}t^{w+h-j}\Bigg]n^{-w} \qquad (6)$$

and

$$m(t) = \sum_{w=1}^{\infty}\sum_{h=0}^{w}\sum_{j=2h}^{h+w}\sum_{i=0}^{[j/2]}\sum_{v=i}^{w+h-j+i}\sum_{\ell=1}^{j-2h+1}\Bigg[2^{\delta_{ij}}\binom{v}{i}\binom{w+h-v}{j-i}$$
$$(-1)^{w+h+j+\ell-1}(\underline{\mathbf{v}}\ \mathbf{w+h-v})^{-1}C_{h,j-2h+1,\ell}\left(Q^{v}+(-1)^{i+1}P^{v}\right)$$
$$\left(Q^{w+h-v}+(-1)^{j-i+1}P^{w+h-v}\right)P^{j-2h+1-\ell-w}$$
$$Q^{\ell-1-w}t^{w+h-j}\Bigg]n^{-w}. \qquad (7)$$

Proof. To derive $b(t)$, consider the difference

$$\begin{aligned} D(t) &= \ln\left(\frac{X+t}{n-X+t}\right) - \ln\left(\frac{P}{Q}\right) \\ &= \ln\left(\frac{nP+\gamma+t}{nP}\right) - \ln\left(\frac{nQ-\gamma+t}{nQ}\right) \\ &= \ln\left(1+\frac{t+\gamma}{nP}\right) - \ln\left(1+\frac{t-\gamma}{nQ}\right). \end{aligned}$$

Using the asymptotic series expansion

$$\sum_{i=1}^{\infty}(-1)^{i+1}\frac{x^i}{i}$$

of the logarithmic function $\ln(1+x)$, $D(t)$ can be expressed as

$$\begin{aligned} D(t) &= \sum_{i=1}^{\infty}(-1)^{i+1}(in^iP^i)^{-1}(t+\gamma)^i - \sum_{i=1}^{\infty}(-1)^{i+1}(in^iQ^i)^{-1}(t-\gamma)^i \\ &= \sum_{i=1}^{\infty}(-1)^{i+1}(in^iP^i)^{-1}\sum_{j=0}^{i}\binom{i}{j}\gamma^j t^{i-j} \\ &\quad - \sum_{i=1}^{\infty}(-1)^{i+1}(in^iQ^i)^{-1}\sum_{j=0}^{i}\binom{i}{j}(-1)^j\gamma^j t^{i-j} \\ &= \sum_{i=1}^{\infty}\sum_{j=0}^{i}\binom{i}{j}(-1)^{i+1}i^{-1}(nPQ)^{-i}t^{i-j}\gamma^j\left(Q^i+(-1)^{j+1}P^i\right) \\ &= \sum_{j=0}^{\infty}\left[\sum_{i=j}^{\infty}\binom{i}{j}(-1)^{i+1}i^{-1}(PQ)^{-i}\left(Q^i+(-1)^{j+1}P^i\right)t^{i-j}n^{-i}\right]\gamma^j \end{aligned} \tag{8}$$

where in the last step, when the summations involving i and j are interchanged a zero term involving $i=0$, $j=0$ has been added.

Taking the expectation of $D(t)$, using equation (1) of Lemma 1, yields

$$\begin{aligned} b(t) = \sum_{j=0}^{\infty}\sum_{i=j}^{\infty}\sum_{h=0}^{[j/2]}\sum_{\ell=1}^{j-2h+1}&\left[\binom{i}{j}(-1)^{i+j+\ell}i^{-1}C_{h,j-2h+1,\ell}\right. \\ &\left.\left(Q^i+(-1)^{j+1}P^i\right)P^{j-h+1-\ell-i}Q^{h-1+\ell-i}t^{i-j}\right]n^{-i+h}. \end{aligned} \tag{9}$$

Setting $w = i - h$, (9) can be expressed as

$$b(t) = \sum_{j=0}^{\infty} \sum_{h=0}^{[j/2]} \sum_{w=j-h}^{\infty} \sum_{\ell=1}^{j-2h+1} \Bigg[\binom{w+h}{j} (-1)^{w+h+j+\ell} \underline{(\mathbf{w}+\mathbf{h})}^{-1} C_{h,j-2h+1,\ell} \left(Q^{w+h} + (-1)^{j+1} P^{w+h}\right) P^{j-2h+1-\ell-w} Q^{\ell-1-w} t^{w+h-j} \Bigg] n^{-w}$$

$$= \sum_{w=0}^{\infty} \sum_{h=0}^{w} \sum_{j=2h}^{h+w} \sum_{\ell=1}^{j-2h+1} \Bigg[\binom{w+h}{j} (-1)^{w+h+j+\ell} \underline{(\mathbf{w}+\mathbf{h})}^{-1} C_{h,j-2h+1,\ell} \left(Q^{w+h} + (-1)^{j+1} P^{w+h}\right) P^{j-2h+1-\ell-w} Q^{\ell-1-w} t^{w+h-j} \Bigg] n^{-w}. \tag{10}$$

If $w = 0$, then $h = 0$ and $j = 0$, and in this case $(Q^{w+h} + (-1)^{j+1} P^{w+h}) = 0$. Removing the zero terms involving $w = 0$ from equation (10), transforms this equation into the form of equation (6).

To derive $m(t)$, consider $D^2(t)$. Using equation (8),

$$D^2(t) = \left\{ \sum_{j_1=0}^{\infty} \Bigg[\sum_{i_1=j_1}^{\infty} \binom{i_1}{j_1} (-1)^{i_1+1} \underline{\mathbf{i}_1}^{-1} (PQ)^{-i_1} \left(Q^{i_1} + (-1)^{j_1+1} P^{i_1}\right) t^{i_1-j_1} n^{-i_1} \Bigg] \gamma^{j_1} \right\}$$
$$\left\{ \sum_{j_2=0}^{\infty} \Bigg[\sum_{i_2=j_2}^{\infty} \binom{i_2}{j_2} (-1)^{i_2+1} \underline{\mathbf{i}_2}^{-1} (PQ)^{-i_2} \left(Q^{i_2} + (-1)^{j_2+1} P^{i_2}\right) t^{i_2-j_2} n^{-i_2} \Bigg] \gamma^{j_2} \right\}$$

$$= \sum_{j=0}^{\infty} \left\{ \sum_{i=0}^{[j/2]} 2^{\delta_{ij}} \Bigg[\sum_{i_1=i}^{\infty} \binom{i_1}{i} (-1)^{i_1+1} \underline{\mathbf{i}_1}^{-1} (PQ)^{-i_1} \left(Q^{i_1} + (-1)^{i+1} P^{i_1}\right) t^{i_1-i} n^{-i_1} \Bigg] \Bigg[\sum_{i_2=j-i}^{\infty} \binom{i_2}{j-i} (-1)^{i_2+1} \underline{\mathbf{i}_2}^{-1} (PQ)^{-i_2} \left(Q^{i_2} + (-1)^{j-i+1} P^{i_2}\right) t^{i_2-j+i} n^{-i_2} \Bigg] \right\} \gamma^j$$

$$= \sum_{j=0}^{\infty} \sum_{i=0}^{[j/2]} 2^{\delta_{ij}} \left\{ \sum_{u=j}^{\infty} \left[\sum_{v=i}^{u-j+1} \left[\binom{v}{i} (-1)^{v+1} \underline{\mathbf{v}}^{-1} (PQ)^{-v} \right.\right.\right.$$
$$\left. (Q^v + (-1)^{i+1} P^v) \, t^{v-i} \right] \left[\binom{u-v}{j-i} (-1)^{u-v+1} (\underline{\mathbf{u}-\mathbf{v}})^{-1}$$
$$\left.\left.\left. (PQ)^{-(u-v)} (Q^{u-v} + (-1)^{j-i+1} P^{u-v}) \, t^{u-v-j+i} \right]\right] n^{-u} \right\} \gamma^j$$
$$= \sum_{j=0}^{\infty} \sum_{u=j}^{\infty} \sum_{i=0}^{[j/2]} \sum_{v=i}^{u-j+i} \left[2^{\delta_{ij}} \binom{v}{i} \binom{u-v}{j-i} (-1)^u (\underline{\mathbf{v}} \, \underline{\mathbf{u}-\mathbf{v}})^{-1} (PQ)^{-u} \right.$$
$$\left. (Q^v + (-1)^{i+1} P^v) (Q^{u-v} + (-1)^{j-i+1} P^{u-v}) \, t^{u-j} \right] n^{-u} \gamma^j. \quad (11)$$

The series for the mean square error is obtained by taking the expectation of $D^2(t)$. Using equation (1) of Lemma 1:

$$m(t) = \sum_{j=0}^{\infty} \sum_{u=j}^{\infty} \sum_{i=0}^{[j/2]} \sum_{v=i}^{u-j+1} \sum_{h=0}^{[j/2]} \sum_{\ell=1}^{j-2h+1} \left[2^{\delta_{ij}} \binom{v}{i} \binom{u-v}{j-i} \right.$$
$$(-1)^{u+j+\ell-1} (\underline{\mathbf{v}} \, \underline{\mathbf{u}-\mathbf{v}})^{-1} C_{h,j-2h+1,\ell}$$
$$(Q^v + (-1)^{i+1} P^v) (Q^{u-v} + (-1)^{j-i+1} P^{u-v})$$
$$\left. P^{j-h+1-\ell-u} Q^{h-1+\ell-u} t^{u-j} \right] n^{-u+h}$$
$$= \sum_{j=0}^{\infty} \sum_{h=0}^{[j/2]} \sum_{w=j-h}^{\infty} \sum_{i=0}^{[j/2]} \sum_{v=i}^{w+h-j+i} \sum_{\ell=1}^{j-2h+1} \left[2^{\delta_{ij}} \binom{v}{i} \binom{w+h-v}{j-i} \right.$$
$$(-1)^{w+h+j+\ell-1} (\underline{\mathbf{v}} \, \underline{\mathbf{w}+\mathbf{h}-\mathbf{v}})^{-1} C_{h,j-2h+1,\ell} (Q^v + (-1)^{i+1} P^v)$$
$$(Q^{w+h-v} + (-1)^{j-i+1} P^{w+h-v})$$
$$\left. P^{j-2h+1-\ell-w} Q^{\ell-1-w} t^{w+h-j} \right] n^{-w}, \quad (12)$$

where $w = u - h$.

When the summations involving j, h and w are interchanged,

$$\sum_{j=0}^{\infty} \sum_{h=0}^{[j/2]} \sum_{w=j-h}^{\infty} \quad \text{becomes} \quad \sum_{w=0}^{\infty} \sum_{h=0}^{w} \sum_{j=2h}^{h+w}.$$

Also, if $w = 0$ then h, j, i and v are all zero. In this case $(Q^v + (-1)^{i+1} P^v) = 0$. Interchanging the summations and removing the zero terms involving

$w = 0$, transforms (12) into the form of equation (7). This completes the proof of the theorem.

The infinite series expansion of the bias and mean square error of $\hat{\lambda}(t)$ established in Theorem 1 can be implemented to calculate these quantities to various orders of n^{-1}. The actual computations required using (6) and (7) are simple but tedious. Corollaries 1 and 2 provide alternative formulations for the bias and mean square error. Although these forms appear more complex, they are more convenient to use since most terms in the expression are constant with respect to n, P and t which need to be computed only once and then used for any future calculations.

In Corollary 1, the computing formula for the bias is expressed in the form given in (4).

Corollary 1. Let X be a random variable distributed as a bionomial with parameters n and P. An asymptotic series in powers of n^{-1} of the bias of the estimator $\hat{\lambda}(t)$ of $\ln(P/Q)$ is

$$b(t) = (Q-P)\sum_{w=1}^{\infty}\sum_{s=0}^{w-1}\sum_{g=2s-w+1}^{w} A_{w,s,g}t^{w-g}(PQ)^{s}(nPQ)^{-w}, \tag{13}$$

where

$$A_{w,s,g} = \sum_{h=0}^{g}\sum_{\ell=s-\left[\frac{g-h}{2}\right]}^{s}\sum_{u=s-\left[\frac{g-h}{2}\right]}^{\min\left(\ell,\left[\frac{w+h-1+\kappa_{g-h}}{2}\right]\right)} \delta_g\delta_\ell\delta_u\binom{w+h}{g+h} (-1)^{w+h+s+1}(w+h)^{-1}C_{h,g-h+1,s-\ell+1}d_{w+h,u+1,\kappa_{g-h}} d_{g-h-2s+2\ell,\ell-u+1,1-\kappa_{g-h}}. \tag{14}$$

Proof. The bias series is obtained by taking the expectation of $D(t)$ in equation (8). This expectation is taken using equation (2) of Lemma 1. That is,

$$b(t) = \sum_{j=0}^{\infty}\sum_{i=j}^{\infty}\sum_{h=0}^{[j/2]}\sum_{\ell=1}^{\left[\frac{j-2h+2}{2}\right]}\Bigg[\binom{i}{j}(-1)^{i+\ell-1}C_{h,j-2h+1,\ell}t^{i-j} (Q^{i}+(-1)^{j+1}P^{i})(Q^{j-2h-2\ell+2}+(-1)^{j}P^{j-2h-2\ell+2}) (PQ)^{h+\ell-i-1}\Bigg]n^{-i+h}$$

$$= \sum_{j=0}^{\infty}\sum_{h=0}^{[j/2]}\sum_{w=j-h}^{\infty}\sum_{\ell=1}^{\left[\frac{j-2h+2}{2}\right]}\Bigg[\binom{w+h}{j}(-1)^{w+h+\ell}\underline{(\mathbf{w}+\mathbf{h})}^{-1}$$

$$C_{h,j-2h+1,\ell} t^{w+h-j}\left(Q^{w+h}+(-1)^{j+1}P^{w+h}\right)$$
$$\left.\left(Q^{j-2h-2\ell+2}+(-1)^{j}P^{j-2h-2\ell+2}\right)(PQ)^{\ell-1-w}\right]n^{-w}$$
$$=\sum_{w=1}^{\infty}\sum_{h=0}^{w}\sum_{j=2h}^{h+w}\sum_{\ell=1}^{\left[\frac{j-2h+2}{2}\right]}\left[\binom{w+h}{j}(-1)^{w+h+\ell}(w+h)^{-1}\right.$$
$$C_{h,j-2h+1,\ell} t^{w+h-j}\left(Q^{w+h}+(-1)^{j+1}P^{w+h}\right)$$
$$\left.\left(Q^{j-2h-2\ell+2}+(-1)^{j}P^{j-2h-2\ell+2}\right)(PQ)^{\ell-1-w}\right]n^{-w}. \tag{15}$$

Lemma 2 is now applied to

$$Q^{w+h}+(-1)^{j+1}P^{w+h} \quad \text{and} \quad Q^{j-2h-2\ell+2}+(-1)^{j}P^{j-2h-2\ell+2}.$$

That is,

$$Q^{w+h}+(-1)^{j+1}P^{w+h}$$
$$=(Q-P)^{1-\kappa_j}\sum_{u=1}^{\left[\frac{w+h+1+\kappa_j}{2}\right]} d_{w+h,u,\kappa_j}(-1)^{u-1}(PQ)^{u-1} \tag{16}$$

and

$$Q^{j-2h-2\ell+2}+(-1)^{j}P^{j-2h-2\ell+2}$$
$$=(Q-P)^{\kappa_j}\sum_{v=1}^{\left[\frac{j-2h-2\ell+4}{2}\right]} d_{j-2h-2\ell+2,v,1-\kappa_j}(-1)^{v-1}(PQ)^{v-1}. \tag{17}$$

The upper limit of the summation in expresssion (17) is simply

$$\left[\frac{j-2h-2\ell+4}{2}\right],$$

since

$$\left[\frac{j-2h-2\ell+4-\kappa_j}{2}\right]=\left[\frac{j-\kappa_j}{2}\right]+(-h-\ell+2)$$

and if j is an even integer or zero then $\kappa_j=0$ and $[(j-\kappa_j)/2]=[j/2]$. On the other hand, if j is an odd integer then $\kappa_j=1$ and $[(j-\kappa_j)/2]=[(j-1)/2]=[j/2]$.

Substituting (16) and (17) into expression (15) yields

$$b(t) = (Q-P)\sum_{w=1}^{\infty}\sum_{h=0}^{w}\sum_{j=2h}^{h+w}\sum_{\ell=1}^{\left[\frac{j-2h+2}{2}\right]}\sum_{u=1}^{\left[\frac{w+h+1+\kappa_j}{2}\right]}\sum_{v=1}^{\left[\frac{j-2h-2\ell+4}{2}\right]}$$
$$\left\{\binom{w+h}{j}(-1)^{w+h+\ell+u+v}(w+h)^{-1}C_{h,j-2h+1,\ell}t^{w+h-j}\right.$$
$$\left. d_{w+h,u,\kappa_j}d_{j-2h-2\ell+2,v,1-\kappa_j}\right\}(PQ)^{\ell+u+v-3}(nPQ)^{-w}. \tag{18}$$

Adjusting the order of the summations, equation (18) is transformed to

$$b(t) = (Q-P)\sum_{w=1}^{\infty}\sum_{s=0}^{w-1}\left\{\sum_{h=0}^{w}\sum_{j=a_1}^{w+h}\sum_{\ell=1}^{a_2}\sum_{u=a_3}^{a_4}\binom{w+h}{j}(-1)^{w+h+s+1}\right.$$
$$(w+h)^{-1}C_{h,j-2h+1,\ell}t^{w+h-j}d_{w+h,u,\kappa_j}$$
$$\left. d_{j-2h-2\ell+2,s-\ell-u+3,1-\kappa_j}\right\}(PQ)^{s}(nPQ)^{-w}, \tag{19}$$

where

$$a_1 = 2h + \max(0, 2s-w-h+1),$$
$$a_2 = \min(s, [j/2]-h)+1,$$
$$a_3 = \max(0, s-[j/2]+h)+1,$$
$$a_4 = \min\left(s-\ell+1, \left[\frac{w+h-1+\kappa_j}{2}\right]\right)+1.$$

Equation (19) is now transformed using two changes of variables. First, j is replaced by $j+2h$. Then u and ℓ are changed to $u+1$ and $\ell+1$ respectively. These changes yield:

$$b(t) = (Q-P)\sum_{w=1}^{\infty}\sum_{s=0}^{w-1}\left\{\sum_{h=0}^{w}\sum_{j=b_1}^{w-h}\sum_{\ell=0}^{b_2}\sum_{u=b_3}^{b_4}\binom{w+h}{j+2h}(-1)^{w+h+s+1}\right.$$
$$\left.(w+h)^{-1}C_{h,j+1,\ell+1}t^{w-j-h}d_{w+h,u+1,\kappa_j}d_{j-2\ell,s-\ell-u+1,1-\kappa_j}\right\}$$
$$(PQ)^{s}(nPQ)^{-w}, \tag{20}$$

where

$$b_1 = \max(0, 2s+1-w-h),$$
$$b_2 = \min(s, [j/2]),$$
$$b_3 = \max(0, s-[j/2]),$$
$$b_4 = \min\left(s-\ell, \left[\frac{w+h-1+\kappa_j}{2}\right]\right).$$

Setting $g = h + j$, and then replacing ℓ by $s - \ell$, (20) can be expressed as

$$b(t) = (Q-P)\sum_{w=1}^{\infty}\sum_{s=0}^{w-1}\Bigg\{\sum_{g=c_1}^{w}\sum_{h=0}^{g}\sum_{\ell=c_2}^{s}\sum_{u=c_3}^{c_4}\binom{w+h}{g+h}(-1)^{w+h+s+1}$$
$$(w+h)^{-1}C_{h,g-h+1,s-\ell+1}\,d_{w+h,u+1,\kappa_{g-h}}\,d_{g-h-2s+2\ell,\ell-u+1,1-\kappa_{g-h}}$$
$$t^{w-g}\Bigg\}(PQ)^{s}(nPQ)^{-w}, \tag{21}$$

where

$$c_1 = \max(0, 2s - w + 1),$$
$$c_2 = \max\left(0, s - \left[\frac{g-h}{2}\right]\right),$$
$$c_3 = \max\left(0, s - \left[\frac{g-h}{2}\right]\right),$$
$$c_4 = \min\left(\ell, \left[\frac{w+h-1+\kappa_{g-h}}{2}\right]\right).$$

Finally, using the indicator variable δ_a, equation (21) becomes

$$b(t) = (Q-P)\sum_{w=1}^{\infty}\sum_{s=0}^{w-1}\Bigg\{\sum_{g=2s-w+1}^{w}\sum_{h=0}^{g}\sum_{\ell=s-\left[\frac{g-h}{2}\right]}^{s}\;\sum_{u=s-\left[\frac{g-h}{2}\right]}^{\min\left(\ell,\left[\frac{w+h-1+\kappa_{g-h}}{2}\right]\right)}$$
$$\delta_g\delta_\ell\delta_u\binom{w+h}{g+h}(-1)^{w+h+s+1}(w+h)^{-1}$$
$$C_{h,g-h+1,s-\ell+1}\,d_{w+h,u+1,\kappa_{g-h}}$$
$$d_{g-h-2s+2\ell,\ell-u+1,1-\kappa_{g-h}}\,t^{w-g}\Bigg\}(PQ)^{s}(nPQ)^{-w}$$
$$= (Q-P)\sum_{w=1}^{\infty}\sum_{s=0}^{w-1}\sum_{g=2s-w+1}^{w} A_{w,s,g}\,t^{w-g}(PQ)^{s}(nPQ)^{-w}$$

and the proof of the corollary is complete.

Using a computer program, the constants $A_{w,s,g}$ have been calculated for various values of w. The constants, or coefficients, calculated have been tabulated in Table 1 for orders up to $w = 7$. As noted by Haldane, the setting of the constant t to $\frac{1}{2}$ leads to an unbiased estimator to order n^{-1}. That is,

Table 1. *$A_{w,s,g}$ Coefficients of the Asymptotic Series $b(t)$ in Powers of n^{-1} of the Bias.*

		g							
w	s	0	1	2	3	4	5	6	7
1	0	1.0000	-5.0000						
2	0	-.50000	1.0000	-.41667					
	1		-1.0000	.50000					
3	0	.33333	-1.5000	2.0000	-.75000				
	1	-.33333	3.0000	-5.0000	2.0000				
	2			1.0000	-.50000				
4	0	-.25000	2.0000	-5.5000	6.0000	-2.0917			
	1	.50000	-6.0000	20.000	-24.000	8.6833			
	2		2.0000	-12.500	19.000	-7.4167			
	3				-1.0000	.50000			
5	0	.20000	-2.5000	11.667	-25.000	24.000	-7.9167		
	1	-.60000	10.000	-55.000	130.00	-132.00	44.667		
	2	.20000	-7.5000	60.000	-172.50	194.00	-68.500		
	3			-8.3333	45.000	-65.000	25.000		
	4					1.0000	-.50000		
6	0	-.16667	3.0000	-21.250	75.000	-137.00	120.00	-37.871	
	1	.66667	-15.000	122.50	-475.00	923.00	-840.00	270.23	
	2	-.50000	18.000	-192.50	880.00	-1893.5	1830.0	-605.82	
	3		-3.0000	70.000	-465.00	1225.0	-1320.0	458.17	
	4				30.000	-150.50	211.00	-80.417	
	5						-1.0000	.50000	
7	0	.14286	-3.5000	35.000	-183.75	541.33	-882.00	720.00	-219.04
	1	-.71429	21.000	-238.00	1365.0	-4281.7	7287.0	-6120.0	1890.5
	2	.85714	-35.000	490.00	-3237.5	11181.	-20300.	17760.	-5603.8
	3	-.14286	14.000	-308.00	2660.0	-10866.	21938.	-20460.	6662.0
	4			28.000	-525.00	3133.7	-7801.5	8162.0	-2798.3
	5					-100.33	483.00	-665.00	252.00
	6							1.0000	-.50000

$$b\left(\frac{1}{2}\right) = \frac{P-Q}{24n^2P^2Q^2} + O(n^{-3})$$

$$= \frac{1}{24}\left[\frac{1}{(nQ)^2} - \frac{1}{(nP)^2}\right] + O(n^{-3}). \tag{22}$$

It is because of this fact that the choice for t suggested is $\frac{1}{2}$.

Due to this property of $t = \frac{1}{2}$, bias coefficients $A_{w,s}\left(\frac{1}{2}\right)$ have been calculated for

$$b(t) = (Q - P) \sum_{w=1}^{\infty} \sum_{s=0}^{w-1} A_{w,s}(t)(PQ)^s(nPQ)^{-w}, \tag{23}$$

where

$$A_{w,s}(t) = \sum_{g=2s-w+1}^{w} A_{w,s,g}t^{w-g}.$$

These coefficients for values of w up to 7 are given in Table 2.

Table 2. *$A_{w,s}\left(\frac{1}{2}\right)$ Coefficients of the Asymptotic Series $b\left(\frac{1}{2}\right)$ in Powers of n^{-1} of the Bias*

	s						
w	0	1	2	3	4	5	6
1	0						
2	−.041667	0					
3	−.083333	.20833	0				
4	−.23229	.96458	−.79167	0			
5	−.85833	4.8979	−7.5875	2.7083	0		
6	−3.9830	28.807	−65.673	50.573	−8.7917	0	
7	−22.319	195.24	−588.65	715.05	−307.85	27.708	0

The formula $b(t)$ in (1) derived for the bias is based on asymptotic series to the logarithmic functions $\ln[1+(t+\gamma)/(nP)]$ and $\ln[1+(t-\gamma)/(nQ)]$. As a result, the finite series expression $b(t)$ does not converge; however, the series is asymptotic and "in such series the terms begin to decrease, and reach a minimum, afterwards increasing" (Bromwich, 1965). The larger the n and the less extreme the P considered, the more the range of possible values are in the interval of convergence of the series expansions of the logarithmic functions. Thus it is anticipated that for such values of n and P, the first few terms of the asymptotic series formulation $b(t)$ will be a good approximation to the bias.

Similar to the bias, a computing formula for the mean square error given in equation (2) can be derived.

Corollary 2. Let X be a random variable distributed as a binomial with parameters n and P. An asymptotic series in powers of n^{-1} of the mean square error of the estimator $\hat{\lambda}(t)$ of $\ln(P/Q)$ is

$$m(t) = \sum_{w=1}^{\infty} \sum_{s=0}^{w-1} \sum_{g=2s-w}^{w} F_{w,s,g} t^{w-g} (PQ)^{s} (nPQ)^{-w}, \tag{24}$$

where

$$\begin{aligned} F_{w,s,g} = {} & \sum_{h=0}^{g} \sum_{i=0}^{\left[\frac{g+h}{2}\right]} \sum_{v=i}^{w-g+i} \sum_{\ell=s-\left[\frac{g-h}{2}\right]}^{s} \sum_{r=s-\left[\frac{g-h}{2}\right]-\left[\frac{w+h-v+1+\kappa_{g-h}-\kappa_i}{2}\right]}^{\min(\ell,\left[\frac{v-1+\kappa_i}{2}\right])} \\ & \sum_{u=s-\left[\frac{g-h}{2}\right]-r-1+\kappa_i-\kappa_i\kappa_{g-h}}^{\min(\ell-r,\left[\frac{w+h-v-1+\kappa_{g-h-i}}{2}\right])} \sum_{f=s-\left[\frac{g-h}{2}\right]-r-u}^{\min(\ell-r-u,1-\kappa_i+\kappa_i\kappa_{g-h})} \delta_g \delta_\ell \delta_r \delta_u \delta_f \\ & 2^{\delta_{i,g+h}} \delta_{v-1} \delta_{w+h-v-1} \delta_{\left[\frac{g-h}{2}\right]+\left[\frac{w+h+\kappa_{g-h}}{2}\right]-s-\kappa_{v-i+1}(1-\kappa_{w+h+\kappa_{g-h}})} \\ & 4^{f} \binom{v}{i} \binom{w+h-v}{g+h-i} (-1)^{w+h+s} (\underline{\mathbf{v}}\ \underline{\mathbf{w+h-v}})^{-1} \\ & C_{h,g-h+1,s-\ell+1} d_{v,r+1,\kappa_i} d_{w+h-v,u+1,\kappa_{g-h-i}} \\ & d_{g-h-2(s-\ell),\ell-r-u-f+1,1-\kappa_{g-h}}. \end{aligned} \tag{25}$$

Proof. The series for the mean square error can be obtained by taking the expected value of (11). Using equation (2) of Lemma 1,

$$\begin{aligned} m(t) = {} & \sum_{j=0}^{\infty} \sum_{u=j}^{\infty} \sum_{i=0}^{[j/2]} \sum_{v=i}^{u-j+i} \sum_{h=0}^{[j/2]} \sum_{\ell=1}^{\left[\frac{j-2h+2}{2}\right]} \Bigg[2^{\delta_{ij}} \binom{v}{i} \binom{u-v}{j-i} (-1)^{u+\ell+1} \\ & (\underline{\mathbf{v}}\ \underline{\mathbf{u-v}})^{-1} C_{h,j-2h+1,\ell} t^{u-j} \left(Q^{v} + (-1)^{i+1} P^{v}\right) \\ & \left(Q^{u-v} + (-1)^{j-i+1} P^{u-v}\right) \left(Q^{j-2h-2\ell+2} + (-1)^{j} P^{j-2h-2\ell+2}\right) \\ & (PQ)^{h+\ell-u-1} \Bigg] n^{-u+h} \\ = {} & \sum_{j=0}^{\infty} \sum_{h=0}^{[j/2]} \sum_{w=j-h}^{\infty} \sum_{i=0}^{[j/2]} \sum_{v=i}^{w+h-j+i} \sum_{\ell=1}^{\left[\frac{j-2h+2}{2}\right]} \Bigg[2^{\delta_{ij}} \binom{v}{i} \binom{w+h-v}{j-i} \end{aligned}$$

$$(-1)^{w+h+\ell+1}(\underline{\mathbf{v}}\ \underline{\mathbf{w}+\mathbf{h}-\mathbf{v}})^{-1}C_{h,j-2h+1,\ell}t^{w+h-j}$$
$$\left(Q^v+(-1)^{i+1}P^v\right)\left(Q^{w+h-v}+(-1)^{j-i+1}P^{w+h-v}\right)$$
$$\left.\left(Q^{j-2h-2\ell+2}+(-1)^jP^{j-2h-2\ell+2}\right)(PQ)^{\ell-1-w}\right]n^{-w}, \qquad (26)$$

where u is replaced by $w+h$ in the last step.

When the order of summation involving j, h, and w are interchanged,

$$\sum_{j=0}^{\infty}\sum_{h=0}^{[j/2]}\sum_{w=j-h}^{\infty} \quad \text{becomes} \quad \sum_{w=0}^{\infty}\sum_{h=0}^{w}\sum_{j=2h}^{h+w}$$

and (26) can be expressed as

$$m(t)=\sum_{w=0}^{\infty}\sum_{h=0}^{w}\sum_{j=2h}^{h+w}\sum_{i=0}^{[j/2]}\sum_{v=i}^{w+h-j+i}\sum_{\ell=1}^{\left[\frac{j-2h+2}{2}\right]}\left[2^{\delta_{ij}}\binom{v}{i}\binom{w+h-v}{j-i}\right.$$
$$(-1)^{w+h+\ell+1}(\underline{\mathbf{v}}\ \underline{\mathbf{w}+\mathbf{h}-\mathbf{v}})^{-1}C_{h,j-2h+1,\ell}t^{w+h-j}$$
$$\left(Q^v+(-1)^{i+1}P^v\right)\left(Q^{w+h-v}+(-1)^{j-i+1}P^{w+h-v}\right)$$
$$\left.\left(Q^{j-2h-2\ell+2}+(-1)^jP^{j-2h-2\ell+2}\right)(PQ)^{\ell-1-w}\right]n^{-w}. \qquad (27)$$

The lower limit of summation for w can be set to 1 since for $w=0$ the corresponding term in (27) is zero.

Using Lemma 2, for positive integers v and $w+h-v$,

$$Q^v+(-1)^{i+1}P^v=(Q-P)^{1-\kappa_i}\sum_{r=1}^{\left[\frac{v+1+\kappa_i}{2}\right]} d_{v,r,\kappa_i}(-1)^{r-1}(PQ)^{r-1}, \qquad (28)$$

and

$$Q^{w+h-v}+(-1)^{j-i+1}P^{w+h-v}$$
$$=(Q-P)^{1-\kappa_{j-i}}\sum_{u=1}^{\left[\frac{w+h-v+1+\kappa_{j-i}}{2}\right]} d_{w+h-v,u,\kappa_{j-i}}(-1)^{u-1}(PQ)^{u-1}. \qquad (29)$$

The indicator functions δ_{v-1} and $\delta_{w+h-v-1}$ can be incorporated, as multiplicative terms, into (28) and (29) in order to extend these expressions to include the cases $v=0$ and $w+h-v=0$. Also, using Lemma 2.2,

$$Q^{j-2h-2\ell+2}+(-1)^jP^{j-2h-2\ell+2}$$
$$=(Q-P)^{\kappa_j}\sum_{g=1}^{\left[\frac{j-2h-2\ell+4}{2}\right]} d_{j-2h-2\ell+2,g,1-\kappa_j}(-1)^{g-1}(PQ)^{g-1}, \qquad (30)$$

where the upper limit of summation in (30) follows for reasons as discussed for equation (17). Substituting (28) to (30) into (33) yields

$$m(t)=\sum_{w=1}^{\infty}\sum_{h=0}^{w}\sum_{j=2h}^{h+w}\sum_{i=0}^{[j/2]}\sum_{v=i}^{w+h-j+i}\sum_{\ell=1}^{\left[\frac{j-2h+2}{2}\right]}\sum_{r=1}^{\left[\frac{v+1+\kappa_i}{2}\right]}\sum_{u=1}^{\left[\frac{w+h-v+1+\kappa_{j-i}}{2}\right]}$$
$$\sum_{g=1}^{\left[\frac{j-2h-2\ell+4}{2}\right]}\left\{2^{\delta_{ij}}\binom{v}{i}\binom{w+h-v}{j-i}(-1)^{w+h+\ell+r+u+g}\right.$$
$$\delta_{v-1}\delta_{w+h-v-1}(\underline{\mathbf{v}}\ \underline{\mathbf{w+h-v}})^{-1}C_{h,j-2h+1,\ell}t^{w+h-j}d_{v,r,\kappa_i}$$
$$\left.d_{w+h-v,u,\kappa_{j-i}}d_{j-2h-2\ell+2,g,1-\kappa_j}\right\}$$
$$(PQ)^{\ell-w+r+u+g-4}n^{-w}(Q-P)^{2(1-\kappa_i+\kappa_i\kappa_j)}, \tag{31}$$

where the power of the $(Q-P)$ term follows from

$$1-\kappa_i+\kappa_j+1-\kappa_{j-i}=2-\kappa_i+\kappa_j-(\kappa_j-\kappa_i)^2=2(1-\kappa_i+\kappa_i\kappa_j).$$

Since

$$\begin{aligned}(Q-P)^{2(1-\kappa_i+\kappa_i\kappa_j)}&=(1-4PQ)^{1-\kappa_i+\kappa_i\kappa_j}\\&=\sum_{f=0}^{1-\kappa_i+\kappa_i\kappa_j}\binom{1-\kappa_i+\kappa_i\kappa_j}{f}(-1)^f4^f(PQ)^f\\&=\sum_{f=0}^{1-\kappa_i+\kappa_i\kappa_j}(-1)^f4^f(PQ)^f,\end{aligned}$$

equation (31) can be expressed as

$$m(t)=\sum_{w=1}^{\infty}\sum_{h=0}^{w}\sum_{j=2h}^{h+w}\sum_{i=0}^{[j/2]}\sum_{v=i}^{w+h-j+1}\sum_{\ell=1}^{\left[\frac{j-2h+2}{2}\right]}\sum_{r=1}^{\left[\frac{v+1+\kappa_i}{2}\right]}\sum_{u=1}^{\left[\frac{w+h-v+1+\kappa_{j-i}}{2}\right]}$$
$$\sum_{f=0}^{1-\kappa_i+\kappa_i\kappa_j}\sum_{g=1}^{\left[\frac{j-2h-2\ell+4}{2}\right]}\left\{4^f2^{\delta_{ij}}\binom{v}{i}\binom{w+h-v}{j-i}(-1)^{w+h+\ell+r+u+g+f}\right.$$
$$\delta_{v-1}\delta_{w+h-v-1}(\underline{\mathbf{v}}\ \underline{\mathbf{w+h-v}})^{-1}C_{h,j-2h+1,\ell}t^{w+h-j}d_{v,r,\kappa_i}$$
$$\left.d_{w+h-v,u,\kappa_{j-i}}d_{j-2h-2\ell+2,g,1-\kappa_j}\right\}(PQ)^{\ell-w+r+u+g+f-4}n^{-w}. \tag{32}$$

Adjusting the order of summations, equation (32) is transformed to

$$
\begin{aligned}
m(t) = \sum_{w=1}^{\infty} \sum_{s=0}^{w-1} \Bigg\{ & \sum_{h=0}^{w} \sum_{j=d_1}^{w+h} \sum_{i=0}^{[j/2]} \sum_{v=i}^{w+h-j+i} \sum_{\ell=1}^{d_2} \sum_{r=d_3}^{d_4} \sum_{u=d_5}^{d_6} \sum_{f=d_7}^{d_8} 2^{\delta_{ij}} \\
& \delta_{[j/2]-h+\left[\frac{w+h+\kappa_j}{2}\right]-s-\kappa_{v-i+1}(1-\kappa_{w+h+\kappa_j})} 4^f \binom{v}{i} \binom{w+h-v}{j-i} \\
& (-1)^{w+h+s} \delta_{v-1} \delta_{w+h-v-1} (\underline{\mathbf{v}}\ \underline{\mathbf{w+h-v}})^{-1} C_{h,j-2h+1,\ell} t^{w+h-j} \\
& d_{v,r,\kappa_i} d_{w+h-v,u,\kappa_{j-i}} d_{j-2h-2\ell+2,s-\ell-r-u-f+4,1-\kappa_j} \Bigg\} \\
& (PQ)^s (nPQ)^{-w},
\end{aligned}
\tag{33}
$$

where

$$
\begin{aligned}
d_1 &= 2h + \max(0, 2s - w - h), \\
d_2 &= \min(s, [j/2] - h) + 1, \\
d_3 &= \max\left(0, s - [j/2] + h - \left[\frac{w+h-v+1+\kappa_j-\kappa_i}{2}\right]\right) + 1, \\
d_4 &= \min\left(s - \ell + 1, \left[\frac{v-1+\kappa_i}{2}\right]\right) + 1, \\
d_5 &= \max(0, s - [j/2] + h - r + \kappa_i - \kappa_i\kappa_j) + 1, \\
d_6 &= \min\left(s - \ell - r + 2, \left[\frac{w+h-v-1+\kappa_{j-i}}{2}\right]\right) + 1, \\
d_7 &= \max(0, s - [j/2] + h - r - u + 2), \\
d_8 &= \min(s - \ell - r - u + 3, 1 - \kappa_i + \kappa_i\kappa_j),
\end{aligned}
$$

and where the upper summation limit of s has been lowered to $w - 1$ since for $s = w$ the indicator function

$$
\delta_{[j/2]-h+\left[\frac{w+h+\kappa_j}{2}\right]-s-\kappa_{v-i+1}(1-\kappa_{w+h+\kappa_j})} = \delta_{-\kappa_1} = \delta_{-1} = 0.
$$

Equation (33) is now transformed as follows: first, j is replaced by $j+2h$; second, r, u and ℓ are replaced by $r+1$, $u+1$ and $\ell+1$ respectively; third, g is set to be $h+j$; and fourth, ℓ is replaced by $s-\ell$. These changes yield:

$$
m(t) = \sum_{w=1}^{\infty} \sum_{s=0}^{w-1} \sum_{g=\max(0,2s-w)}^{w} \Bigg\{ \sum_{h=0}^{g} \sum_{i=0}^{\left[\frac{g+h}{2}\right]} \sum_{v=i}^{w-g+i} \sum_{\ell=e_1}^{s} \sum_{r=e_2}^{e_3} \sum_{u=e_4}^{e_5} \sum_{f=e_6}^{e_7} 2^{\delta_{i,g+h}}
$$

$$\delta_{v-1}\delta_{w+h-v-1}\delta_{\left[\frac{g-h}{2}\right]+\left[\frac{w+h+\kappa_{g-h}}{2}\right]-s-\kappa_{v-i+1}(1-\kappa_{w+h+\kappa_{g-h}})}$$

$$4^f\binom{v}{i}\binom{w+h-v}{g+h-i}(-1)^{w+h+s}(\underline{\mathbf{v}}\ \underline{\mathbf{w}+\mathbf{h}-\mathbf{v}})^{-1}C_{h,g-h+1,s-\ell+1}$$

$$\left.d_{v,r+1,\kappa_i}d_{w+h-v,u+1,\kappa_{g-h-i}}d_{g-h-2(s-\ell),\ell-r-u-f+1,1-\kappa_{g-h}}\right\}$$

$$t^{w-g}(PQ)^s(nPQ)^{-w}, \tag{34}$$

where

$$e_1 = \max\left(0, s-\left[\frac{g-h}{2}\right]\right),$$

$$e_2 = \max\left(0, s-\left[\frac{g-h}{2}\right]-\left[\frac{w+h-v+1+\kappa_{g-h}-\kappa_i}{2}\right]\right),$$

$$e_3 = \min\left(\ell, \left[\frac{v-1+\kappa_i}{2}\right]\right),$$

$$e_4 = \max\left(0, s-\left[\frac{g-h}{2}\right]-r-1+\kappa_i-\kappa_i\kappa_{g-h}\right),$$

$$e_5 = \min\left(\ell-r, \left[\frac{w+h-v-1+\kappa_{g-h-i}}{2}\right]\right),$$

$$e_6 = \max\left(0, s-\left[\frac{g-h}{2}\right]-r-u\right),$$

$$e_7 = \min\left(\ell-r-u, 1-\kappa_i+\kappa_i\kappa_{g-h}\right).$$

Using the indicator function, more specifically, δ_g, δ_ℓ, δ_r, δ_u and δ_f, equation (34) is transformed into equation (24) and the proof is complete.

The constants $F_{w,s,g}$ have been calculated for various values of w using a computer program. The calculated constants are presented in Table 3 for values of w up to and including 7.

Disregarding terms of order higher than n^{-1}, the familiar expression for the mean square error,

$$m(t) = \frac{1}{nPQ} = \frac{1}{nP} + \frac{1}{nQ},$$

derived, for example by Woolf (1955), follows from using the coefficients in Table 3.

For $t = 1/2$, the mean square error coefficients $F_{w,s}(1/2)$ have been calculated for

$$m(t) = \sum_{w=1}^{\infty}\sum_{s=0}^{w-1} F_{w,s}(t)(PQ)^s(nPQ)^{-w}, \tag{35}$$

Table 3. *$F_{w,s,g}$ Coefficients of the Asymptotic Series $m(t)$ in Powers of n^{-1} of the Mean Square Error*

		g							
w	*s*	0	1	2	3	4	5	6	7
1	0		1.0000						
2	0	1.0000	−3.0000	1.7500					
	1	−4.0000	8.0000	−5.0000					
3	0	−1.0000	5.5000	−8.8333	4.000				
	1	4.0000	−22.000	37.333	−17.667				
	2		12.000	−24.000	13.000				
4	0	.91667	−8.3333	25.917	−32.000	12.757			
	1	−4.3333	42.667	−142.17	184.50	−75.778			
	2	2.6667	−45.333	181.00	−260.33	113.42			
	3			−28.000	56.000	−29.000			
5	0	−.83333	11.417	−58.278	136.67	−114.18	52.861		
	1	4.6667	−70.500	387.72	−960.00	1051.10	−393.97		
	2	−5.3333	111.00	−716.78	1959.0	−2284.1	887.73		
	3		−26.667	313.33	−1090.0	1446.7	−603.33		
	4				60.000	−120.00	61.000		
6	0	.76111	−14.700	111.63	−422.50	829.43	−784.75	270.46	
	1	−4.9667	105.90	−866.83	3470.8	−7099.6	6911.1	−2423.6	
	2	8.2000	−220.60	2077.8	−9138.0	19935.	−20259.	7289.3	
	3	−2.0444	116.80	−1559.3	8294.0	−20318.	22204.	−8329.3	
	4			173.33	−1746.7	5641.0	−1749.7	2905.1	
	5					−124.00	248.00	−125.00	
7	0	−.70000	18.150	−192.00	1066.6	−3329.3	5762.9	−5024.2	1649.3
	1	5.2333	−149.20	1698.2	−9981.7	32526.	−58135.	51845.	−17263.
	2	−11.267	386.33	−4976.7	31944.	−110960.	207660.	−191180.	64919.
	3	6.1333	−323.47	5347.6	−40171.	155010.	−311720.	301080.	−105220.
	4		42.933	−1519.5	16506.	−78474.	179420.	−187820.	68762.
	5				−933.33	8670.7	−26782.	32984.	−13181.
	6						252.00	−504.00	253.00

where

$$F_{w,s}(t) = \sum_{g=2s-w}^{w} F_{w,s,g} t^{w-g}.$$

These values are presented in Table 4 for w from 1 to 7. When $w = 1$, of course, there is no change in $m(t)$ since it is independent of this parameter.

Similar to the discussion on the bias, the formula for mean square error given by (5) is an asymptotic series and it is anticipated that for large n and

Table 4. *$F_{w,s}\left(\frac{1}{2}\right)$ Coefficients of the Asymptotic Series $m\left(\frac{1}{2}\right)$ in Powers of n^{-1} of the Mean Square Error*

	s						
w	0	1	2	3	4	5	6
1	1.0000						
2	.50000	–2.0000					
3	.83333	–4.0000	4.0000				
4	2.2517	–14.007	23.000	–8.0000			
5	8.3389	–64.212	152.59	–115.00	16.000		
6	39.162	–360.05	1124.2	–1364.1	533.00	–32.000	
7	222.73	–2381.5	9220.0	–15585	10884	–2359.0	64.000

P values in the midrange that the first few terms of the formulation will be a good approximation to the mean square error.

Corollary 2 provides an asymptotic series approximation of the mean square error of $\hat{\lambda}(t)$. Using this result and $b(t)$ of Corollary 1, a similar asymptotic series in powers of n^{-1} of the variance of this esimator can be derived.

3. DISCUSSION

Asymptotic series in powers of n^{-1} were developed in regards to the bias and mean square error for logit estimation. Such series are characterized by terms that begin to decrease, reach a minimum, and increase thereafter. For large n and moderate P, the first several terms of each series will provide an adequate approximate to the quantity it represents. Even for small values of n, the performance of these series are often satisfactory by utilizing the first few terms. The one notable exception is the asymptotic series for the bias of the logit estimator at small values of P. The optimal number of terms to retain depends on the parameters n and P.

The series for the bias illustrates the reasoning for the customary 1/2 cell adjustment. However, the boundary problem of the estimator and the mean square error criterion raise issues in regards to the "optimality" of 1/2. The asymptotic series $b(t)$ and $m(t)$ for the bias and mean square error of the logit estimator, respectively, have been developed as functions of t. They will provide an analytical basis for a more thorough investigation of cell adjustment and its ramifications on the estimation of the log odds ratio and common odds ratio.

The asymptotic series have been mathematically expressed in such a way that general coefficients can be developed. This makes the application of the formulas to the log odds ratio a simple step. Based on these formulations, a thorough investigation of the bias and mean square error of the log odds ratio estimator can be investigated. In particular, as noted above, an analytical evaluation of the effect of the cell adjustments can be examined.

REFERENCES

Anscombe, F. J. (1956), "On estimating binomial response relations". *Biometrika* **43**, 461–464.

Bromwich, T. J. I'A. (1965). *An Introduction to the Theory of Infinite Series*, Second edition. London: Macmillan.

Gart, J. J., and J. R. Zweifel (1967), "On the bias of various estimators of the logit and its variance with application to quantal bioassay". *Biometrika* **54**, 181–187.

Gart, J. J., H. M. Pettigrew, and D. G. Thomas (1985), "The effect of bias, variance estimation, skewness and kurtosis of the empirical logit on weighted least squares analysis". *Biometrika* **72**, 179–190.

Haldane, J. B. S. (1955), "The estimation and significance of the logarithm of a ratio of frequencies". *Annals of Human Genetics* **20**, 309–311.

Hotelling, H. (1953), "New light on the correlation coefficient and its transform". *Journal of the Royal Statistical Society*, Series B **15**, 193–232.

Naylor, A. F. (1967), "Small sample considerations in combining 2 × 2 tables". *Biometrics* **23**, 349–356.

Woolf, B. (1955), "On estimating the relation between blood group and disease". *Annals of Human Genetics* **19**, 251–253.

Walter W. Hauck [1]

ESTIMATION OF A COMMON ODDS RATIO

ABSTRACT

For comparing two proportions from stratified samples, one approach is via inference on the odds ratio. The various point and interval estimators of a common odds ratio from multiple 2×2 tables are reviewed in this paper, with particular emphasis on the point estimators, their asymptotic properties, and what is known about finite-sample properties. Based on research to date, the conditional maximum likelihood and Mantel-Haenszel estimators are recommended as the point estimators of choice. Confidence interval methods have not been studied as well, but there is a method associated with the Mantel-Haenszel estimator that is a good choice. There is the possibility of improvement in the finite-sample properties of these estimators. However, further work is needed before one of these modifications can be recommended for general use.

1. INTRODUCTION

One of the common problems of statistical inference is that of comparing two proportions adjusting for stratification. This problem arises in nonrandomized studies where the data may be stratified according to values of a confounding variable in order to eliminate the bias due to confounding (Anderson *et al.*, 1980, Chapter 7). In randomized studies, stratification may be employed to increase the precision of inferences, particularly if stratified randomization had been used (Green and Byar, 1978).

One approach to the comparison of two proportions is by inference regarding the corresponding odds ratio. In this paper, the various point and interval estimators of the odds ratio are reviewed. Asymptotic properties under both asymptotic conditions, fixed number of strata and increasing

[1] Department of Epidemiology and International Health, HSW 1699, University of California, San Francisco, California 94143

I. B. MacNeill and G. J. Umphrey (eds.), Biostatistics, 125–149.

number of strata, are presented as each estimator is introduced in Section 4. Finite-sample comparisons of the estimators are presented in Section 5.

2. NOTATION AND ASSUMPTIONS

After stratification, the data fall into K 2×2 strata as follows for stratum k:

		Group		
		1	2	
Outcome	Success	X_{1k}	X_{2k}	T_k
	Fail	$N_{1k} - X_{1k}$	$N_{2k} - X_{2k}$	$N_k - T_k$
		N_{1k}	N_{2k}	N_k

Each X_{ik} is assumed to be binomial with parameters P_{ik} and N_{ik}, $i = 1, 2$ and $k = 1, \ldots, K$, and to be independent of all the other binomial variables. Implicit in the binomial assumption is an assumption of within-stratum homogeneity. In particular, within each stratum, the same probability of success is assumed to apply to all units belonging to each group. Consequently, results based on this assumption will not apply in the case of stratification of a numerical variable such as age, though one can argue that they ought to be a reasonable approximation if the number of strata is large enough to ensure approximate homogeneity.

The odds ratio in the kth stratum is

$$\psi_k = P_{1k}Q_{2k}/(Q_{1k}P_{2k}),$$

where $Q_{ik} = 1 - P_{ik}$, $i = 1, 2$ and $k = 1, \ldots, K$, and $0 < \psi_k < \infty$ is assumed throughout. In this review, we consider only the case where the odds ratio is the same in all strata:

$$\psi_k = \psi \qquad \text{for all } k = 1, \ldots, K.$$

The natural logarithm of the odds ratio will be denoted by $\gamma_k = \ln \psi_k$ and $\gamma = \ln \psi$, and the total sample size will be denoted by $N = \sum_{k=1}^{K} N_k$.

3. ASYMPTOTIC CASES

Consideration of the asymptotic properties of estimators of ψ is complicated by the existence of two appropriate asymptotic cases, referred to here as the fixed-strata and increasing-strata cases. In the fixed-strata case, the number of strata is fixed, and the sample sizes within each stratum become large. In the increasing-strata case, the converse holds; the stratum sizes are fixed, but the number of strata becomes large. More formally,

Fixed-strata asymptotic conditions:

a) K is fixed; and
b) $N_{ik} = NA_{ik}$, where $A_{ik} > 0$ for all i and k, $\sum_{k=1}^{K}\sum_{i=1}^{2} A_{ik} = 1$, and $N \to \infty$.

Increasing-strata asymptotic conditions:

a) The $N_{ik}, i = 1, 2$ and $k = 1, \ldots, K$ are fixed; and
b) $K \to \infty$.

For the fixed-strata case, inference is based on the binomial distributions introduced in Section 2. When working with the increasing-strata case, the principal distribution of interest will be that of X_{ik} given T_k, N_{1k}, and N_{2k}. As originally noted by Fisher (1935), this distribution is the extended hypergeometric (Harkness, 1965):

$$Pr[X_{1k} = x \mid T_k, N_{1k}, N_{2k}] = \frac{\binom{N_{1k}}{x}\binom{N_{2k}}{T_k - x}\psi^x}{\sum_{u=a_k}^{b_k}\binom{N_{1k}}{u}\binom{N_{2k}}{T_k - u}\psi^u}, \qquad (3.1)$$

where $a_k = \max(0, T_k - N_{2k})$ and $b_k = \min(T_k, N_{1k})$. When inference is based on the distribution (3.1), asymptotic results for the increasing-strata case will depend on the marginal joint distribution of (T_k, N_{1k}, N_{2k}). Following Breslow (1981), it is convenient to index possible values of this triplet (possible marginal configurations) by $j = 1, \ldots, J$, where J is assumed to be finite, and let $\pi_j = \lim_{k \to \infty} K_j/K$ where K_j is the number of tables with configuration j.

4. THE ESTIMATORS

We will consider five odds ratio estimators (Table 1), two estimators appropriate in the fixed-strata case and three that are appropriate in both asymptotic cases. Lastly, some modifications to the primary estimators are

Table 1. *Point Estimators of the Odds Ratio from Multiple 2×2 Tables*

A. Appropriate in fixed–strata case only	
Unconditional maximum likelihood	Gart (1962)
Woolf's	Woolf (1955), Gart (1966)
B. Appropriate in both asymptotic cases	
Conditional maximum likelihood	Birch (1964)
Empirical Bayes	Pierce and Sands (1975)
Mantel-Haenszel	Mantel and Haenszel (1959)

introduced. Confidence interval methods will be introduced along with their corresponding point estimator.

4.1 Maximum Likelihood (Unconditional)

The first estimator is maximum likelihood based directly on the binomial distributions introduced in Section 2. The corresponding likelihood, dropping the combinatorial terms, is

$$L = \prod_{i=1}^{2} \prod_{k=1}^{K} P_{ik}^{X_{ik}} Q_{ik}^{N_{ik}-X_{ik}}. \tag{4.1}$$

Since

$$P_{2k} = P_{1k}/(\psi Q_{1k} + P_{1k}), \qquad k = 1, \ldots, K,$$

the likelihood is actually a function of only $K+1$ parameters $P_{11}, \ldots, P_{1k}, \psi$. The maximum likelihood estimate of ψ, $\hat{\psi}_{ML}$, is then found by maximizing (4.1) with respect to the $K+1$ parameters simultaneously. This is the odds ratio estimate obtained by fitting the no-three-factor-interaction model in log-linear or logit analyses, and any of the standard methods for obtaining estimates for those models could be used here. This may require working with γ and

$$\rho_k = \ln(P_{1k}/Q_{1k}), \qquad k = 1, \ldots, K$$

to avoid boundary problems if an iterative maximization process is used. If ψ (or γ) is the only parameter of interest, the maximization problem can be reduced to that for the single parameter; see Section 4.3.

In the fixed-strata asymptotic case, the usual optimality results for maximum likelihood estimation hold, and $\hat{\psi}_{ML}$ has the following asymptotic distribution (Gart, 1962):

$$N^{1/2}(\hat{\psi}_{ML} - \psi) \underset{N\to\infty}{\overset{d}{\longrightarrow}} \mathcal{N}(0, \psi^2/V), \tag{4.2}$$

where $V = \sum_{k=1}^{K} v_k$ and

$$v_k^{-1} = (A_{1k}P_{1k}Q_{1k})^{-1} + (A_{2k}P_{2k}Q_{2k})^{-1}, \qquad k = 1, \ldots, K. \tag{4.3}$$

In the increasing-strata asymptotic case, however, $\hat{\psi}_{ML}$ need not even be consistent. In the case of matched pairs for example (the extreme case of $N_{1k} = N_{2k} = 1$ for all k), $\hat{\psi}_{ML}$ converges to ψ^2 as the number of pairs (K) becomes large (Andersen, 1973). Pike *et al.* (1980) demonstrated the considerable asymptotic bias, as $K \to \infty$, of $\hat{\psi}_{ML}$ when the sample sizes per table are small and Breslow (1981) has demonstrated that the N_k must be large for the asymptotic bias to be small. The problem is that the usual maximum likelihood properties require a parameter space that remains fixed as $N \to \infty$. This is violated in the increasing-strata case, since the number of parameters increases with K.

To construct confidence intervals based on $\hat{\psi}_{ML}$ or $\hat{\gamma}_{ML}$ is straightforward. Intervals for γ, for example, are of the form $\hat{\gamma}_{ML} \pm z_{\alpha/2}\hat{V}^{-1/2}$, where $z_{\alpha/2}$ is the upper $\alpha/2$ percentage point of the standard normal distribution and $\hat{V}$ is an estimator of NV in (4.2). That is, we have

$$\hat{v}_k^{-1} = (N_{1k}\hat{P}_{1k}\hat{Q}_{1k})^{-1} + (N_{2k}\hat{P}_{2k}\hat{Q}_{2k})^{-1}, \tag{4.4}$$

where the $\hat{P}_{1k}$ and $\hat{P}_{2k}$ are based on the maximum likelihood estimates for all $K+1$ parameters:

$$\hat{P}_{1k} = [1 + \exp(-\hat{\rho}_k)]^{-1}$$

and

$$\hat{P}_{2k} = \hat{P}_{1k}/(\hat{\psi}_{ML}\hat{Q}_{1k} + \hat{P}_{1k}).$$

Clearly this approach to confidence intervals will be valid only in the fixed-strata case.

Intervals for ψ can be constructed by exponentiating the γ intervals or directly using the large-sample variance of $\hat{\psi}_{ML}$. The exponentiation route is preferable here (and for the other interval methods of similar form) as it avoids the problem of negative values in an interval for ψ.

4.2 Woolf's Estimator

Woolf (1955) proposed a noniterative estimator of γ:

$$\hat{\gamma}_W = \sum_{k=1}^{K} w_k \hat{\gamma}_k \Big/ \sum_{k=1}^{K} w_k,$$

where

$$\begin{aligned}\hat{\gamma}_k &= \ln\left[p_{1k}q_{2k}/(q_{1k}p_{2k})\right] \\ &= \ln\{X_{1k}(N_{2k} - X_{2k})/[(N_{1k} - X_{1k})X_{2k}]\}\end{aligned}$$

is the estimator of γ from the kth stratum,

$$\begin{aligned}p_{ik} &= X_{ik}/N_{ik}, \\ q_{ik} &= 1 - p_{ik},\end{aligned}$$

and

$$\begin{aligned}w_k^{-1} &= (N_{1k}p_{1k}q_{1k})^{-1} + (N_{2k}p_{2k}q_{2k})^{-1} \\ &= 1/X_{1k} + 1/(N_{1k} - X_{1k}) + 1/X_{2k} + 1/(N_{2k} - X_{2k}). \qquad (4.5)\end{aligned}$$

Since

$$N^{1/2}(\hat{\gamma}_k - \gamma) \underset{N\to\infty}{\overset{d}{\longrightarrow}} \mathcal{N}(0, 1/v_k),$$

w_k^{-1} can be seen as an estimate of the large-sample variance, $(Nv_k)^{-1}$, of $\hat{\gamma}_k$ and thus $\hat{\gamma}_W$ is based on the principle of weighting inversely proportional to the variance. (Note that the variance estimators (4.4) and (4.5) differ only on how the P_{ik} are estimated.) The Woolf estimator can also be viewed as a special case of the minimum logit chi-square estimator, proposed for quantal-response bioassay by Berkson (1944).

Modifications to Woolf's estimator have been considered since $\hat{\gamma}_W$ cannot be calculated if there is a zero in any cell of any of the K tables. Modification to $\hat{\gamma}_k$ and w_k will be considered separately.

Haldane (1955) and Anscombe (1956) proposed a modification to the estimator of the log odds that corresponds to

$$\hat{\gamma}'_k = \ln\{(X_{1k} + .5)(N_{2k} - X_{2k} + .5)/\left[(N_{1k} - X_{1k} + .5)(X_{2k} + .5)\right]\}$$

as the estimator of the logarithm of the odds ratio. Haldane showed that the choice of 0.5 eliminates the N^{-1} terms in the bias and that therefore

$$E(\hat{\gamma}'_k) = \gamma_k + O(N^{-2}).$$

The modified weights, w'_k, are defined analogously to the $\hat{\gamma}'_k$:

$$(w'_k)^{-1} = 1/(X_{1k} + .5) + 1/(N_{1k} - X_{1k} + .5) \\ + 1/(X_{2k} + .5) + 1/(N_{2k} - X_{2k} + .5), \qquad (4.6)$$

as proposed by Gart (1966). This "modified Woolf" estimator, $\hat{\gamma}_{MW}$, is then calculated exactly as the original Woolf estimator after the addition of 0.5 to every cell. This estimator is also known as the empirical logit estimator.

Gart (1962) showed that, in the fixed-strata case, $\hat{\psi}_W = \exp(\hat{\gamma}_W)$, has the same asymptotic distribution (4.2) as $\hat{\psi}_{ML}$ and, in particular, that $\hat{\psi}_W$ is asymptotically efficient. Since the 0.5 corrections for the modified Woolf estimator are asymptotically negligible in the fixed-strata case, this asymptotic optimality property will be shared by $\hat{\psi}_{MW}$.

The nature of the Woolf estimator is that it requires sufficiently large-sample sizes in each table in order for each $\hat{\gamma}_k$ to be at least reasonable in terms of bias. In particular, Breslow (1981) has shown that both N_1 and N_2 must be large in order to reduce the asymptotic bias. Consequently, the Woolf estimators should not be expected to perform well in the increasing-strata case.

Confidence intervals based on Woolf's estimator are of the same form as for unconditional maximum likelihood. For γ we have $\hat{\gamma}_{MW} \pm z_{\alpha/2} W^{-1/2}$, where $W = \sum w'_k$. W^{-1} is an estimate of the large-sample variance of $\hat{\gamma}_{MW}$.

4.3 Conditional Maximum Likelihood

In the increasing-strata asymptotic case, as mentioned earlier, the usual maximum likelihood estimator breaks down because there are $K+1$ parameters to be estimated and K is not fixed. The K parameters corresponding to the K tables, $\rho_1, \ldots, \rho_K$, are nuisance parameters in this problem in the sense that ψ is the primary parameter of interest. An estimation approach that avoids the difficulty due to an increasing number of parameters is conditional maximum likelihood (CML). The principle of this approach is to determine the likelihood for the parameter(s) of interest, ψ in this case, conditional on minimal sufficient statistics for the nuisance parameters and then to estimate the target parameter by maximizing this conditional likelihood.

For the problem of interest here, $T_1, \ldots, T_K$ are the minimally sufficient statistics for $\rho_1, \ldots, \rho_K$ and the consequent conditional likelihood is the product of the extended hypergeometric likelihoods (3.1) from each table:

$$L' = \prod_{k=1}^{K} \frac{\binom{N_{1k}}{X_{1k}}\binom{N_{2k}}{T_k - X_{1k}}\psi^{X_{1k}}}{\sum_{u=a_k}^{b_k}\binom{N_{1k}}{u}\binom{N_{2k}}{T_k - u}\psi^{u}}. \qquad (4.7)$$

Birch (1964) has shown that the value of ψ that maximizes L', $\hat{\psi}_{CML}$, is the solution to

$$\sum_{k=1}^{K} X_{1k} = \sum_{k=1}^{K} E(X_{1k} \mid T_k, \psi). \tag{4.8}$$

Miettinen (1970) gives the somewhat simpified form of (4.8) for M:1 matching. Thomas (1975) gives a computer program for finding $\hat{\psi}_{CML}$.

Due to the relative difficulty of solving (4.8), particularly in the days before high-speed computers were commonly available, a number of closed-form approximations to $\hat{\psi}_{CML}$ were proposed. [Birch (1964) and Goodman (1969) are notable examples.] These approximations share the property of being consistent only at $\psi = 1$. They are therefore of limited usefulness and are not considered further in this paper.

One approximation worth mention is based on the asymptotic normality of the extended hypergeometric distribution (Cornfield, 1956; Hannan and Harkness, 1963). Specifically, as the N_{ik} become large, the conditional distribution of X_{ik} given T_k becomes normal with mean

$$\tilde{E}[X_{1k} \mid T_k, \psi] = \tilde{X}_k$$

and variance

$$\tilde{V}[X_{1k} \mid T_k, \psi] = \left(\frac{1}{\tilde{X}_k} + \frac{1}{T_k - \tilde{X}_k} + \frac{1}{N_{1k} - \tilde{X}_k} + \frac{1}{N_{2k} - T_k + \tilde{X}_k}\right)^{-1},$$

where $\tilde{X}_k$ is the solution, such that $a_k \leq \tilde{X}_k \leq b_k$, to

$$\tilde{X}_k(N_{2k} - T_k + \tilde{X}_k)/\left[(T_k - \tilde{X}_k)(N_{1k} - \tilde{X}_k)\right] = \psi.$$

(Fleiss (1979) gives the closed-form solution to $\tilde{X}_k$ as a function of ψ.) $X_+ = \sum X_{1k}$ is then asymptotically normal with mean

$$\tilde{E}[X_+ \mid T_1, \ldots, T_k, \psi] = \sum_{k=1}^{K} \tilde{E}[X_{1k} \mid T_k, \psi]$$

and variance

$$\tilde{V}[X_+ \mid T_1, \ldots, T_k, \psi] = \sum_{k=1}^{K} \tilde{V}[X_{1k} \mid T_k, \psi].$$

Gart (1970) proposed replacing the right side (4.8) by its asymptotic normal approximation and then finding the solution. Breslow (1976) showed that the resultant estimator ($\hat{\psi}_{AML}$ in Gart's notation) is actually the unconditional maximum likelihood estimator and, hence, that the usual and conditional maximum likelihood estimators of ψ are asymptotically equivalent in the *fixed-strata* asymptotic case (a result that was also noted in the errata to Gart, 1971). A program for finding $\hat{\psi}_{AML}$, such as that of Thomas (1975), is then an alternative to direct maximization of the likelihood (4.1) with respect to the $K+1$ parameters, as long as the ρ_k are not of interest.

Andersen (1970) determined the properties of conditional maximum likelihood estimators in the increasing-strata asymptotic case. Applied to estimation of the odds ratio, Andersen's results yield

$$K^{1/2}(\hat{\psi}_{CML} - \psi) \underset{K\to\infty}{\overset{d}{\longrightarrow}} \mathcal{N}\left(0, \psi^2 \Big/ \sum_{j=1}^{J} \pi_j \mathrm{Var}_j(X)\right),$$

where $\mathrm{Var}_j(X)$ denotes the exact variance of the extended hypergeometric distribution for marginal configuration j, and π_j is defined at the end of Section 3. The issue of the asymptotic efficiency of $\hat{\psi}_{CML}$ in the increasing-strata asymptotic case is unresolved. Andersen shows that the Cramér-Rao lower bound for the asymptotic variance is not attained by $\hat{\psi}_{CML}$, so there may be an estimator that is consistent and asymptotically normal with a lower variance. Lindsay (1983) and Davis (1985) have partly resolved this. Lindsay's results show, for the case $N_{1k} = N_1$ and $N_{2k} = N_2$ for all k, that the CML estimator is asymptotically efficient if the ρ_k are from a distribution with more than $(N_1+N_2)/2$ points of support. In addition, Davis has shown that, except for matched pairs, no weighted average of the stratum-specific odds ratio is as efficient asymptotically as CML in increasing-strata cases.

Once we allow that the ρ_k are from some distribution, we could try empirical Bayes techniques. Though there has been considerable work on random effects logistic models, little of this seems applicable to the common odds ratio problem. (See Stiratelli *et al.* (1984) for references in this area and a model applicable if the γ_k are also assumed random.) Ejigou and McHugh (1981) did develop an empirical Bayes estimator for the case of $M:1$ matching, but then showed that it was the CML estimator. The approach of Pierce and Sands (1975) is applicable to the odds ratio problem but has not been tried (to this author's knowledge), except for matched pairs (Pierce, 1976). From Lindsay's results, we know that the empirical Bayes approach will not improve on CML asymptotically, but there is the possibility it may do so in finite samples.

There are two approaches to confidence intervals that are associated with conditional maximum likelihood. The first is the so-called exact method, based on the hypergeometric distribution (4.7). Gart (1970) gives the exact interval for ψ with confidence coefficient at least $1-\alpha$ as $(\hat{\psi}_1, \hat{\psi}_2)$, found as the solution to

$$\sum_{J\in R_1} \prod_{k=1}^{K} P[j_k \mid T_k, \hat{\psi}_1] = \alpha/2$$

and

$$\sum_{J\in R_2} \prod_{k=1}^{K} P[j_k \mid T_k, \hat{\psi}_2] = \alpha/2,$$

where $J = (j_1, \ldots, j_k)$. The set R_1 consists of all J such that $\sum_{k=1}^{K} j_k \geq X_+$; R_2 consists of all J such that $\sum_{k=1}^{K} j_k \leq X_+$. In words, the lower (upper) limit, $\hat{\psi}_1$ ($\hat{\psi}_2$), is the value of ψ such that the probability of the observed or larger (smaller) value of X_+ is $\alpha/2$.

Determination of the exact limits can be very time-consuming, particularly if the N_{ik} are large. One approximation is to replace the extended hypergeometric distribution by its asymptotic normal approximation, as described above. This was done for a single 2×2 table by Cornfield (1956) and extended to multiple 2×2 tables by Gart (1970, 1971). Gart's method is that $1-\alpha$ confidence intervals for ψ, $(\hat{\psi}_1, \hat{\psi}_2)$, are found as the solutions to

$$[X_+ - \tilde{E}(X_+ \mid T_1, \ldots, T_k, \hat{\psi}_2)]/\tilde{V}(X_+ \mid T_1, \ldots, T_k, \hat{\psi}_2)^{1/2} = -z_{\alpha/2}$$

and

$$[X_+ - \tilde{E}(X_+ \mid T_1, \ldots, T_k, \hat{\psi}_1)]/\tilde{V}(X_+ \mid T_1, \ldots, T_k, \hat{\psi}_1)^{1/2} = +z_{\alpha/2}. \quad (4.9)$$

Thomas' (1975) program includes the exact and Cornfield-Gart interval methods. Mantel (1977) has proposed intervals of the form (4.9) except using the exact hypergeometric mean and variance for X_+ and using a continuity correction.

A second approach based on conditional maximum likelihood would be to estimate the variance of $\hat{\gamma}_{CML}$ from the inverse of the negative second derivative matrix of the conditional likelihood in the sample. Such an approach is directly analogous to that for unconditional maximum likelihood. Unlike unconditional maximum likelihood, however, most currently available programs for solving for $\hat{\gamma}_{CML}$ do not involve the second derivatives of the likelihood and so this approach has not seen application. One computer program that does compute the inverse second derivatives is that of Breslow and Day (1980, Appendix V; also Smith *et al.*, 1981) though that program is for a more general problem.

4.4 Mantel-Haenszel Estimator

Mantel and Haenszel (1959) proposed what is likely the most commonly used odds ratio estimator, though its properties have been determined analytically only in recent years. The Mantel-Haenszel estimator is

$$\hat{\psi}_{MH} = \sum_{k=1}^{K} X_{1k}(N_{2k} - X_{2k})/N_k \Big/ \sum_{k=1}^{K}(N_{1k} - X_{1k})X_{2k}/N_k$$
$$= \sum_{k=1}^{K} S_k\hat{\psi}_k \Big/ \sum_{k=1}^{K} S_k, \tag{4.10}$$

where

$$\hat{\psi}_k = X_{1k}(N_{2k} - X_{2k})/\left[(N_{1k} - X_{1k})X_{2k}\right]$$

and

$$S_k = (N_{1k} - X_{1k})X_{2k}/N_k$$
$$= (1/N_{1k} + 1/N_{2k})^{-1}(1 - p_{1k})p_{2k}. \tag{4.11}$$

The first form of $\hat{\psi}_{MH}$ given in (4.10) is most convenient for calculation, while the second is most convenient for some analytic purposes.

Hauck (1979) obtained the asymptotic distribution of $\hat{\psi}_{MH}$ for the fixed-strata case (though the derivation for the case of heterogeneous odds ratios was corrected by Guilbaud, 1983):

$$N^{1/2}(\hat{\psi}_{MH} - \psi) \underset{N\to\infty}{\overset{d}{\longrightarrow}} \mathcal{N}\left(0, \psi^2 \sum_{k=1}^{K}(M_k^2/v_k) \Big/ \left(\sum_{k=1}^{K} M_k\right)^2\right), \tag{4.12}$$

where

$$M_k = (1/A_{1k} + 1/A_{2k})^{-1}(1 - P_{1k})P_{2k},$$

and v_k is given by (4.3). By applying the Cauchy-Schwarz inequality, $\hat{\psi}_{MH}$ can be seen to be asymptotically inefficient for all values of the odds ratio except $\psi = 1$ (except for a few degenerate cases; see Tarone *et al.*, 1983).

The asymptotic distribution in the increasing-strata case was found by Breslow (1981):

$$K^{1/2}(\hat{\psi}_{MH} - \psi) \underset{K\to\infty}{\overset{d}{\longrightarrow}} \mathcal{N}\left(0, \frac{\sum_{j=1}^{J} \pi_j \mathrm{Var}_j(R - S)}{\sum_{j=1}^{J} \pi_j [E_j(S)]^2}\right), \tag{4.13}$$

where $R = X_1(N_2 - X_2)/N$, $S = (N_1 - X_1)X_2/N$, and E_j and Var_j denote the mean and variance given the jth marginal configuration. When $\psi = 1$, the asymptotic variance of $\hat{\psi}_{MH}$ is equal to that of $\hat{\psi}_{CML}$, analogous to the relation to $\hat{\psi}_{ML}$ in the fixed-strata asymptotic case.

The basic form for confidence intervals based on the Mantel-Haenszel estimator is $\hat{\gamma}_{MH} \pm z_{\alpha/2}\hat{V}^{1/2}$, where $\hat{V}$ is an estimate of the variance of $\hat{\gamma}_{MH}$. The presence of two different asymptotic variance formulas for the Mantel-Haenszel estimator complicates the problem of choosing an estimator of the variance. There have been a number of proposals each of which could be used for confidence intervals; see Table 2. If we were to set criteria that an ideal $\hat{V}$ would satisfy, we would ask that $\hat{V}$ be appropriate in both asymptotic cases and that $\hat{V}$ not be appreciably more difficult to calculate than $\hat{\psi}_{MH}$ itself. Of the nine variance estimators listed in Table 2, only $\hat{V}_U$ and $\hat{V}_{US}$ proposed by Robins *et al.* (1986) satisfy both these criteria. Of these two $\hat{V}_{US}$ is the variance estimator of choice, since it, but not $\hat{V}_U$, is invariant under interchange of rows.

Robins *et al.* (1986) worked with the variance in (4.13), that is actually applicable in both asymptotic cases (Breslow and Liang, 1982), and found estimators of the numerator and denominator that were consistent in both asymptotic cases. The first proposal, $\hat{V}_C$, is to evaluate the hypergeometric means and variances in (4.13) at $\psi = \hat{\psi}_{MH}$. $\hat{V}_C$ is appropriate in both asymptotic cases, but is computationally burdensome. In the second estimator of Robins *et al.*, $\hat{V}_U$, the numerator and denominator of $\hat{V}_U$ are consistent estimators, in both asymptotic cases, of the numerator and denominator of the variance (4.13). However, $\hat{V}_U$ is not invariant under interchange of rows or columns. For a symmetrized version, $\hat{V}_{US}$, they considered the arithmetic average of $\hat{V}_U$ with $\hat{V}_U$ recomputed after interchange of rows. (Breslow and Liang had similarly proposed symmetrizing $\hat{V}'_H$ by taking the geometric mean of $\hat{V}'_H$ with $\hat{V}'_H$ recomputed after interchange of rows.) The resulting estimator is

$$\hat{V}_{US}(\hat{\gamma}_{MH}) = \frac{\sum_k F_k R_k}{2R_+^2} + \frac{\sum_k (F_k S_k + G_k R_k)}{2R_+ S_+} + \frac{\sum_k G_k S_k}{2S_+^2},$$

where $F_k = (X_{1k} + N_{2k} - X_{2k})/N_k$, $G_k = (N_{1k} - X_{1k} + X_{2k})/N_k$, $R_+ = \sum_k R_k$, and $S_+ = \sum_k S_k$.

Robins *et al.* also reported a simulation study that compared $\hat{V}_U$, $\hat{V}_{US}$, $\hat{V}_C$, $\hat{V}_B$, $\hat{V}_{BL}$, $\hat{V}'_H$, and $\hat{V}_G$ in terms of agreement with the actual variance and validity of confidence intervals for situations corresponding to both asymptotic cases. As would be expected from the above discussion, only $\hat{V}_U$, $\hat{V}_{US}$, and $\hat{V}_C$ consistently gave confidence interval coverage rates close to nominal. They recommend use of one of these three variance estimators for all types of data. By the criteria set earlier, $\hat{V}_{US}$ is the method of choice. An important caveat is that there is no reason to expect any of these variance estimators

Table 2. *Estimators of the Variance of the Mantel-Haenszel Estimator*

Estimator	*Principal Reference*	*Comments*
	A. Valid in Fixed-Strata Case Only	
$\hat{V}_H$	Hauck (1979)	An overestimator (Robins *et al.*, 1986); not invariant under interchange of rows or columns (Ury, 1982).
$\hat{V}'_H$	Breslow and Liang (1982)	Modification to $\hat{V}_H$ that is less of an estimator (Robins *et al.*, 1986); not invariant under interchange of rows or columns.
$\hat{V}_G$	Gart, reported by Ury (1982)	Modification to $\hat{V}_H$ using estimates of the P_{ik} consistent with the estimated odds ratio and thus is invariant under interchange of rows or columns [analogous to the relation between (4.4) and (4.5)].
	B. Valid in Increasing-Strata Case Only	
$\hat{V}_B$	Breslow (1981)	An underestimator (Robins *et al.*, 1986); simplified form for matched studies given by Connett *et al.* (1982) and Fleiss (1984).
	C. Valid in Both Asymptotic Cases	
$\hat{V}_{BL}$	Breslow and Liang (1982)	Weighted combination of $\hat{V}'_H$ and $\hat{V}_B$; not invariant under interchange of rows or columns.
$\hat{V}_C$	Robins *et al.* (1986)	Evaluate hypergeometric means and variances in (4.13) at $\psi = \hat{\psi}_{MH}$.
$\hat{V}_U$	Robins *et al.* (1986)	Based on consistent estimation of the numerator and denominator of (4.13); not invariant under interchange of rows and columns.
$\hat{V}_{US}$	Robins *et al.* (1986)	Symmetrized form of $\hat{V}_U$.

to perform well in the case of small total sample size, i.e., small number of tables each with small sample sizes.

In addition to the choices for $\hat{V}$ listed in Table 2, confidence intervals can be obtained from the test-based method proposed by Miettinen (1976). Applied in this context, Miettinen's proposal is to determine a variance, $\hat{V}_M$, from

$$\chi^2 = \hat{\gamma}^2/\hat{V}_M,$$

where χ^2 is a one degree-of-freedom chi-square statistic for testing the null hypothesis $\gamma = 0$. The choice of χ^2 is left open. As one example, Hauck and Wallemark (1983), in their comparative study, used the test proposed by Mantel and Haenszel (1959), but without the continuity correction.

It should be noted that the test-based principle is not valid (Greenland, 1984) and so problems are to be expected. One problem discussed by Halperin (1977) (with rejoinder by Miettinen, 1977), is that the method requires that the variance of the point estimator not depend on the value of the parameter. For estimation of the odds ratio, this can be, at best, an approximation near the null value ($\gamma = 0$). We thus have a theoretical basis for expecting that the test-based intervals may not do well for γ away from zero. Regardless of these considerations, the test-based method is mentioned here because it has received consideration as a confidence interval method for odds ratios from single 2×2 tables (Brown, 1981; Gart and Thomas, 1982), and because it has been used for odds ratios from multiple tables (Centers for Disease Control, 1983, for example).

4.5 Modifications to the Mantel-Haenszel, Maximum Likelihood, and Woolf Estimators

Hauck *et al.* (1982) proposed modifying the Mantel-Haenszel and (unconditional) maximum likelihood estimators in a manner analogous to Gart's (1966) modification to the Woolf estimator. The modified Woolf estimator can be viewed as applying Woolf's original formula to tables that had 0.5 added to each cell. Hauck *et al.'s* proposal was to apply the Mantel-Haenszel and maximum likelihood formulas to such modified tables, after adding some suitable constant to each cell.

This proposal had two motivations. First, the modified Woolf estimator had been shown to have a smaller variance than the Mantel-Haenszel and maximum likelihood estimators (McKinlay, 1975; Hauck *et al.*, 1980) and it was conjectured that at least some of this increase in precision was due to the larger effective sample size for the Woolf estimator ($N + 2K$ instead of N). Second, the preliminary study by Hauck *et al.* found that Mantel-Haenszel and maximum likelihood tended to be overestimators (for $\psi > 1$). Since the modification would tend to bring the estimates of ψ closer to 1, it was possible that modification might improve the bias properties of the estimators simultaneous with a decrease in variance.

Hauck *et al.* (1980) also suggested that, in terms of bias, a 0.5 correction was too great for the Woolf estimator. Hitchcock (1962) had suggested 0.25 as a correction in a related situation, so 0.25 and 0.5 corrections were considered for the Mantel-Haenszel and maximum likelihood estimators and a 0.25 correction for the Woolf estimator. Additional related modifications to the Woolf estimator have been considered by Wells (1982). All these estimators are asymptotically equivalent, in the fixed-strata case, to the corresponding original estimator.

Pseudocount methods (Fienberg and Holland, 1970) would let the data determine the appropriate constant to add to each cell, instead of sticking with constants 0.5 and 0.25. The rationale is that the more the data were consistent with ψ near 1.0, the larger the modification would be, thus pulling the estimators in towards 1.0. Hauck *et al.* (1982) considered two such variants of the Mantel-Haenszel, unconditional maximum likelihood, and Woolf estimators and found that, away from $\psi = 1$, these pseudocount variants became much more biased than the other modifications. These two variants will not be considered further.

An alternative modification to the Mantel-Haenszel estimator, based on the jackknife principle, was proposed by Breslow and Liang (1982). Letting $\hat{\gamma}_{MH}^{(k)}$ denote the Mantel-Haenszel estimator of γ after deleting the kth table, the jackknife estimator proposed by Breslow and Liang is

$$\hat{\gamma}_{JK} = K\hat{\gamma}_{MH} - (K-1)\sum_{k=1}^{K} \hat{\gamma}_{HM}^{(k)}/K,$$

and $\hat{\psi}_{JK} = \exp(\hat{\gamma}_{JK})$.

Jewell (1984) also considered jackknifing as one of many possibilities he considered for reducing the bias of the Mantel-Haenszel estimator and its logarithm in matched samples and of the conditional maximum likelihood estimator for matched pairs. Jewell's basic approach was to eliminate the K^{-1} term in the bias. For example, for matched pairs, direct adjustment for this term yields

$$\hat{\psi}_{ADJ} = 2R_+/(2S_+ + 1)$$

and

$$\hat{\gamma}_{ADJ} = \ln\left\{R_+(2R_+ + 1)/\left[S_+(2S_+ + 1)\right]\right\}$$

compared to $\hat{\psi}_{CML} = \hat{\psi}_{MH} = R_+/S_+$.

5. FINITE-SAMPLE COMPARISONS

5.1 Point Estimators

In recent years, a number of simulation studies have been conducted for the purpose of determining and then comparing the properties (bias and precision) of the various odds ratio point estimators (Table 3). The first of these studies was done by McKinlay (1975), who considered cases corresponding to the fixed-strata asymptotic case. McKinlay found that the Woolf estimator was consistently the most precise, but that the bias of the Woolf estimator increased rapidly with increased number of strata for a fixed sample size. Of the three estimators she considered, only the Mantel-Haenszel estimator consistently removed the bias in all the cases she considered.

Table 3. *Finite-Sample Studies of the Point Estimators of ψ and γ*

Reference	*Relevant Asymptotic Case*	*Estimators Compared**
McKinlay (1975)	Fixed-Strata	Birch, MH, MW
McKinlay (1978)	Fixed-Strata	Birch, Goodman, MH, MW, UML
Lubin (1981)	Increasing-Strata	CML, UML
Hauck *et al.* (1982)	Fixed-Strata	MH, MW, UML, (+4 variants of each)
Jewell (1984)	Matched Pairs	CML (=MH) (+4 variants)
Hauck (1984)	Fixed-Strata	CML, MH, UML
Liang (1984)	Both	CML, MH, MW, UML

*CML = conditional maximum likelihood; MH = Mantel-Haenszel; MW = modified Woolf; UML = unconditional maximum likelihood.

In her second simulation study, McKinlay (1978) was interested primarily in the effect of heterogeneity of the strata-specific odds ratios (the ψ_k) on the properties of the estimators of the log odds ratio, but some results were for the homogeneous case of interest in this review. As in McKinlay (1975), she found the modified Woolf estimator to be usually most precise but with marked increases in bias for larger numbers of strata. Between the Mantel-Haenszel and maximum likelihood estimators, they were either similar or the Mantel-Haenszel estimator was less biased and more precise.

As noted by Hauck *et al.* (1982), when considered as a comparison of odds ratio estimators from stratified samples (rather than as a comparison of matching to stratification as was McKinlay's primary intention), McKinlay's (1975) work did not provide sufficient information to choose a stratified estimator—she did not consider either of the maximum likelihood estimators. Also, one could not determine to what extent her results were due to the approximation inherent in stratifying a continuous variable. In McKinlay (1978), the (unconditional) maximum likelihood estimator was considered, but the approximation inherent in stratifying a numerical variable remained.

The studies of Hauck *et al.* (1980) and Hauck *et al.* (1982) were intended to correct for these problems for cases corresponding to fixed-strata asymptotics and, in the 1982 paper, to study the modifications introduced in Section 4.5. Results for the Woolf estimator with both the 0.25 and 0.5 modifications confirmed McKinlay's work—it was more precise than the corresponding modification to the Mantel-Haenszel and unconditional maximum likelihood estimators, but more biased, sometimes considerably so. Comparing the Mantel-Haenszel and maximum likelihood estimators, the Mantel-Hasenszel estimator was the *more* precise (when compared to the maximum likelihood estimator with the same modification), as McKinlay (1978) had found. The choice of estimator depended on whether one wished to estimate ψ or γ. The original Mantel-Haenszel and maximum likelihood estimators overestimated ψ by an amount that increased with ψ ($\psi > 1$). Use of the 0.25 correction for Mantel-Haenszel and the 0.5 correction for maximum likelihood considerably reduced the overestimation. For estimating ψ, the original Mantel-Haenszel estimator and the maximum likelihood estimator with the 0.25 correction had the best bias properties.

The omission by Hauck *et al.* (1982) of the conditional maximum likelihood estimator was rectified by Hauck (1984) who compared the conditional maximum likelihood estimator to its two common competitors, the Mantel-Haenszel and unconditional maximum likelihood estimators for fixed-strata cases. (Similar comparisons were also included by Liang (1984), who also considered the modified Woolf estimator and found similar results.) Of these three estimators, the conditional maximum likelihood is superior in terms of both bias and precision. The properties of the Mantel-Haenszel estimator were similar; it was less biased as an estimator of ψ than conditional maximum likelihood in 11 of the 36 cases considered by Hauck, but in all 11 of these cases, both estimators were essentially unbiased (absolute bias < 0.012). Otherwise conditional maximum likelihood was the equal or slightly superior for both γ and ψ. Unconditional maximum likelihood was uniformly inferior. However, compared to the modified versions of the Mantel-Haenszel and unconditional maximum likelihood estimators, the situation is not so clear. For estimating ψ, the 0.25-modified Mantel-Haenszel and 0.5-modified unconditional maximum likelihood estimators were superior in terms of both bias and precision for the cases considered (minimum of 20% decrease in average absolute bias and 10% decrease in average variance, with the exception of one case of the 12 where the average variances were essentially equal). For estimating γ, the conditional maximum likelihood estimator is slightly superior to the best of the other estimators (Mantel-Haenszel and 0.25-modified maximum likelihood) in terms of bias and comparable to the Mantel-Haenszel estimator in precision.

Maximum likelihood estimation in situations corresponding to the increasing-strata case (small numbers per table) was considered by Lubin (1981) and Liang (1984). Lubin compared conditional to unconditional maximum likelihood and Liang compared those two methods and the Mantel-Haenszel and Woolf estimator for $1 : m$ matching and other cases of small numbers per table. Both found that unconditional maximum likelihood was a consistent overestimator of γ ($\gamma > 0$), as also noted by Jewell (1984) for matched samples. Conditional maximum likelihood, while still a consistent overestimator, had a much smaller bias (estimated to be up to 1/50 that of the unconditional maximum likelihood estimator in one of Lubin's cases) and was more precise. Liang found that the Mantel-Haenszel estimator was only slightly inferior to CML, as in the fixed-strata cases.

For matched pairs, Jewell (1984) found that the jackknife was less effective at bias reduction than methods that adjusted the Mantel-Haenszel estimator for the K^{-1} term in the expansion for the bias. Jewell prefers $\hat{\psi}_{ADJ}$ for estimating ψ and different choices for estimating γ depending on the underlying parameter values.

5.2 Interval Estimators

There has been considerably less attention given to evaluation of the properties of the various odds ratio interval estimators. Methods for a single 2×2 table were compared by Gart and Thomas (1972, 1982), Fleiss (1979), and Brown (1981). Seven methods for multiple tables have been considered by Hauck and Wallemark (1983). However, the results on the three Mantel-Haenszel methods have been superseded by the work of Robins *et al.* (1986), who compared six Mantel-Haenszel variances in confidence intervals. Both Hauck and Wallemark, and Robins *et al.* considered both asymptotic cases.

In the fixed strata cases, we can consider the six methods considered by Hauck and Wallemark as falling into four groups. Two methods, namely Gart's asymptotic approximation to the exact and the Mantel-Haenszel with Hauck's variance formula were conservative (that is, coverage probabilities greater than nominal) for all investigated combinations. The result for Hauck's formula agrees with Robins *et al.*'s finding that $\hat{V}_H'$ overestimates the variance of $\hat{\gamma}_{MH}$. The asymptotic exact method is more conservative in multiple 2×2 tables than for a single table (Gart and Thomas, 1982).

Next is a method whose coverage probability falls below the nominal level for all investigated combinations, namely the Mantel-Haenszel method with Breslow's variance formula. This variance formula is intended for large numbers of tables and would not be expected to do well in the large sample cases. Again, the results here agree with those of Robins *et al.*, who found $\hat{V}_B$ to be an underestimator of the variance of $\hat{\gamma}_{MH}$.

The third group consists of two methods that were inconsistent in the sense that their coverage probabilities were not consistently above or consistently below nominal. Coverage probabilities for the Woolf method ranged from the most conservative to the most liberal of all methods. Hauck *et al.* (1982) had found that the Woolf point estimator underestimates the magnitude of γ for $\gamma \neq 0$, and that this bias become larger as γ increases. This bias is reflected in the confidence intervals; where the coverage probability falls below nominal, ψ tends to lie above the intervals. For example, for $K = 10$, $N = 500$, $\psi = 5.0$, the estimated coverage probability for nominal 95% intervals was .897, and the estimated coverage probability of the interval lying below ψ is .102. Gart and Thomas (1982) found a similar bias for a single table, though of small magnitude. It appears that all cell frequencies must be large for this method to be reasonable.

The second inconsistent method is the Mantel-Haenszel estimator with standard error determined by Miettinen's test-based method. For values of ψ at or near 1, its coverage probabilities are near nominal. However as ψ increases the estimated coverage probability eventually falls below nominal. For a single table, Brown (1981) and Gart and Thomas (1982) found similar results and, by considering values of ψ larger than those considered by Hauck and Wallemark, obtained coverage probabilities much lower than nominal than those reported by Hauck and Wallemark.

Lastly, Hauck and Wallemark reported a method that performed consistently well, in the large-sample cases: the Mantel-Haenszel estimator with Breslow and Liang's compromise variance formula. Its estimated coverage probability was never far from nominal. In Robins *et al.*'s study, the coverage probabilities using $\hat{V}_{US}$ were generally slightly closer to nominal than those using $\hat{V}_{BL}$.

The seven methods considered by Hauck and Wallemark for the matched sample cases fell into three groups. Three methods, the exact and two Mantel-Haenszel methods, with Hauck's and BL's compromise variance formulas, were consistently conservative. The compromise formula was the least conservative of the three, and, in general, Hauck's formula was the most conservative of the Mantel-Haenszel methods. As in the large-sample cases, the confidence interval results agreed with Robins *et al.*'s results that $\hat{V}'_H$ tended to overestimate the variance of $\hat{\gamma}_{MH}$. Occasionally, though, the coverage probability was slightly lower for $\hat{V}'_H$ than for $\hat{V}_B$ or $\hat{V}_C$. In contrast to the fixed-strata cases, Robins *et al.* found that the standard deviation of $\hat{V}'_H$ was much larger than that of $\hat{V}_B$ in matched sample cases so that $\hat{V}'_H < \hat{V}_B$ is possible.

Three methods were inconsistent in the same sense as above. The inappropriateness of the Woolf method for matched samples was confirmed by Hauck and Wallemark's simulation; it was erratic to say the least, with coverage probabilities for the nominal 95% intervals ranging from 7% to 99.9%. The test-based and asymptotic exact methods were also inconsistent. Both were conservative for ψ at or near one. For larger ψ, the estimated coverage probabilities dropped well below nominal, particularly for the asymptotic approximation to the exact method. The test-based method behaved as in the fixed-strata cases in terms of coverage probabilities; it has coverage probabilities near nominal for near one, but is unreliable away from one. Unlike the large-sample cases, however, there is a caveat to its behavior near one. The average interval length and, particularly, the standard deviation of the interval length are inflated when $\psi = 1$, relative to the other reasonable methods. A value of chi-square near zero must occasionally grossly inflate $\hat{V}_M$ and hence the interval length.

The method that was most consistently near nominal of those considered by Hauck and Wallemark was the Mantel-Haenszel estimator with Breslow's variance formula. Of the seven methods they considered, this was the method of choice. The only caution is that the estimated coverage probabilities tended to drop slightly below nominal for some of the 1:8 matching cases. Robins *et al.* also found coverage probabilities below nominal for tables with $N_1 = N_2 = 5$. As for the fixed-strata cases, confidence intervals using the variances proposed by Robins *et al.* had coverage probabilities consistently near nominal.

6. CONCLUSIONS

The primary motivation for considering and reviewing the various odds ratio estimators has been the desire to reach a reasonable recommendation as to which estimator to use in the different situations that can be encountered. If, to begin with, we leave aside consideration of the modifications presented in Section 4.6 which have been proposed recently, the choice of estimator is reasonably clear. The Mantel-Haenszel and conditional maximum likelihood estimators share the property of being robust in the sense that they behave reasonably in both the fixed-strata and increasing-strata asymptotic cases. The Woolf and unconditional maximum likelihood estimators, on the other hand, are badly biased in the increasing-strata case. These asymptotic considerations have been found to apply also in various finite-sample studies. It thus seems reasonable to restrict consideration to the Mantel-Haenszel and CML estimators.

Considerations of asymptotic efficiency and minimum bias in finite-samples give an advantage to conditional maximum likelihood over the Mantel-Haenszel estimator. Conversely, there are three advantages to the Mantel-Haenszel method. First, it is considerably simpler, since it is noniterative, with bias and variance properties close to conditional maximum likelihood. Second, with the work of Robins *et al.* (1986), we have an associated confidence interval procedure that is also noniterative and applicable in both asymptotic cases. Third, Donald (1984) for the fixed-strata case and Liang (1985) for the increasing-strata case, have shown that the Mantel-Haenszel estimator remains consistent and asymptotically normal when the binomial assumption is violated and observations within a stratum are allowed to be dependent. In the fixed-strata case, Donald showed that unconditional maximum likelihood remains consistent, but Liang showed that the CML estimator does not remain consistent in the increasing-strata case. In summary, the Mantel-Haenszel estimator is a very reasonable choice.

Since all the estimators considered here are biased in finite samples, there is the opportunity to modify the estimators to simultaneously improve (or at least not worsen) their bias and precision properties. Modifications of the type proposed by Hauck *et al.* (1982), analogous to the modified Woolf estimator, appear reasonable in the case of moderate numbers in each table, but their motivation in the increasing-strata case is lacking. The jackknifed Mantel-Haenszel estimator proposed by Breslow and Liang (1982) and also considered by Jewell (1984) also appears reasonable, but here, since the estimator is defined by deletion of tables, the motivation in the case of a small number of tables is lacking. Breslow and Liang note, with respect to the jackknifed variance, further work is needed to determine the proper unit for deletion and the optimum weights for combining the pseudovalues. Other possibilities are the bias-adjusted estimators developed by Jewell.

"Further work is needed" is actually a good closing. While the conditional maximum likelihood and Mantel-Haenszel are clearly the point estimators of choice among estimators that have been reasonably well studied to date, it is also clear that improvement of finite-sample properties is possible.

REFERENCES

Andersen, E. B. (1970), "Asymptotic properties of conditional maximum-likelihood estimators". *Journal of the Royal Statistical Society,* Series B **32,** 283–301.

Andersen, E. B. (1973), *Conditional Inference and Models for Measuring.* Copenhagen: Mentalhygienisk Forlag.

Anderson, S., A. Auquier, W. W., Hauck, D. Oakes, W. Vandaele, and H. Weisberg (1980), *Statistical Methods for Comparative Studies*. New York: Wiley and Sons.

Anscombe, F. J. (1956), "On estimating binomial response relations". *Biometrika* **43**, 461–464.

Berkson, J. (1944), "Application of the logit function to bioassay". *Journal of the American Statistical Association* **39**, 357–365.

Birch, M. W. (1964), "The detection of partial association, I: the 2 × 2 case". *Journal of the Royal Statistical Society*, Series B **26**, 313–324.

Breslow, N. E. (1976), "Regression analysis of the log odds ratio: A method for retrospective studies". *Biometrics* **32**, 409–416.

Breslow, N. E. (1981), "Odds ratio estimators when the data are sparse". *Biometrika* **68**, 73–84.

Breslow, N. E., and N. E. Day (1980), *Statistical Methods in Cancer Research, Volume 1—The Analysis of Case-Control Studies*. Lyon: International Agency for Research on Cancer.

Breslow, N. E., and K. Y. Liang (1982), "The variance of the Mantel-Haenszel estimator". *Biometrics* **38**, 943–952; Errata: *Biometrics* **40**, 1217 (1984).

Brown, C. C. (1981), "The validity of approximation methods for interval estimation of the odds ratio". *American Journal of Epidemiology* **113**, 474–480.

Centers for Disease Control Cancer and Steroid Hormone Study (1983), "Long-term oral contraceptive use and the risk of breast cancer." *Journal of the American Medical Association* **249**, 1591–1595.

Connett, J., A. Ejigou, R. McHugh, and N. Breslow (1981), "The precision of the Mantel-Haenszel estimator in case-control studies with multiple matching". *American Journal of Epidemiology* **116**, 875–877.

Cornfield, J. (1956), "A statistical problem arising from retrospective studies". *Proceedings of the Third Berkeley Symposium on Mathematical Statistics and Probability* **4**, 135–148.

Davis, L. J. (1985), "Weighted averages of the observed odds ratios when the number of tables is large". *Biometrika* **72**, 203–205.

Donald, A. (1984), "The analysis of clustered data in sets of 2 × 2 contingency tables." Ph.D. thesis, The University of Western Ontario.

Ejigou, A., and R. McHugh (1981), "Relative risk estimation under multiple matching". *Biometrika* **68**, 85–91.

Fienberg, S. E., and P. Holland (1970), "Methods for eliminating zero counts in contingency tables". In *Random Counts in Models and Structures*, ed. G. P. Patil, pp. 233–260. University Park: Pennsylvania State University Press.

Fisher, R. A. (1935), "The logic of inductive inference". *Journal of the Royal Statistical Society*, Series A **98**, 39–54.

Fleiss, J. L. (1979), "Confidence intervals for the odds ratio in case-control studies: the state of the art". *Journal of Chronic Diseases* **32**, 69–77.

Fleiss, J. L. (1984), "The Mantel-Haenszel estimator in case-control studies with varying numbers of controls matched to each case". *American Journal of Epi-*

demiology **120,** 1–3.

Gart, J. J. (1962), "On the combination of relative risks". *Biometrics* **18,** 601–610.

Gart, J. J. (1966), "Alternative analyses of contingency tables". *Journal of the Royal Statistical Society,* Series B **28,** 164–179.

Gart, J. J. (1970), "Point and interval estimation of the common odds ratio in the combination of 2 × 2 tables with fixed marginals". *Biometrika* **57,** 471–475.

Gart, J. J., (1971), "The comparison of proportions: a review of significance tests, confidence intervals and adjustments for stratification". *Review of the International Statistical Institute* **39,** 148–169; errata: **40,** 221 (1972).

Gart, J. J., and D. G. Thomas (1972), "Numerical results on approximate confidence limits for the odds ratio". *Journal of the Royal Statistical Society,* Series B **34,** 441–447.

Gart, J. J., and D. G. Thomas (1982), "The performance of three approximate confidence limit methods for the odds ratio". *American Journal of Epidemiology* **115,** 453–470.

Goodman, L. A. (1969), "On partitioning χ^2 and detecting partial association in three-way contingency tables". *Journal of the Royal Statistical Society,* Series B **31,** 486–498.

Green, S. B., and D. P. Byar (1978), "The effect of stratified randomization on size and power of statistical tests in clinical trials". *Journal of Chronic Diseases* **31,** 445–454.

Greenland, S. (1984), "A counter example to the test-based principle of setting confidence limits". *American Journal of Epidemiology* **120,** 4–7.

Guilbaud, O. (1983), "On the large-sample distribution of the Mantel-Haenszel odds-ratio estimator". *Biometrics* **39,** 523–525.

Haldane, J. B. S. (1955), "The estimation and significance of the logarithm of a ratio of frequencies". *Annals of Human Genetics* **20,** 309–311.

Halperin, M. (1977), "Letter to the editor re: Miettinen (1976)". *American Journal of Epidemiology* **105,** 496–498.

Hannan, J., and W. Harkness (1963), "Normal approximation to the distribution of two independent binomials, conditional on fixed sum". *Annals of Mathematical Statistics* **34,** 1593–1595.

Harkness, W. L. (1965), "Properties of the extended hypergeometric distribution". *Annals of Mathematical Statistics* **36,** 938–945.

Hauck, W. W. (1979), "The large-sample variance of the Mantel-Haenszel estimator of a common odds ratio". *Biometrics* **35,** 817–819.

Hauck, W. W. (1984), "A comparative study of conditional maximum likelihood estimation of a common odds ratio". *Biometrics* **40,** 1117–1123.

Hauck, W. W., S. Anderson, and F. J. Leahy (1982), "Finite-sample properties of some old and some new estimators of a common odds ratio from multiple 2 × 2 tables". *Journal of the American Statistical Association* **77,** 145–152.

Hauck, W. W., F. J. Leahy, and S. Anderson (1980), "Finite-sample properties of estimators of a common odds ratio from multiple 2 × 2 tables". Presented at the August, 1980 Joint Statistical Meetings in Houston, Texas.

Hauck, W. W., and C.-B. Wallemark (1983), "A comparison of confidence interval methods for a common odds ratio". Presented at the August, 1983 Joint Statistical Meetings in Toronto, Ontario.

Hitchcock, S. E. (1962), A note on the estimation of the parameters of the logistic function, using the minimum logit χ^2 method". *Biometrika* **49**, 250–252.

Jewell, N. P. (1984), "Small-sample bias of point estimators of the odds ratio from matched sets". *Biometrics* **40**, 421–435.

Liang, K. Y. (1984), "The comparison of four estimators of a common odds ratio in 2 × 2 tables". Johns Hopkins University, Department of Biostatistics, paper No. 544.

Liang, K. Y. (1985), "Odds ratio reference with dependent data". *Biometrika* **72**, 678–682.

Lindsay, B. G. (1983), "Efficiency of the conditional score in a mixture setting". *Annals of Statistics* **11**, 486–497.

Lubin, J. H. (1981), "An empirical evaluation of the use of conditional and unconditional likelihoods for case-control data". *Biometrika* **68**, 567–571.

Mantel, N. (1977), "Tests and limits for the common odds ratio of several 2 × 2 contingency tables: methods in analogy with the Mantel-Haenszel procedure". *Journal of Statistical Planning and Inference* **1**, 179–189.

Mantel, N., and W. M. Haenszel (1959), "Statistical aspects of the analysis of data from retrospective studies of disease". *Journal of the National Cancer Institute* **22**, 719–748.

McKinlay, S. M. (1975), "The effects of bias on estimators of relative risk for pair-matched and stratified samples". *Journal of the American Statistical Association* **70**, 859–864.

McKinlay, S. M. (1978), "The effect of nonzero second-order interaction on combined estimators of the odds ratio". *Biometrika* **65**, 191–202.

Miettinen, O. S. (1970), "Estimation of relative risk from individually matched series". *Biometrics* **26**, 75–86.

Miettinen, O. (1976), "Estimability and estimation in case-referent studies". *American Journal of Epidemiology* **103**, 226–235.

Miettinen, O. (1977), "Reply to Halperin (1977)". *American Journal of Epidemiology* **105**, 498–502.

Pierce, D. A. (1976), "A random effects model for matched pairs of binomial data". Technical Report No. 55, Department of Statistics, Oregon State University, Corvallis, Oregon.

Pierce, D. A., and B. R. Sands (1975), "Extra-Bernoulli variation in binary data". Technical Report No. 46, Department of Statistics, Oregon State University, Corvallis, Oregon.

Pike, M. C., A. P. Hill, and P. G. Smith (1980), "Bias and efficiency in logistic analyses of stratified case-control studies". *International Journal of Epidemiology* **9**, 89–95.

Robins, J., N. Breslow, and S. Greenland (1986), "Estimators of the Mantel-Haenszel variance consistent in both sparse-data and large-strata limiting models". *Biometrics,* in press.

Smith, P. G., M. C. Pike, A. P. Hill, N. E. Breslow, and N. E. Day (1981), "Multivariate conditional logistic analysis of stratum-matched case-control studies (Algorithm AS 162)". *Applied Statistics* **30**, 190–197.

Stiratelli, R., N. Laird, and J. H. Ware (1984), "Random-effects models for serial observations with binary response". *Biometrics* **40**, 961–971.

Tarone, R. E., J. J. Gart, and W. W. Hauck (1983), "On the asymptotic inefficiency of certain noniterative estimators of a common relative risk or odds ratio". *Biometrika* **70**, 519–522.

Thomas, D. G. (1975), "Exact and asymptotic methods for the combination of 2×2 tables". *Computers and Biomedical Research* **8**, 423–446.

Ury, H. K. (1982), "Hauck's approximate large-sample variance of the Mantel-Haenszel estimator". *Biometrics* **38**, 1094–1095.

Wells, G. A. (1982), "Contributions to the estimation of the logit, log odds and common odds ratio". Ph.D. thesis, The University of Western Ontario.

Woolf, B. (1955), "On estimating the relation between blood group and disease". *Annals of Human Genetics* **19**, 251–253.

Alan Donald and Allan Donner [1]

THE EFFECT OF CLUSTERING ON THE ANALYSIS OF SETS OF 2×2 CONTINGENCY TABLES

ABSTRACT

The usual methods of analyzing data in sets of 2×2 contingency tables are invalid when observations are correlated. This happens when, for example, families rather than individuals are assigned to treatments. This paper compares two models of clustering in dichotomous data: the beta-binomial model and a model developed by Cohen and Altham. The beta-binomial distribution is then used to model responses in K 2×2 contingency tables. This leads to an expression for the variance of the maximum likelihood estimator of the common odds ratio, which is compared to the usual MLE of the odds ratio. The performance of the Mantel-Haenszel estimator of the odds ratio under cluster sampling is assessed using a simulation study.

1. INTRODUCTION

Two problems that often complicate the analysis of dichotomous data in comparative studies are the necessity for stratification and the occurrence of dependent responses. Each of these problems has been solved to some extent. The stratification of fourfold tables may be dealt with using the familiar Mantel-Haenszel procedures, which have been extensively studied in the past 25 years. The phenomenon of dependent responses—which we call the clustering problem—has recently begun to receive more attention. Cohen (1976) and Altham (1976) suggested a heuristic model for clustering in categorical data; Brier (1980) used the Dirichlet multinomial distribution. This paper explores what occurs when the two problems coincide. Detailed proofs, which are omitted, are given by Donald (1985) and Donald and Donner (1985).

[1] Department of Epidemiology and Biostatistics, The University of Western Ontario, London, Ontario N6A 5C1

I. B. MacNeill and G. J. Umphrey (eds.), Biostatistics, 151–167.

2. REVIEW

2.1 The Stratification Problem

Consider a set of K 2×2 contingency tables, each consisting of 2 binomial distributions. The underlying probabilities of the tth table may be written as in Table 1 and the responses as in Table 2. We assume that the odds ratio, ψ, is common to all tables. That is,

$$\psi = \frac{p_{1t}q_{2t}}{p_{2t}q_{1t}}. \tag{1}$$

Table 1. *Underlying Probabilities of Success in the tth of K 2×2 Contingency Tables. $q_{jt} = 1 - p_{jt}$.*

	Response		
Group	Positive	Negative	Sum
1	p_{1t}	q_{1t}	1
2	p_{2t}	q_{2t}	1

Mantel and Haenszel (1959) suggested the following estimate of ψ:

$$\hat{\psi}_{MH} = \frac{\sum_t x_{1t}y_{2t}/n_t}{\sum_t x_{2t}y_{1t}/n_t} \tag{2}$$

and the test of $H_0 : \psi = 1$ by the statistic

$$X^2_{MH} = \frac{\sum_t x_{1t} - E(x_{1t})^2}{\sum_t \text{var}(x_{1t})} \tag{3}$$

which is approximately chi-square with one degree of freedom. Hauck (1979) showed that if K is fixed and the number of observations per table increases, then the asymptotic variance of $\hat{\psi}_{MH}$ is

$$\text{var}(\hat{\psi}_{MH}) = \psi^2 \frac{\sum_t w_t v_t}{(\sum_t w_t)^2}, \tag{4}$$

where

$$w_t = \frac{p_{2t}q_{1t}n_t}{n_{1t}n_{2t}}$$

and

$$vt = \frac{1}{n_{1t}p_{1t}q_{1t}} + \frac{1}{n_{2t}p_{2t}q_{2t}}.$$

The Mantel-Haenszel procedures, because they employ closed rather than iterative estimators, are common in epidemiological research.

Table 2. *Data Array in the tth of K 2×2 Contingency Tables.*

	Response		
Group	Positive	Negative	Sum
1	x_{1t}	y_{1t}	n_{1t}
2	x_{2t}	y_{2t}	n_{2t}
Sum	x_t	y_t	n_t

Maximum likelihood estimation, developed by Gart (1962), requires iterative techniques. The variance of the maximum likelihood estimator of Ψ, $\hat{\psi}_{ML}$, is

$$\text{var}(\hat{\psi}_{ML}) = \psi^2 \sum_t \frac{n_{1t}n_{2t}p_{1t}q_{1t}p_{2t}q_{2t}}{n_{1t}p_{1t}q_{1t} + n_{2t}p_{2t}q_{2t}}. \quad (5)$$

2.2 The Clustering Problem

Correlated responses in categorical data manifest themselves in the tendency of clusters to respond in similar ways. In an independent binomial sample of 100 with an underlying probability parameter of .1, for example, the expected 10 positive responses will be distributed randomly throughout the sample. If, however, the sample consists of 20 clusters of size five with high correlation within clusters, it is more likely that the positive responses will be concentrated in one or two clusters and that the remaining clusters will display only negative responses.

Cohen (1976) and Altham (1976) suggested the following model for this phenomenon for multinomial distributions with r response categories in clusters of size m. Let p_i be the probability of a randomly selected subject displaying response i ($i = 1, \ldots, r$) and let p_{ij} be the probability that the second member of a cluster is in category j, given that the first member is in category i. Then

$$p_{ij} = \begin{cases} ap_i + (1-a)p_i & \text{if } i = j \\ (1-a)p_i p_j & \text{if } i \neq j, \end{cases}$$

where $0 \leq a \leq 1$. Under this model, the density function for a cluster of size m may be written as:

$$f(x; p, a, m) = \begin{cases} ap + (1-a)p^m & \text{if } x = m \\ aq + (1-a)q^m & \text{if } x = 0 \\ (1-a)\binom{m}{x} p^x q^{m-x} & \text{otherwise.} \end{cases}$$

It is straightforward to see that if a is 0, the responses are independent, and if a is large there is high correlation. In fact, if $\hat{p}_i = x_i/n$ is the usual estimator of p_i, then $\text{var}(\hat{p}_i) = Cp_i(1-p_i)$, where $C = 1 + (m-1)a$. The correction factor C is identical to the sample size correction factor derived for clustered continuous data with $a = \rho$. (See Snedecor and Cochran, 1980, pp. 240)

Brier (1980) used the Dirichlet multinomial distribution (Mossimann, 1962) to model clustered categorical data. This distribution is based on the assumption that the probability parameters vary between clusters according to a multivariate beta distribution. In the case of a dichotomous response, the Dirichlet multinomial distribution becomes the beta-binomial distribution studied by Skellam (1948), Griffiths (1973), Plackett and Paul (1978) and Paul and Plackett (1978). The probability of x positive responses in a cluster of size m is

$$f(x; p, k, m) = \binom{m}{x} \frac{\Gamma(m)\Gamma(x+kp)\Gamma(m-x+kq)}{\Gamma(m+k)\Gamma(kp)\Gamma(kq)},$$

where $0 \leq k \leq \infty$. A large k indicates low clustering. The variance of the maximum likelihood estimator of p, $\hat{p} = x/m$, is Cpq, where $C = (m+k)/(1+k)$. It may be shown that the intracluster correlation coefficient (Cochran, 1980, p. 241) is

$$\rho = \frac{1}{1+k}; \tag{6}$$

thus, $C = 1 + (m-1)\rho$.

In addition to the useful and satisfying qualities of (6), the beta-binomial model holds two advantages over the Cohen-Altham model. First, it is founded on a more heuristically reasonable model: clustering in categorical data is, by definition, the tendency of all members of a cluster to act alike, and this effect is fully accounted for by the variation of probabilities between clusters. Second, the parameter a of the Cohen-Altham model possesses an intuitive appeal only when it is 0 or 1. To see this, consider Figure 1, which consists of graphs of the two dichotomous density functions for $p = .5$, $a = \rho = .25$ and cluster size $m = 5$. The dotted line, connecting density values for the Cohen-Altham model, is in fact a weighted combination of the Bernoulli distribution (shown by the high probabilities for 0 or 5 positive responses) and the binomial distribution (shown by the central peak). The beta distribution (solid line) displays a shape that is more consistent with intuition: one expects that clustering would force the high probabilities away from the central values of 2 and 3, leaving a dip, rather than a peak, in the centre. In the following, therefore, we use the beta-binomial model as a description of the clustering phenomenon.

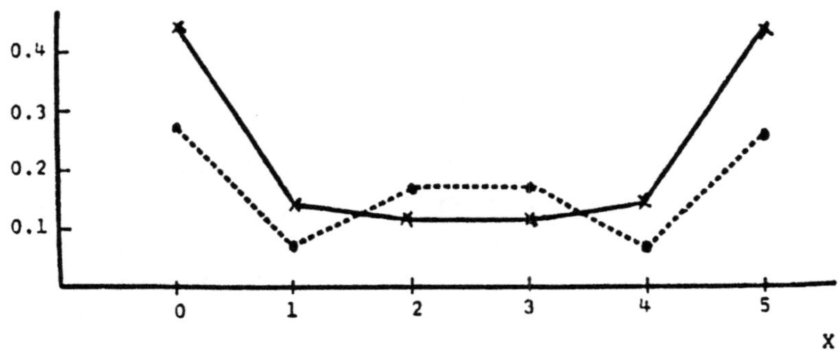

Figure 1. *Comparison of the Cohen-Altham (dotted line) and beta-binomial (solid line) models for clustered data in a cluster of size 5 with* $p = .5$ *and* $a = \rho = .25$.

3. MAXIMUM LIKELIHOOD THEORY

Consider a set of K 2×2 contingency tables in which N_{jt} clusters of size m are assigned to the jth row of the tth table. Using the beta-binomial density to describe the response, x_{jti}, of the ith cluster in this row and expression (1) to write in terms of p_{jt} and q_{jt}, we may write the likelihood function as:

$$L(p_{11}, p_{12}, \ldots, p_{1k}, k; x_{111}, \ldots, x_{2kN})$$
$$= \sum_{t=1}^{K} \sum_{i=1}^{N_{jt}} \sum_{j=1}^{2} \binom{m}{x} \frac{G(x_{jti}, kp_{jt}), G(m - x_{jti}, kq_{jt})}{G(m,k)},$$

where $G(n,z)$ is the gamma function ratio

$$G(n,z) = \frac{\Gamma(n+z)}{\Gamma(z)}$$
$$= \begin{cases} (n-1+z)(n-2+z)\ldots(z) & \text{if } n > 0 \\ 1 & \text{if } n = 0 \end{cases}$$

for integer n and real z. Maximization of $\log L$ requires iterative techniques. Details of the derivation and inversion of the information matrix were given by Donald (1984).

The variance of the maximum likelihood estimator of the common odds ratio is

$$V_{ML} = \frac{\psi^2}{k^2} \frac{R}{RS - T^2}, \tag{7}$$

where R, S and T are as in Table 3.

Table 4 contains sample variances of the MLE for various sets of parameters. The clusters are assumed to be evenly distributed among the tables. As with the usual (non-clustered) MLE, the variance increases as the common odds ratio increases. Increasing the intracluster correlation serves to effectively decrease the sample size and thus inflate the variance; a similar effect occurs when the cluster size is increased (without increasing the total sample size). The limiting expression of (7) as $k \to \infty$ ($\rho \to 0$) is identical to Gart's expression (5) for the variance when there is no clustering. A similar situation occurs when $m = 1$. On the other hand as $k \to 0$ ($\rho \to 1$), each cluster tends to act as a single individual, which cuts the effective sample size by a factor of $1/m$.

These phenomena may be seen in Figure 2, which contains graphs of the ratio of the variance of the usual MLE (5) divided by the variance of the clustered MLE (7). For values of the common odds ratio close to 1, this ratio is approximately equal to $1/(1+(m-1)\rho)$. It can be shown, using methods developed by Brillinger (1975), that under beta-binomial cluster sampling the usual (non-clustered) maximum likelihood estimator is consistent. Thus, the usual maximum likelihood techniques, with variances inflated by $C_m = 1 + (m-1)\rho$, are dependable under cluster sampling if the odds ratio is not

Table 3. *Expressions contained in formula (7) for the variance of the maximum likelihood estimated of the common odds ratio under cluster sampling.*

$$R = -NH''(m,k) - \sum_t \left\{ (C_{lt} + C_{2t}) - \frac{(p_{lt}q_{lt}B_{lt} + p_{2t}q_{2t}B_{2t})^2}{(p_{lt}q_{lt})^2 A_{lt} + (p_{2t}q_{2t})^2 A_{2t}} \right\}$$

$$S = \sum_t \left\{ \frac{1}{(p_{lt}q_{lt}^2 A_{lt}} + \frac{1}{(p_{2t}q_{2t}^2 A_{2t}} \right\}$$

$$T = \sum_t \frac{(p_{lt}q_{lt}p_{2t}q_{2t}^2(p_{lt}q_{lt}A_{lt}B_{2t} - p_{2t}q_{2t}A_{2t}B_{lt})}{(p_{lt}q_{lt})^2 A_{lt} + (p_{2t}q_{2t})^2 A_{2t}}$$

$$A_{jt} = -N_{jt}\sum_{x=1}^{m} \binom{m}{x} \frac{G(x,kp_{jt})G(m-x,kq_{jt})}{G(m,k)} \sum_{t=0}^{x-1} \frac{1}{(t+kp_{jt})^2} + \frac{G(x,kq_{jt})G(m-x,kp_{jt})}{G(m,k)} \sum_{t=0}^{x-1} \frac{1}{(t+kq_{jt})^2}$$

$$B_{jt} = -N_{jt}\sum_{x=1}^{m} \binom{m}{x} \frac{p_{jt}G(x,kp_{jt})G(m-x,kq_{jt})}{G(m,k)} \sum_{t=0}^{x-1} \frac{1}{(t+kp_{jt})^2} - \frac{q_{jt}G(x,kq_{jt})G(m-x,kp_{jt})}{G(m,k)} \sum_{t=0}^{x-1} \frac{1}{(t+kq_{jt})^2}$$

$$C_{jt} = -N_{jt}\sum_{x=1}^{m} \binom{m}{x} \frac{p_{jt}G(x,kp_{jt})G(m-x,kq_{jt})}{G(m,k)} \sum_{t=0}^{x-1} \frac{1}{(t+kp_{jt})^2} + \frac{q_{jt}G(x,kq_{jt})G(m-x,kp_{jt})}{G(m,k)} \sum_{t=0}^{x-1} \frac{1}{(t+kq_{jt})^2}$$

$$H''(m,k) = \frac{d^2}{dk^2} G(m,k)$$

Table 4. *Sample variances of the maximum likelihood estimator of the common odds ratio under cluster sampling. Probability bands consist of the probability parameters associated with the first row of each table. The sample size in each case is 240.*

1. Number of tables = 5, Common odds ratio = 1
   Probability band = (.05, .10, .15, .20, .25)

	CLUSTER SIZE				
ρ	2	4	6	8	12
.05	.9556	.8779	.8120	.7555	.6634
.20	.8540	.6660	.5489	.4684	.3643
.40	.7474	.5089	.3914	.3204	.2376
.60	.6562	.4029	.2962	.2362	.1702
.80	.5745	.3201	.2252	.1750	.1222
.95	.5180	.2666	.1805	.1368	.0925

2. Number of tables = 5, Common odds ratio = 1
   Probability band = (.10, .30, .50, .70, .90)

	CLUSTER SIZE				
ρ	2	4	6	8	12
.05	.9539	.8735	.8058	.7478	.6539
.20	.8435	.6457	.5250	.4434	.3394
.40	.7307	.4839	.3657	.2956	.2153
.60	.6405	.3836	.2779	.2195	.1560
.80	.5650	.3100	.2164	.1672	.1159
.95	.5154	.2642	.1784	.1350	.0911

Table 4. (continued)

3. Number of tables = 10, Common odds ratio = 5
Probability band = (.04, .07, .10, .13, .19, .22, .25, .28, .31)

	CLUSTER SIZE				
ρ	2	3	4	6	12
.05	.9609	.9235	.8890	.8275	.6867
.20	.8712	.7739	.6982	.5871	.4051
.40	.7693	.6326	.5411	.4247	.2668
.60	.6752	.5191	.4258	.3177	.1870
.80	.5861	.4214	.3318	.2355	.1295
.95	.5218	.3550	.2698	.1831	.0942

4. Number of tables = 10, Common odds ratio = 5
Probability band = (.05, .15, .25, .35, .45, .55, .65, .75, .85, .95)

	CLUSTER SIZE				
ρ	2	3	4	6	12
.05	.9591	.9174	.8792	.8117	.6604
.20	.8503	.7386	.6539	.5335	.3471
.40	.7383	.5887	.4920	.3732	.2211
.60	.6473	.4843	.3898	.2833	.1598
.80	.5700	.4031	.3139	.2195	.1179
.95	.5185	.3511	.2660	.1798	.0918

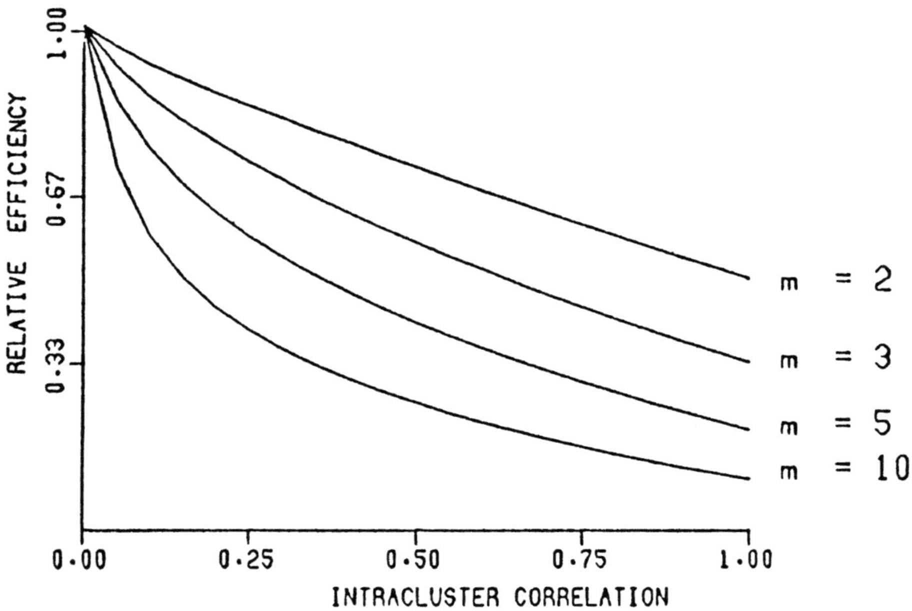

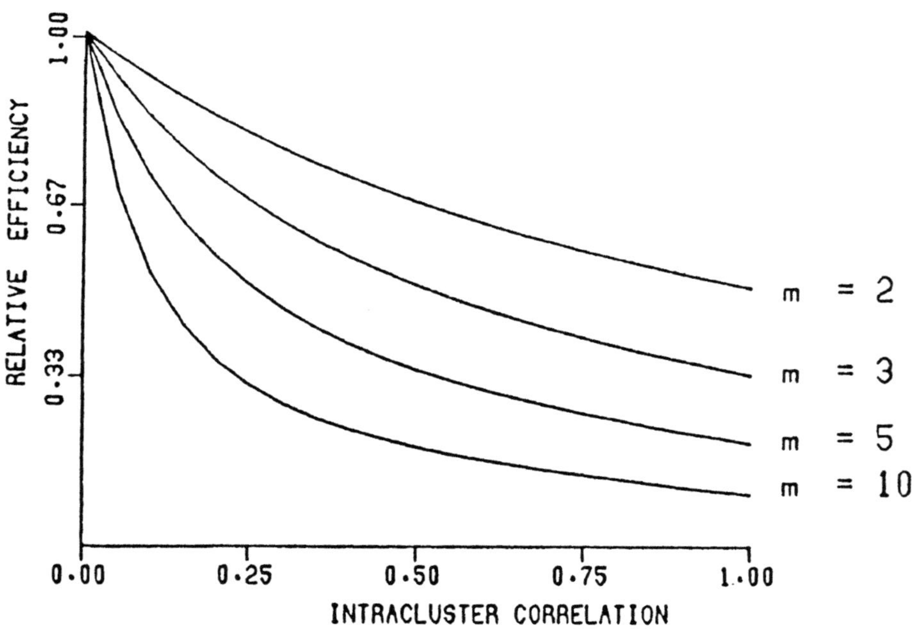

Figure 2. *Graphs of the variance ratio V_G/V_{ML} for 10 tables; probability band* (.05, .15, .25, .35, .55, .65, .75, .85, .95)*; sample size 240; odds ratios 5 (upper graph), 1 (lower graph).*

large. For larger values of ψ, the graphs tend to flatten, as can be seen for $\psi = 5$ in Figure 2.

If the cluster sizes are mixed, let a_m be the proportion of clusters of size m. The results of Bradley and Gart (1962) on associated populations may then be applied to obtain

$$V_{ML} = \sum a_m V_m,$$

where V_m is the variance of the MLE for cluster size m.

4. MANTEL-HAENSZEL PROCEDURES

4.1 Variance of the Mantel-Haenszel Estimator

The Hauck variance (4) of $\hat{\psi}_{MH}$ may be adjusted to compensate for cluster sampling. Let N_{jtm} be the number of clusters of size m in row j of table t and a_{jt} be the proportion of clusters in row j of table t. Let

$$c_{jt} = \frac{\sum\limits_m C_m m N_{jtm}}{\sum\limits_m m N_{jtm}}$$

and

$$V_t^* = \frac{c_{1t}}{a_{1t} p_{1t} q_{1t}} + \frac{c_{2t}}{a_{2t} p_{2t} q_{2t}}.$$

Then

$$\text{var}(\hat{\psi}_{MH}) = \frac{\psi^2}{n} \frac{\sum\limits_t w_t^2 V_t^*}{(\sum\limits_t w_t^2)^2},$$

where w_t is as in Hauck's formula (4).

4.2 Adjustment to the Mantel-Haenszel Test

The Mantel-Haenszel test statistic (3) may be adjusted using the weighted adjustment factors c_{jt} defined above. Under beta-binomial sampling, the adjusted Mantel-Haenszel chi-square statistic, X^2_{MHC}, is asymptotically chi-square with one degree of freedom:

$$X^2_{MH} = \left(\sum_t \frac{| x_{1t} y_{2t} - x_{2t} y_{1t} |}{n_{1t} c_{1t} + n_{2t} c_{2t}} \right)^2 \Big/ \sum_t \frac{n_{1t} n_{2t} x_t y_t}{n_t^2 (n_{1t} c_{1t} + n_{2t} c_{2t} - 1)}.$$

5. SIMULATION STUDY

Since the results above are valid asymptotically only, we designed a simulation study that, in part, examined the performance of the Mantel-Haenszel procedures in a sample of size 240 under clustering modelled by the beta-binomial distribution. Full details of this study, which also assessed maximum likelihood techniques and methods of estimating intraclass correlation, were given by Donald (1985).

The parameters used for the simulation were:

Number of tables: 5

Probability bands:

Narrow band—(0.5, .10, .15, .20, .25)

Wide band—(.10, .30, .50, .70, .90)

Cluster sizes: 2, 4, 6

Odds ratios: 1, 3, 5

Correlation coefficients 0, .05, .10, .20, .50, .80

For each set of parameters, 300 iterations were performed.

Compared with the MLE, the Mantel-Haenszel estimator of the odds ratio performed well in terms of variance and bias for small values of the intracluster correlation coefficient and for small cluster size. As Table 5 indicates, increasing the value of $C_m = 1 + (m-1)\rho$ diminishes the efficiency of the Mantel-Haenszel estimator. As expected, increasing the value of the common odds ratio inflated the variance of both the Mantel-Haenszel and the maximum likelihood estimators. Both estimators displayed consistently positive bias for all parameter combinations (Tables 6 and 7). This bias tended to rise for larger values of ψ and ρ, but not with respect to the cluster size.

Table 8 contains the observed rejection rates of the Mantel-Haenszel test of $H_0 : \psi = 1$ at $\alpha = .05$. For values of the intracluster correlation coefficient less than .20 and cluster size 4 or fewer, the test performed acceptably; larger values of ρ and m led to high rejection rates.

In conclusion, the Mantel-Haenszel procedures perform well if the intracluster correlation coefficient is expected to be less than .1 and the clusters are less than 4. In other circumstances, the Mantel-Haenszel estimate of the common odds ratio has positive bias and a large variance and the Mantel-Haenszel test of significance is likely to yield a spuriously significant result.

Table 5. *Observed variance relative efficiencies of odds ratio estimators. The ratio is the variance of the maximum likelihood estimator divided by the variance of the Mantel-Haenszel estimator. Hence, an entry greater than 1 indicates that the Mantel-Haenszel estimator has lower variance.*

Probability band = (.05, .10, .15, .20, .25)

		INTRACLUSTER CORRELATION					
ψ	m	.00	.05	.10	.20	.50	.80
	2	1.044	1.032	1.053	1.029	.985	1.029
1	4	1.039	1.083	1.063	1.039	.765	.883
	6	1.063	1.089	1.065	1.097	.981	.508
	2	.921	.985	1.092	.903	1.039	.901
3	4	1.057	.906	.877	.971	.588	.439
	6	1.019	1.013	.724	.944	.609	.470
	2	1.023	.883	1.021	.790	1.113	.705
5	4	.985	1.183	.909	1.035	.798	.171
	6	1.013	1.244	.875	.759	.331	.796

Probability band = (.10, .30, .50, .70, .90)

		INTRACLUSTER CORRELATION					
ψ	m	.00	.05	.10	.20	.50	.80
	2	1.034	1.035	1.062	1.095	1.074	1.094
1	4	1.044	1.034	1.064	1.040	.993	.953
	6	1.033	1.076	1.068	1.083	1.072	1.016
	2	1.036	1.086	1.087	1.176	1.099	1.173
3	4	1.065	1.035	1.080	1.067	1.191	1.174
	6	1.077	1.060	1.029	1.206	.554	.152
	2	1.014	1.069	1.020	1.033	1.086	.900
5	4	1.070	1.029	1.091	1.375	.998	.464
	6	1.075	1.073	1.154	.902	.257	.197

Table 6. *Observed biases of the Mantel-Haenszel estimator of the common odds ratio under cluster sampling.*

Probability band = (.05, .10, .15, .20, .25)

		INTRACLUSTER CORRELATION					
ψ	m	.00	.05	.10	.20	.50	.80
	2	.070	.075	.028	.044	.033	.092
1	4	.095	.102	.108	.136	.209	.185
	6	.075	.103	.091	.093	.196	.501
	2	.536	.780	.516	.553	.600	.498
3	4	.790	1.140	1.055	1.168	2.016	1.158
	6	.554	.866	.978	1.064	1.237	.911
	2	1.339	2.036	1.322	1.654	1.209	1.251
5	4	1.804	1.949	2.255	2.599	2.773	4.640
	6	1.618	1.662	2.083	2.086	2.589	.091

Probability band = (.10, .30, .50, .70, .90)

		INTRACLUSTER CORRELATION					
ψ	m	.00	.05	.10	.20	.50	.80
	2	.029	.051	.078	.033	.073	.048
1	4	.053	.071	.072	.098	.104	.096
	6	.020	.017	.037	.081	.051	.182
	2	.136	.122	.132	.113	.301	.338
3	4	.187	.277	.375	.488	.505	.706
	6	.162	.188	.271	.477	1.030	1.334
	2	.272	.295	.531	.221	.727	.899
5	4	.336	.397	.430	.642	1.208	2.731
	6	.275	.535	.595	1.094	3.223	3.733

Table 7. *Observed biases of the Gart (non-clustered unconditional maximum likelihood) estimator of the common odds ratio under cluster sampling.*

Probability band = (.05, .10, .15, .20, .25)

		INTRACLUSTER CORRELATION					
ψ	m	.00	.05	.10	.20	.50	.80
	2	.072	.077	.026	.043	.040	.100
1	4	.097	.107	.112	.143	.226	.206
	6	.077	.108	.096	.098	.212	.541
	2	.591	.834	.582	.629	.738	.583
3	4	.888	1.140	1.168	1.354	2.228	1.520
	6	.626	.989	1.079	1.230	1.471	1.304
	2	1.474	2.109	1.454	1.810	1.614	1.384
5	4	1.804	1.949	2.255	2.599	2.773	4.640
	6	1.816	1.934	2.282	2.416	3.047	.702

Probability band = (.10, .30, .50, .70, .90)

		INTRACLUSTER CORRELATION					
ψ	m	.00	.05	.10	.20	.50	.80
	2	.030	.053	.081	.035	.077	.053
1	4	.055	.073	.076	.103	.115	.107
	6	.021	.018	.040	.087	.062	.228
	2	.195	.189	.211	.193	.405	.478
3	4	.245	.347	.460	.592	.727	.998
	6	.216	.252	.346	.657	1.337	1.790
	2	.406	.469	.699	.406	.967	1.242
5	4	.505	.562	.645	.949	1.713	3.286
	6	.435	.719	.849	1.466	3.565	4.311

Table 8. *Observed rejection rates for the Mantel-Haenszel test of* $H_0 : \psi = 1$ *under the null hypothesis. Asterisks indicate the significance of the difference between the observed rate and* .05 (*, $p < .05$; **, $p < .01$; ***, $p < .001$).

1. Probability band = (.05, .10, .15, .20, .25)

	INTRACLUSTER CORRELATION					
m	.00	.05	.10	.20	.50	.80
2	.048	.019*	.030	.049	.082**	.124***
4	.056	.072	.069	.100***	.200***	.267***
6	.055	.093**	.106***	.116***	.223***	.306***

2. Probability band = (.10, .30, .50, .70, .90)

	INTRACLUSTER CORRELATION					
m	.00	.05	.10	.20	.50	.80
2	.034	.041	.037	.085**	.090**	.069
4	.050	.057	.096***	.110***	.140***	.179***
6	.027	.074	.131***	.181***	.191***	.304***

REFERENCES

Altham, P. M. E. (1976), "Discrete variable analysis for individuals grouped into families." *Biometrika* **63**, 263–269.

Bradley, R. A., and J. J. Gart (1962), "The asymptotic properties of ML estimators when sampling from associated populations." *Biometrika* **49**, 205–214.

Brier, S. (1980), " Analysis of contingency tables under cluster sampling." *Biometrika* **67**, 591–596.

Brillinger, D. R. (1975), "Statistical inference for stationary point processes." In *Stochastic Processes and Related Topics*, ed. M. L. Puri. New York: Academic Press.

Cochran, W. G. (1980), *Sampling Techniques*. New York: Wiley and Sons.

Cohen, J. E. (1976), "The distribution of the chi-squared statistic under clustered sampling from contingency tables." *Journal of the American Statistical Association* **71**, 665–670.

Donald, A. (1984), "The analysis of clustered data in sets of 2 × 2 contingency tables." Ph.D. Thesis, The University of Western Ontario.

Donald, A., and A. Donner (1985), "Effects of clustering on the analysis of sets of 2 × 2 contingency tables: maximum likelihood theory and adjustments to the Mantel-Haenszel procedures." In preparation.

Gart, J. J. (1962), "On the combination of relative risks." *Biometrics* **18**, 601–610.

Griffiths, D. A. (1973), "Maximum likelihood estimation for the beta-binomial distribution and an application to the household distribution of the total number of cases of a disease." *Biometrics* **29**, 637–648.

Hauck, W. W. (1981), "The large-sample variance of the Mantel-Haenszel estimator of a common odds ratio." *Biometrics* **35**, 817–819.

Mantel, N., and W. Haenszel (1959), "Statistical aspects of the analysis of data from retrospective studies of disease." *Journal of the National Cancer Institute* **22**, 719–748.

Mossimann, J. E. (1962), "On the compound multinomial distribution, the multivariate beta-distribution and correlation among proportions." *Biometrika* **49**, 65–82.

Paul, S. R., and R. L. Plackett (1978), "Inference sensitivity for Poisson mixtures." *Biometrika* **65**, 591–602.

Plackett, R. L., and S. R. Paul (1978), "Dirichlet models for square contingency tables." *Communications in Statistics A, Theory and Methods* **10**, 939–952.

Skellam, J. G. (1948), "A probability distribution derived from the binomial distribution regarding the probability of success as variable between sets of trials." *Journal of the Royal Statistical Society*, Series B **10**, 257–261.

Snedecor, G. W., and W. G. Cochran (1980), *Statistical Methods*, Seventh Edition. Ames, Iowa: Iowa State University Press.

Nathan Mantel [1] and S. R. Paul [2]

GOODNESS-OF-FIT ISSUES IN TOXICOLOGICAL EXPERIMENTS INVOLVING LITTERS OF VARYING SIZE

1. INTRODUCTION

Data on fetal deaths in litters of varying sizes were used by Haseman and Soares (1976) and re-used by Kupper and Haseman (1978) to exemplify the fitting of certain models for explaining the variation between litters in the occurrence of such deaths. Model fitting plays an important role, for then any treatment effects, litter-size effects, or other possible causes of variation can be studied by determining whether estimated model parameters are significantly affected by such causes of variation.

In the work of Hasemen and Soares (1976), the principal models fitted, but found wanting, were a Poisson model and a binomial model. Clear improvements in fit were shown by Kupper and Haseman (1978) by either a beta-binomial (BB) model or a correlated-binomial (CB) model. Since neither of these two models nested into the other and differences in fit were not extreme, no formal test of which was the more appropriate model was feasible.

Actually, both of these models take intra-litter correlation into account. By the beta-binomial model, a form of compound binomial, a positive intra-litter correlation would result because of litter members sharing a common binomial parameter. With the correlated binomial model, the intra-litter correlation could be either positive, because of the community of the prenatal exposures of litter members, or negative, because of competition between litter members.

Paul (1985) derived a model for combining the features of both of these

[1] Department of Mathematics, Statistics and Computer Science, The American University, 4900 Auburn Avenue, Bethesda, Maryland 20814

[2] Department of Mathematics and Statistics, University of Windsor, Windsor, Ontario N9B 3P4

I. B. MacNeill and G. J. Umphrey (eds.), Biostatistics, 169–176.

models into a beta-correlated-binomial (BCB) model. A particular situation for which applicability was seen was in the distribution of α grades among students, barring detailed data on separate examination results. A BCB model could take into account simultaneously the variation in difficulty among examinations and the variation in ability among students.

Since a hierarchical relationship existed between the BCB model and either the BB or the CB model, the significance of any apparent improvement in fit could be evaluated. One could determine whether the BB aspect of the BCB model was needed for a good fit, or if the CB aspect of BCB model was needed or if both aspects were needed.

When all three models were fitted and goodness-of-fit chi squares calculated, a puzzling result was found. For BB and CB, those chi squares were the same as Kupper and Haseman had reported, verifying that their calculating procedures had been followed. But the BCB chi square in some very few instances exceeded either the BB or the CB chi square.

This anomaly we found, on consideration, to be related to the non-constancy of litter size in the data. Further, such a variety of improprieties was found to attend the calculation of goodness-of-fit chi square as to render that statistic unsuitable for its apparent purpose in the circumstance of varying litter size. While difficulties can be avoided through use of the likelihood chi square, some difficulties remain.

In what follows, we will describe the problems we found to be associated with the goodness-of-fit chi squares. Though our investigation had involved an abundance of calculations, we will not be giving the results of those calculations here. The case is that once each difficulty is identified, the unsuitability of the goodness-of-fit chi square becomes more and more manifest.

Aside from questions of the suitability of the goodness-of-fit chi square, we plan to discuss some other aspects of toxicological experiments relating to litter size.

2. SOME BACKGROUND

Essentially the data considered by Haseman and Soares (1976) and Kupper and Haseman (1978) consist of paired observations of the form (r_i, n_i), denoting that litter i is of size n_i and contains r_i affected individuals. In Haseman and Soares (1976), data are presented in the form $f(r, n)$, the frequency of litters in which $r_i = r$, $n_i = n$.

In both papers, in those tables where observed and fitted frequencies are shown and compared via chi square, what is given is $f(r)$, the number of litters in which $r_i = r$ irrespective of n_i, i.e. $\sum_{n=1} f(r, n)$, summation being up through the largest litter size in the data. Chi square is then calculated as $\sum_{r=1}[f(r) - E(f(r))]^2/E(f(r))$. In the examples given r values in excess of 3 have been lumped. Note that the $E(f(r))$ are obtained by maximum likelihood fitting.

For future reference we define s_i as $n_i - r_i$, so that in parallel with $f(r,n)$ and $f(r)$ we could have $g(s,n)$ and $g(s)$.

3. PARTICULAR DIFFICULTIES

A. Non-correspondence of the data being fitted and the data used in assessing goodness-of-fit

The actual data being fitted are the assemblage of (r_i, n_i) paired observations or, equivalently, the set of $f(r,n)$ frequencies. But the set of frequencies used in calculating chi square consists of the $f(r)$ values, the n values being ignored. Conceivably, the $f(r)$ values could be well fitted, but not the $f(r,n)$ values. For the goodness-of-fit chi square to show proper behavior, our maximum likelihood fitting should have been to the set of $f(r)$ values, given the assemblage of n_i values. Of course, it would have been foolish and inelegant to fit the $f(r)$ data when we had the $f(r,n)$ data.

But, note that were all the n_i equal, the impropriety of considering $f(r)$ rather than $f(r,n)$ would not occur.

B. Unsuitability of use of a $\sum(O-E)^2/E$-type formula for calculating chi square (O = observed, E = expected, e.g. $f(r)$ and $E(f(r))$)

In many instances, consideration of $(O-E)^2/E$ properly takes into account the covariances amongst the O values, but not generally. The general approach is to get the variance-covariance matrix of the O_j values, then, after deleting the last row and column (or several last rows and columns where data have been lumped) inverting so as to get a matrix with elements C_{jk}. Chi square is then calculated as $\sum c_{jk} d_j d_k$, where $d_j = O_j - E_j$, $d_k = O_k - E_k$. (In a birth order problem involving sibships of varying sizes, Mantel and Halperin (1963) demonstrated a proper calculation of chi square, but took advantage of a particular circumstance so as to avoid matrix inversion.)

Computation of the variance-covariance matrix to be inverted is straightforward. We will consider the individual litter with outcome (r_i, n_i), though adaptation for $f(r,n)$ values is simple.

For litter i, suppose that the fitted probability that $r_i = j$ is $P_i(j)$, i.e. $P_i(0)$, $P_i(1), \ldots, P_i(n_i)$ with $P_i(j) = 0$ for $j > n_i$. These $P_i(j)$ values are the contributions of litter i to the E_j values, while the contribution to the O_j values is 1 for $j = r_i$, 0 for all other j. The variance of the contribution to any O_j value is $P_i(j)(1 - P_i(j))$ while the covariance of the contribution to O_j and to O_k is $-P_i(j)P_i(k)$. The overall variance-covariance matrix of the O_j is simply the sum of the individual litter contributions to the matrix.

C. The need to modify chi square when comparing alternative fittings

Essentially, when we have computed our chi square properly, it tells us how much worse our fitted values match the data than would the results of a perfectly fitting model. But, how should we judge the comparison of two imperfectly-fitting models?

The chi squares for two fits could vary because of differences in their (d_j) vectors or differences in their (c_{jk}) inverse matrices. To make proper comparison, the (c_{jk}) inverse matrix must be kept constant between any two fits. Mantel and Patwary (1961) have described the need to keep constant the weighting function in assessing alternative fittings. An illustration was given by Mantel and Brown (1973) of data fitted by a variety of models. For any two models which nested, then the simpler model (the one requiring the fewer fitted parameters) would be considered the null model to provide the common weighting function, i.e. the c_{jk} values or, when appropriate, the $1/E_j$ values. The difference in the chi squares is then influenced only by the difference in the (d_j) vectors.

D. Do we count the responders (e.g. defectives, deaths, etc.) or the non-responders (e.g. normal fetuses, survivors, etc.)

Suppose we had turned our data around so as to focus on the normal fetuses, i.e. the (s_i, n_i) data and the $g(s,n)$ frequencies. Instead of our considering a litter of size 10 to have 2 defectives, we would consider it to have 8 non-defectives. Gross *et al.* (1970) in studying rat reproduction emphasized the number in the litter surviving through weaning, ignoring the losses at earlier stages.

Considering the s_i rather than the r_i would make no difference were the n_i all equal. And even with the n_i unequal, there is really no change in the fitted parameter values or in the true fit to the data. Where the change comes about is in the calculation of chi square relative to the $g(s)$ frequencies. The chi square relative to the $g(s)$ frequencies may not accord with the chi square relative to the $f(r)$ frequencies, let alone with the $g(s,n)$ or equivalent $f(r,n)$ frequencies. Compensating errors could keep either or both of these chi squares small even if the more detailed data are fit poorly.

Summarizing A – D

With varying litter size, the data used in calculating the goodness-of-fit chi square are not the same as the data actually fitted. Apart from that, the chi-square computation formula becomes inappropriate and, instead, a matrix inversion approach would be needed. Litter size differences can also result in different assessments of goodness-of-fit according to whether one looks at the marginal distributions of defectives or deaths in a litter or at the marginal distribution of normals or survivors in a litter. Apart from litter size variation, if alternative models are being compared via goodness-of-fit chi squares, then the same weighting function must be used for both fittings.

4. THE LIKELIHOOD CHI SQUARE

The difficulties associated with use of the goodness-of-fit chi square under circumstances of varying litter size largely disappear when it is the likelihood value which provides the test statistic, i.e. the likelihood chi square. But, an important caution remains.

First off, the calculated likelihood for the fitted model relates to the initial (r_i, n_i) paired values, or the equivalent $f(r, n)$ frequencies. The clumping into $f(r)$ frequencies plays no role in calculating the likelihood. Next, it is immaterial to the fitting, and to the calculated likelihood, whether we consider our observations to consist of (r_i, n_i) pairs and $f(r, n)$ frequencies, or if we considered instead the (s_i, n_i) pairs and the $g(s, n)$ frequencies. But caution is necessary because our data could very readily be insufficient for proper asymptotic behavior of our likelihood chi square—some other way of evaluating the likelihood could be necessary.

Consider the case of a single fixed litter size, say $n = 10$. There would be 11 possible outcomes for a litter with r_i ranging from 0 to 10. Some of those outcomes could have extremely low probabilities. Unless the total number of litters were rather large, asymptotic behavior for the likelihood chi square could not be expected. The goodness-of-fit chi square, acceptable in the fixed n case, might be applied by collapsing outcomes.

But with varying n, the possible number of outcomes gets very much larger. In an example given by Haseman and Soares (1976), n ranged from 1 to 20, so that the total number of cells would be $2 + 3 + \ldots + 21 = 230$. The apparent degrees of freedom would be 210 minus the number of fitted parameters. In even a study of large size, the largest number of cells would have trivially small fitted expectations, and actual frequencies of zero.

It would appear that the likelihood chi square would be unsuitable for gauging goodness-of-fit, yet it should still be suitable for comparing alternative fittings by models with fewer or more fitted parameters, with one model a special case of the other, i.e. a nesting or hierarchical relationship exists between the models. The many cells with zero observations would play no role in distinguishing between the likelihoods. Likelihood chi square for distinguishing between fits would be with limited degrees of freedom.

Illustration of how the likelihood chi square, as well as the goodness-of-fit chi square, can be used to distinguish between alternative models is provided by Mantel and Brown (1973).

5. DEFINITIONS OF r_i AND n_i

Simple fitting of the (r_i, n_i), testing of alternative models, and even assessing treatment effects based on such data can be an incomplete analysis.

Consider that the relationship between r_i and n_i can be altered by non-specific effects of treatment on the response of interest. If treatment increases, or decreases, the viability of defective fetuses through some early stage of gestation, it will increase, or alternatively decrease, the relative frequency of such defective fetuses at a later stage. More generally, differential viability of defective and normal fetuses resulting from treatment will influence the relative frequency of defective fetuses apart from any specific teratogenic or anti-teratogenic effect (or other specific adverse or protective effect) of treatment.

This difficulty can be addressed by studying the influence of treatment on r_i alone, on $s_i = n_i - r_i$ alone, and even on n_i alone, i.e., that treatment may affect r_i, s_i and n_i, separate analysis of treatment effect on each of these indices would be helpful. Also, quite apart from treatment effects, the data can be analyzed for effects of litter size itself on the governing model parameters. Thus, litters can be grouped by broad litter-size categories, with parameters fitted overall and also separately by broad category.

In their work, Haseman and Soares (1976) quite reasonably took for n_i the number of implants in a pregnancy, r_i being the number of dead implants—in this case, interest was in dominant lethal effects. But in other situations, n_i would necessarily have to be either the number of unresorbed fetuses or live unresorbed fetuses, with r_i the number of such fetuses with some specific defect. Where pregnancies have to be carried to term, n_i might be either the number of fetuses carried to term (in which case a candidate r_i could be the number of stillbirths), or n_i might be the number of live births in the litter.

A somewhat comprehensive view of this kind was taken by Gross *et al.* (1970) in their study relating to rat reproduction. While their interest centered on the numbers per litter surviving to weaning, more specifically 21 days post birth, they considered that there were various stages in the reproduction process.

Thus, drug treatment might influence the likelihood of conception at a mating. Next, it might influence the number of implants occurring in a successful mating. Successive inroads on surviving pregnancy would occur as there were embryonic deaths, stillbirths, perinatal deaths in the first four days after deaths, and finally, deaths between 4 days and weaning at 21 days. Since their interest was on rat reproduction generally, not on specific aspects, Gross *et al.* (1970) focused on the number of 21-day survivors. In a sense, they were analyzing only the s_i values, without regard to the n_i values. But, if the separate stages were considered, the s_i value at one stage would become the n_i value for the next stage.

6. SUMMARY

When litters vary in size certain goodness-of-fit chi-square tests for adequacy of statistical models, e.g. a binomial model, a beta-binomial model or a correlated beta binomial model can become unsuitable. Reasons include that the data by which the goodness-of-fit is gauged are not identical to the data actually fitted. Also, the blindly followed computation of chi square does not properly take into account the variance-covariance structure of the data. In any case, the goodness-of-fit chi square, even were the variance-covariance structure taken into account, would differ according to whether one were considering responders or non-responders. In comparing alternative models, changes in the variance-covariance matrix should not be allowed. While a likelihood chi-square approach could bypass most of these difficulties, in the case of varying litter sizes, the likelihood chi square could be unsuitable for assessing goodness-of-fit, but suitable for distinguishing between alternative fittings. Concerns for other aspects of toxicological experiments involving litters must be carefully considered in any proper evaluation.

ACKNOWLEDGMENT

The research for this paper was partially supported by NSERC of Canada.

REFERENCES

Gross, M. A., O. G. Fitzhugh, and N. Mantel (1970), "Evaluation of safety for food additives; an illustration involving the influence of methyl salicylate on rat reproduction." *Biometrics* **26**, 181–194.

Haseman, J. E., and E. R. Soares (1976), "The distribution of fetal death in control mice and its implications on statistical tests for dominant lethal effects."

Mutation Research **41**, 277–288.

Kupper, L. L., and J. K. Haseman (1978), "The use of a correlated binomial model for the analysis of certain toxicological experiments." *Biometrics* **34**, 69–76.

Mantel, N., and C. Brown (1973), "A logistic reanalysis of Ashford and Sowden's data on respiratory symptoms in British coal miners." *Biometrics* **29**, 649–665.

Mantel, N., and M. Halperin (1963), "Analysis of birth-rank data". *Biometrics* **19**, 324–340.

Mantel, N., and K. M. Patwary (1961), "Interval estimation of single parametric functions". *Proceedings of the 32nd Session of the International Statistical Institute, Tokyo, 1960. Bulletin of the International Statistical Institute* **38**, 227–240.

Paul, S. R. (1985), "A three parameter generalization of the binomial distribution". *Communications in Statistics A, Theory and Methods* **14**, 1497–1506.

Shelley B. Bull and Allan Donner [1]

DERIVATION OF LARGE SAMPLE EFFICIENCY OF MULTINOMIAL LOGISTIC REGRESSION COMPARED TO MULTIPLE GROUP DISCRIMINANT ANALYSIS

ABSTRACT

The comparison of logistic regression and normal discrimination (Efron, 1975) is extended to the case of more than two response groups. The large sample distribution of the maximum likelihood estimate of the log odds ratio is derived for each of the two procedures, using matrix calculus methods, and relative estimating efficiency (RE) is then defined as the inverse of the variance ratio. It is shown that Efron's efficiency measure, asymptotic relative efficiency, is a special case of the measure given here. We evaluate two cases in which RE is parameterized only by the response group frequencies and the log odds ratio parameters. The results are presented in two-dimensional contour diagrams.

1. INTRODUCTION

Cornfield (1962) developed the binomial logistic risk function in the context of a two-group discriminant function analysis; one group being individuals with clinically manifest heart disease, and the other, those without. He sought a parsimonious model of the joint dependence of coronary heart disease risk on serum cholesterol and systolic blood pressure in data from the Framingham follow-up study (Greenhouse, 1982). In discussing this model, Cornfield noted that "our present interest is not in discrimination . . . , but . . . to study the quantitative nature of the dependence of risk on serum cholesterol and systolic blood pressure levels."

The analogous multiple group problem assumes that a vector of explanatory variables has a multivariate normal distribution with equal covariance

[1] Department of Epidemiology and Biostatistics, The University of Western Ontario, London, Ontario N6A 5C1 (both authors)

I. B. MacNeill and G. J. Umphrey (eds.), Biostatistics, 177–197.

matrices, but different mean vectors, in each of several populations. For example, in Cornfield's problem, the coronary heart disease group might be divided into two mutually exclusive groups: individuals with myocardial infarction or individuals with angina pectoris. Indeed, this approach might be more informative since "the combination of these two conditions into a single category is not, of course, intended to imply that they necessarily have the same etiology" (Cornfield, 1962). Expressing the logistic parameters as functions of the normal parameters, application of Bayes theorem in the discrimination model yields the multinomial logistic probability functions, (Lachenbruch, 1975).

There are two different procedures for estimating the logistic parameters: (1) substitution of the unconditional maximum likelihood estimates of the normal distribution parameters into the logistic parameter expressions, and (2) maximum likelihood estimation based on the conditional probability of being in a particular group given the explanatory variables. The former procedure is referred to as multiple group discriminant analysis and the latter, multinomial logistic regression. Details are given in Section 2.

Efron (1975) compared the efficiency of the two group linear discrimination and binomial logistic regression procedures with respect to classification error rate and found that logistic regression is between one half and two thirds as effective as normal discrimination when the normality assumptions are satisfied. He used asymptotic relative efficiency, a measure of relative test efficiency, to compare the two procedures.

As Cornfield's (1962) comments indicate, it is typical in epidemiological research to be interested in estimating the effect of explanatory factors on the risk of an individual being in one of two or more response groups. We therefore consider the relative efficiency of the two procedures in terms of coefficient estimation and extend the comparison to the case of multiple groups. The large sample distribution of the estimate of the log odds ratio is derived for each of the procedures, using matrix calculus methods. Relative estimating efficiency is then defined as the inverse of the asymptotic variance ratio. It is shown that Efron's efficiency measure, asymptotic relative efficiency, is a special case of the measure given here.

We evaluate relative estimating efficiency, focussing on its relationship to the value of the log odds ratio for three response groups in the case of two explanatory variables. Of secondary interest is the effect of response group frequencies on relative efficiency.

2. ASSUMPTIONS, DEFINITIONS AND NOTATION

A random variable $\mathbf{X}$ can arise from one of $J+1$ p-dimensional populations such that:

$$\mathbf{X} \sim N_p(\boldsymbol{\mu}_j, \Sigma) \text{ with probability } \pi_j \quad (j = 0, 1, \ldots, J).$$

The normal populations thus differ in mean but not in variance-covariance structure. The space is exhausted by the populations in that $\pi_j > 0$ for all j and $\sum_{j=0}^{J} \pi_j = 1$, so $\pi_0 = 1 - \sum_{j=1}^{J} \pi_j$. We observe n independent realizations of the random variable $\mathbf{X}$ denoted by the pairs $(z_i, \mathbf{x}_i)$ $i = 1, 2, \ldots, n$, where z_i indicates from which population $\mathbf{x}_i$ came and $\mathbf{x}_i$ is a vector of measurements. We can define indicators δ_{ij} where

$$\delta_{ij} = \begin{cases} 1 & \text{if } \mathbf{x}_i \text{ is from population } j \\ 0 & \text{otherwise.} \end{cases}$$

We also define $\lambda_j \equiv \log(\pi_j/\pi_0)$, $j = 1, 2, \ldots, J$.

Given these assumptions, we consider two approaches to the estimation of the logistic parameters β_{0j} and $\boldsymbol{\beta}'_j = (\beta_{1j}, \beta_{2j}, \ldots, \beta_{pj})$.

The normal discrimination procedure, based on the full likelihood $L_1 = \prod_{i=1}^{n} f(z_i, \mathbf{x}_i)$, produces maximum likelihood estimates of the normal parameters π_j, $\boldsymbol{\mu}_j$ $(j = 1, \ldots, J)$, $\boldsymbol{\mu}_0$ and Σ. The parameters β_{0j} and $\boldsymbol{\beta}_j$ are then estimated by:

$$\hat{\beta}_{0j} = \hat{\lambda}_j - \frac{1}{2}\left(\hat{\boldsymbol{\mu}}'_j \hat{\Sigma}^{-1} \hat{\boldsymbol{\mu}}_j - \hat{\boldsymbol{\mu}}'_0 \hat{\Sigma}^{-1} \hat{\boldsymbol{\mu}}_0\right),$$
$$\hat{\boldsymbol{\beta}}_j = \hat{\Sigma}^{-1}\left(\hat{\boldsymbol{\mu}}_j - \hat{\boldsymbol{\mu}}_0\right).$$

The logistic regression procedure is based on the conditional likelihood $L_2 = \prod_{i=1}^{n} f(z_i|\mathbf{x}_i)$, where, given the values $\mathbf{x}_1, \ldots, \mathbf{x}_n$, the z_i are conditionally independent multinomial random variables with parameters

$$\theta_0(\mathbf{x}_i) = \Pr(z_i = 0|\mathbf{x}_i) = \left[1 + \sum_{k=1}^{J} \exp\{\beta_{0k} + \boldsymbol{\beta}'_k x_i\}\right]^{-1}$$

and

$$\theta_j(\mathbf{x}_i) = \Pr(z_i = j|\mathbf{x}_i) = \exp\{\beta_{0j} + \boldsymbol{\beta}'_j \mathbf{x}_i\} \Big/ \left[1 + \sum_{k=1}^{J} \exp\{\beta_{0k} + \boldsymbol{\beta}'_k \mathbf{x}_i\}\right].$$

The maximum likelihood estimates $\tilde{\beta}_{0j}$ and $\tilde{\boldsymbol{\beta}}_j$ are obtained by maximizing L_2 with respect to β_{0j} and $\boldsymbol{\beta}_j$ (see Cox, 1970; Mantel, 1966).

We express the estimates $\hat{\beta}_{0j}$, $\hat{\boldsymbol{\beta}}_j$ and $\tilde{\beta}_{0j}$, $\tilde{\boldsymbol{\beta}}_j$ more concisely using the matrix notation $\hat{B}$ and $\tilde{B}$, where $B = (B_0|B_*)$,

$$B_0' = (\beta_{01}, \ldots, \beta_{0J}) \quad \text{and} \quad B_* = (\boldsymbol{\beta}_1, \ldots, \boldsymbol{\beta}_J).$$

A comparison of the two approaches to estimation is made in terms of large sample variance. This is appropriate, since the logistic regression and normal discrimination estimators have asymptotic normal distributions (see Theorems 1 and 2). Since we are comparing estimators with variances of order $(1/n)$, we measure the relative estimating efficiency of the logistic regression estimator, t_2, to the normal discrimination estimator, t_1, by the variance ratio: $V(t_1)/V(t_2)$, where $V(t_1)$ and $V(t_2)$ are asymptotic variances. In particular, Theorem 3 gives the relative efficiency of the logistic regression estimator of the parameter β_{11}, where β_{11} is interpreted as the increase in the log odds ratio of an observation being in population 1 relative to population 0, for a unit increase in x_1.

We define the following vectors and matrices:

$\mathbf{e}_s$ is a $p \times 1$ vector with 1 in the sth position and zero elsewhere,

$\mathbf{i}_j$ is a $J \times 1$ vector with 1 in the jth position and zero elsewhere,

$\mathbf{1}_m$ is an $m \times 1$ vector of 1's,

I_m is the $m \times m$ identity matrix,

$0_{m \times n}$ is an $m \times n$ matrix of zeros, and

$\text{diag}[d_1, \ldots, d_m]$ is a diagonal (or block diagonal) matrix constructed from the elements (or matrices) $d_1, \ldots, d_m$.

Also useful are the three matrix operators: $\otimes$, the Kronecker product; vec and vech. The Kronecker product of matrices is defined by $A \otimes B = \{a_{ij}B\}_{pr \times qs}$ $(i = 1, \ldots, p;\ j = 1, \ldots, q)$, where we have $A_{p\times q} = \{a_{ij}\}$ and $B_{r\times s}$. The vec and vech operators, discussed by Henderson and Searle (1979) and by McCulloch (1982), convert matrices into column vectors. The operator vec(X) takes the columns $\mathbf{x}_1, \mathbf{x}_2, \ldots, \mathbf{x}_q$ of a $p \times q$ matrix X and stacks the columns so that vec(X) is a vector of length pq. For $m\times m$ symmetric X, vech(X) is defined similarly including only the distinct elements of X, that is, the elements on or above the diagonal. Vech(X) is thus a column vector of length $\frac{1}{2}m(m+1)$. It can also be shown that $\text{vec}(ABC) = (C' \otimes A)\text{vec}(B)$.

Three special matrices may be defined in terms of relationships among the column vectors vec(X), vec(X') and vech(X). The $pq \times pq$ vec-permutation matrix $I_{(p,q)}$, defined by $\text{vec}(X_{p\times q}) = I_{(p,q)}\text{vec}(X'_{p\times q})$, is related to Kronecker products by the identity $B_{r\times s} \otimes A_{p\times q} = I_{(p,r)}(A_{p\times q} \otimes B_{r\times s})I_{(s,q)}$. For $m \times m$ symmetric X, we can define the matrices G and H such that $\text{vec}(X) = G\text{vech}(X)$ and $\text{vech}(X) = H\text{vec}(X)$. The dimensions of G are m^2 by $\frac{1}{2}m(m+1)$, while those of H are $\frac{1}{2}m(m+1)$ by m^2. H is a non-unique left inverse of G, one choice being $H = (G'G)^{-1}G'$. Magnus and

Neudecker (1979) discussed the definition and results stated here and also reviewed earlier literature on the subject.

3. LARGE SAMPLE DISTRIBUTION OF DISCRIMINANT ESTIMATES

Lemma 1. Assuming that a random variable $\mathbf{X}$ can arise from one of $J+1$ p-dimensional normal populations such that

$$\mathbf{X} \sim N_p(\boldsymbol{\mu}_j, \Sigma) \text{ with probability } \pi_j \ (j = 0, 1, 2, \ldots, J),$$

the normal discrimination procedure produces the following maximum likelihood estimates of the normal parameters: $\tilde{\boldsymbol{\lambda}} = \{\hat{\lambda}\} = \{(\hat{\pi}_j/\hat{\pi}_0)\}$, $\hat{\boldsymbol{\mu}}_j = \sum_{i=1}^n \delta_{ij}\mathbf{x}_i/n_j$, and $\hat{\boldsymbol{\sigma}} = \text{vech}(\hat{\Sigma})$, where $n_j = \sum_{i=1}^n \delta_{ij}$, $\hat{\pi}_j = n_j/n$, and $\hat{\Sigma} = \sum_{i=1}^n \sum_{j=0}^J \delta_{ij}(\mathbf{x}_i - \hat{\boldsymbol{\mu}}_j)(\mathbf{x}_i - \hat{\boldsymbol{\mu}}_j)'/n$. Using the usual methods based on Fisher's expected information matrix, one can derive asymptotic distributions for the maximum likelihood estimates:

(a) $L[\sqrt{n}(\hat{\boldsymbol{\lambda}} - \boldsymbol{\lambda})] \longrightarrow N_J(\mathbf{0}, V_{\boldsymbol{\lambda}})$,

(b) $L[\sqrt{n}(\hat{\boldsymbol{\mu}}_j - \boldsymbol{\mu}_j)] \longrightarrow N_p(\mathbf{0}, V_{\boldsymbol{\mu}_j})$,

(c) $L[\sqrt{n}(\hat{\boldsymbol{\sigma}} - \boldsymbol{\sigma})] \longrightarrow N_q(\mathbf{0}, V_\Sigma)$, where $q = \frac{1}{2}p(p+1)$,

with $\hat{\boldsymbol{\lambda}}$, $\hat{\boldsymbol{\mu}}_j$ $(j = 0, 1, \ldots, J)$ and $\hat{\boldsymbol{\sigma}}$ asymptotically uncorrelated, where $V_{\boldsymbol{\lambda}} = \text{diag}[\pi_1^{-1}, \ldots, \pi_J^{-1}] + \pi_0^{-1}\mathbf{1}_J\mathbf{1}'_J$, $V_{\boldsymbol{\mu}_j} = \pi_j^{-1}\Sigma$, and $V_\Sigma = 2H(\Sigma \otimes \Sigma)H'$. The matrix H is of order $p(p+1)/2$ by p^2 and may be defined as $H = (G'G)^{-1}G'$. McCulloch (1982) has given an analytic definition of G.

Theorem 1. The logistic parameter estimates, $\hat{B}_0$ and $\hat{B}_*$, produced by the normal discrimination procedure have the asymptotic distribution:

$$L[\sqrt{n}\ \text{vec}(\hat{B} - B)] \longrightarrow N(\mathbf{0}, \Sigma_1(B)),$$

where

$$\frac{1}{n}\Sigma_1(B) = \begin{bmatrix} \text{Var}(\hat{B}_0) & \text{Cov}(\hat{B}_0, \text{vec}\hat{B}_*) \\ \text{Cov}(\text{vec}\hat{B}_*, \hat{B}_0) & \text{Var}(\text{vec}\hat{B}_*) \end{bmatrix}$$

with

$$\text{Var}(\hat{B}_0) = \frac{1}{n}\Big[Q + \frac{1}{\pi_0}(1 + \boldsymbol{\mu}_0'\Sigma^{-1}\boldsymbol{\mu}_0)\mathbf{1}_J\mathbf{1}'_J + S(U'\Sigma^{-1}U \otimes Q)S'$$
$$+ \frac{1}{2}S(U'\Sigma^{-1}U \otimes U'\Sigma^{-1}U - U_0'\Sigma^{-1}U_0 \otimes U_0'\Sigma^{-1}U_0)S'\Big],$$

$$\begin{aligned}\mathrm{Cov}(\hat{B}_0, \mathrm{vec}\hat{B}_*) = \frac{1}{n}\Big[\frac{1}{\pi_0}(\boldsymbol{\mu}_0'\Sigma^{-1} \otimes \mathbf{1}_J\mathbf{1}_J') + S(U'\Sigma^{-1} \otimes Q) \\ - S(U'\Sigma^{-1} \otimes U'\Sigma^{-1}(U - U_0) \\ - U_0'\Sigma^{-1} \otimes U_0'\Sigma^{-1}(U - U_0))\Big],\end{aligned}$$

$$\begin{aligned}\mathrm{Var}(\mathrm{vec}\hat{B}_*) = \frac{1}{n}\Big[\Sigma^{-1} \otimes (Q + \frac{1}{\pi_0}\mathbf{1}_J\mathbf{1}_J' + \Sigma^{-1} \otimes (U - U_0)'\Sigma^{-1}(U - U_0) \\ + \left(\Sigma^{-1}(U - U_0) \otimes (U - U_0)'\Sigma^{-1}\right) I_{(J,p)}\Big],\end{aligned}$$

and

$$\begin{aligned}Q &= \mathrm{diag}[\pi_1^{-1}, \ldots, \pi_J^{-1}], \\ U &= (\boldsymbol{\mu}_1|\boldsymbol{\mu}_2|\ldots|\boldsymbol{\mu}_J), \\ U_0 &= \boldsymbol{\mu}_0\mathbf{1}_J', \\ S &= \sum_{j=1}^{J}\mathbf{i}_j(\mathbf{i}_j' \otimes \mathbf{i}_j').\end{aligned}$$

Corollary 1.1. The large sample variances and covariances of the logistic parameter estimates, $\hat{\beta}_{sj}$ $(s = 1, 2, \ldots, p;\ j = 1, 2, \ldots, J)$, produced by the normal discrimination procedure are:

$$\begin{aligned}\mathrm{Var}(\hat{\beta}_{sj}) =& \frac{1}{n}\Big\{\sigma^{ss}\left(\frac{1}{\pi_0} + \frac{1}{\pi_j} + (\boldsymbol{\mu}_j - \boldsymbol{\mu}_0)'\Sigma^{-1}(\boldsymbol{\mu}_j - \boldsymbol{\mu}_0)\right) \\ &+ (\boldsymbol{\mu}_j - \boldsymbol{\mu}_0)'\Sigma^{-1}\mathbf{e}_s\mathbf{e}_s'\Sigma^{-1}(\boldsymbol{\mu}_j - \boldsymbol{\mu}_0)\Big\}, \\ \mathrm{Cov}(\hat{\beta}_{sj}, \hat{\beta}_{sk}) =& \frac{1}{n}\Big\{\sigma^{ss}\left(\frac{1}{\pi_0} + (\boldsymbol{\mu}_j - \boldsymbol{\mu}_0)'\Sigma^{-1}(\boldsymbol{\mu}_k - \boldsymbol{\mu}_0)\right) \\ &+ (\boldsymbol{\mu}_j - \boldsymbol{\mu}_0)'\Sigma^{-1}\mathbf{e}_s\mathbf{e}_s'\Sigma^{-1}(\boldsymbol{\mu}_k - \boldsymbol{\mu}_0)\Big\}, \\ \mathrm{Cov}(\hat{\beta}_{sj}, \hat{\beta}_{tj}) =& \frac{1}{n}\Big\{\sigma^{st}\left(\frac{1}{\pi_0} + \frac{1}{\pi_j} + (\boldsymbol{\mu}_j - \boldsymbol{\mu}_0)'\Sigma^{-1}(\boldsymbol{\mu}_j - \boldsymbol{\mu}_0)\right) \\ &+ (\boldsymbol{\mu}_j - \boldsymbol{\mu}_0)'\Sigma^{-1}\mathbf{e}_t\mathbf{e}_s'\Sigma^{-1}(\boldsymbol{\mu}_j - \boldsymbol{\mu}_0)\Big\},\end{aligned}$$

where $\sigma^{st} = \mathbf{e}_s'\Sigma^{-1}\mathbf{e}_t$.

Proof of Theorem 1. The normal discrimination procedure produces estimates of the logistic model parameters that are functions of the normal parameter estimates. From Lemma 1, we have the asymptotic distributions of $\hat{\lambda}_j$, $\hat{\boldsymbol{\mu}}_j$ and $\hat{\Sigma}$. The asymptotic distributions of $\hat{\beta}_{0j}$, $\hat{\beta}_{sj}$, can then be obtained by means of the multivariate "delta" method (see Bishop *et al.*, 1975, Theorem 14.6-2):

$$L[\sqrt{n}\ \mathrm{vec}(\hat{B}-B)] \longrightarrow N\left(\mathbf{0}, \left[\frac{\partial \mathrm{vec} B}{\partial \boldsymbol{\theta}'}\right] \mathrm{Var}\ (\hat{\boldsymbol{\theta}}) \left[\frac{\partial \mathrm{vec} B}{\partial \boldsymbol{\theta}'}\right]'\right),$$

where $\boldsymbol{\theta}' = (\boldsymbol{\lambda}', \boldsymbol{\mu}_0', \boldsymbol{\mu}_*', \boldsymbol{\sigma}')$, $\boldsymbol{\mu}_*' = \mathrm{vec}'U$, and the relationship between B and $\boldsymbol{\theta}$ can be expressed in matrix notation by:

$$B_0 = \boldsymbol{\lambda} - \frac{1}{2}\sum_{j=1}^{J} \mathbf{i}_j \otimes \left\{\mathbf{i}_j'[U'\Sigma^{-1}U - U_0'\Sigma^{-1}U_0]\mathbf{i}_j\right\},$$

$$B_* = (U - U_0)'\Sigma^{-1}.$$

Let

$$M = \left[\frac{\partial \mathrm{vec} B}{\partial \boldsymbol{\theta}'}\right] = [M_{\boldsymbol{\lambda}} | M_{\boldsymbol{\mu}_0} | M_{\boldsymbol{\mu}_*} | M_{\Sigma}]$$

and

$$\mathrm{Var}(\hat{\boldsymbol{\theta}}) = \frac{1}{n}\ \mathrm{diag}\left[V_{\boldsymbol{\lambda}}, V_{\boldsymbol{\mu}_0}, V_{\boldsymbol{\mu}_*}, V_{\Sigma}\right].$$

Then

$$\begin{aligned}\mathrm{Var}(\mathrm{vec}\hat{B}) &= M[\mathrm{Var}(\hat{\boldsymbol{\theta}})]M' \\ &= \frac{1}{n}\left[M_{\boldsymbol{\lambda}}V_{\boldsymbol{\lambda}}M_{\boldsymbol{\lambda}}' + M_{\boldsymbol{\mu}_0}V_{\boldsymbol{\mu}_0}M_{\boldsymbol{\mu}_0}' + M_{\boldsymbol{\mu}_*}V_{\boldsymbol{\mu}_*}M_{\boldsymbol{\mu}_*}' + M_{\Sigma}V_{\Sigma}M_{\Sigma}'\right],\end{aligned}$$

where

$$M_{\boldsymbol{\lambda}} = \left[\frac{\partial \mathrm{vec} B}{\partial \boldsymbol{\lambda}'}\right] = \left[\frac{I_J}{0_{pJ\times J}}\right],$$

$$M_{\boldsymbol{\mu}_0} = \left[\frac{\partial \mathrm{vec} B}{\partial \boldsymbol{\mu}_0'}\right] = \left[\frac{I_J \otimes \boldsymbol{\mu}_0'}{I_{(p,J)}}\right](\mathbf{1}_J \otimes \Sigma^{-1}),$$

$$M_{\boldsymbol{\mu}_*} = \left[\frac{\partial \mathrm{vec} B}{\partial \mathrm{vec}'\boldsymbol{\mu}_*}\right] = \left[\frac{-S(I_J \otimes U')}{I_{(p,J)}}\right](I_J \otimes \Sigma^{-1}),$$

$$M_{\Sigma} = \left[\frac{\partial \mathrm{vec} B}{\partial \mathrm{vech}'\Sigma}\right] = \left[\frac{\frac{1}{2}S(U'\otimes U' - U_0'\otimes U_0')}{-(I_p \otimes (U-U_0)')}\right](\Sigma^{-1}\otimes\Sigma^{-1})G,$$

and $V_{\boldsymbol{\mu}_*} = \mathrm{diag}[\pi_1^{-1}, \ldots, \pi_J^{-1}] \otimes \Sigma$. The other variances are as stated in Lemma 1. The derivation of the M matrices follows from the matrix calculus results of Dwyer and MacPhail (1948), MacRae (1974) and McCulloch (1982). Details are given by Bull (1983).

Evaluating Var(vec$\hat{B}$) we consider:

$$M_{\boldsymbol{\lambda}} V_{\boldsymbol{\lambda}} M'_{\boldsymbol{\lambda}} = \begin{bmatrix} V_{\boldsymbol{\lambda}} & 0_{J\times pJ} \\ 0_{pJ\times J} & 0_{pJ\times pJ} \end{bmatrix}_{(p+1)J\times(p+1)J} = \begin{bmatrix} 1 & 0_{1\times p} \\ 0_{p\times 1} & 0_{p\times p} \end{bmatrix} \otimes V_{\boldsymbol{\lambda}},$$

$$M_{\boldsymbol{\mu}_0} V_{\boldsymbol{\mu}_0} M'_{\boldsymbol{\mu}_0} = \frac{1}{\pi_0} \begin{bmatrix} (\mathbf{1}\ \mathbf{1}' \otimes \boldsymbol{\mu}_0'\Sigma^{-1}\boldsymbol{\mu}_0) & (\mathbf{1}\ \mathbf{1}' \otimes \boldsymbol{\mu}_0'\Sigma^{-1}) I_{(J,p)} \\ I_{(p,J)}(\mathbf{1}\ \mathbf{1}' \otimes \Sigma^{-1}\boldsymbol{\mu}_0) & I_{(p,J)}(\mathbf{1}\ \mathbf{1}' \otimes \Sigma^{-1}) I_{(J,p)} \end{bmatrix}$$

$$= \frac{1}{\pi_0} \begin{bmatrix} \boldsymbol{\mu}_0'\Sigma^{-1}\boldsymbol{\mu}_0 & \boldsymbol{\mu}_0'\Sigma^{-1} \\ \Sigma^{-1}\boldsymbol{\mu}_0 & \Sigma^{-1} \end{bmatrix} \otimes \mathbf{1}\ \mathbf{1}', \quad \text{where } \mathbf{1} = \mathbf{1}_J,$$

$$M_{\boldsymbol{\mu}_*} V_{\boldsymbol{\mu}_*} M'_{\boldsymbol{\mu}_*} = \begin{bmatrix} S(Q \otimes U'\Sigma^{-1}U)S' & S(Q \otimes U'\Sigma^{-1}) I_{(J,p)} \\ I_{(p,J)}(Q \otimes \Sigma^{-1}U)S' & I_{(p,J)}(Q \otimes \Sigma^{-1}) I_{(J,p)} \end{bmatrix}$$

$$= \begin{bmatrix} S(U'\Sigma^{-1}U \otimes Q)S' & S(U'\Sigma^{-1} \otimes Q) \\ (\Sigma^{-1}U \otimes Q)S' & \Sigma^{-1} \otimes Q \end{bmatrix},$$

and

$$M_\Sigma V_\Sigma M'_\Sigma = \begin{bmatrix} (1) & (2) \\ (2)' & (3) \end{bmatrix},$$

where

$$\begin{aligned} (1) &= S(U'\Sigma^{-1}U \otimes U'\Sigma^{-1}U - U_0'\Sigma^{-1}U_0 \otimes U_0'\Sigma^{-1}U_0)S'/2, \\ (2) &= -S(U'\Sigma^{-1} \otimes U'\Sigma^{-1}(U - U_0) - U_0'\Sigma^{-1} \otimes U_0'\Sigma^{-1}(U - U_0)), \\ (3) &= (\Sigma^{-1} \otimes (U - U_0)'\Sigma^{-1}(U - U_0)) \\ &\quad + (\Sigma^{-1}(U - U_0) \otimes (U - U_0)'\Sigma^{-1}) I_{(J,p)}. \end{aligned}$$

Therefore,

$$n\mathrm{Var}(\hat{B}_0) = V_\lambda + \frac{1}{\pi_0}\boldsymbol{\mu}_0'\Sigma^{-1}\boldsymbol{\mu}_0 \mathbf{1}\ \mathbf{1}' + S(U'\Sigma^{-1}U \otimes Q)S' + (1),$$

$$n\mathrm{Cov}(\hat{B}_0, \mathrm{vec}\hat{B}_*) = \frac{1}{\pi_0}(\boldsymbol{\mu}_0'\Sigma^{-1} \otimes \mathbf{1}\ \mathbf{1}') + S(U'\Sigma^{-1} \otimes Q) + (2),$$

and

$$n\mathrm{Var}(\mathrm{vec}\hat{B}_*) = \frac{1}{\pi_0}\Sigma^{-1} \otimes \mathbf{1}\ \mathbf{1}' + \Sigma^{-1} \otimes Q + (3).$$

Furthermore, one obtains the result of Corollary 1.1 by evaluating

$$\mathrm{Var}(\hat{\beta}_{sj}) = (\mathbf{e}'_s \otimes \mathbf{i}'_j)\mathrm{Var}(\mathrm{vec}\hat{B}_*)(\mathbf{e}_s \otimes \mathbf{i}_j),$$
$$\mathrm{Cov}(\hat{\beta}_{sj}, \hat{\beta}_{tj}) = (\mathbf{e}'_s \otimes \mathbf{i}'_j)\mathrm{Var}(\mathrm{vec}\hat{B}_*)(\mathbf{e}_t \otimes \mathbf{i}_j),$$

and

$$\mathrm{Cov}(\hat{\beta}_{sj}, \hat{\beta}_{sk}) = (\mathbf{e}'_s \otimes \mathbf{i}'_j)\mathrm{Var}(\mathrm{vec}\hat{B}_*)(\mathbf{e}_s \otimes \mathbf{i}_k).$$

4. LARGE SAMPLE DISTRIBUTION OF LOGISTIC REGRESSION ESTIMATES

In finding the large sample distribution of the logistic regression estimates, it is easier to work in the so-called standard or canonical situation in which the distributions are in a simpler form. The standard situation is specified by standard population mean vectors which are related to general population mean vectors by a linear, possibly non-homogeneous, transformation represented by the matrix A. The asymptotic variance of an estimator in the general model can then be expressed as a function of the variance-covariance matrix of the estimators in the standard situation; this is stated without proof in Lemma 2. Lemma 3 defines precisely the standard situation and the transformation matrix A for any particular case. We show that A is a function of the population means and variances in the general model.

Lemma 2. We assume that a random variable $\mathbf{X}$ can arise from one of $J+1$ p-dimensional normal populations such that

$$\mathbf{X} \sim N_p(\boldsymbol{\mu}_j, \Sigma_x) \text{ with probability } \pi_j \quad (j = 0, 1, \ldots, J).$$

By means of a linear transformation $\mathbf{Y} = A\,\mathbf{X} + \mathbf{a}$, we obtain the random variable

$$\mathbf{Y} \sim N_p(\mathbf{v}_j, \Sigma_y) \text{ with probability } \pi_j,$$

where $\mathbf{v}_j = A\boldsymbol{\mu}_j + \mathbf{a}$ and $\Sigma_y = A\Sigma_x A'$.

If $\tilde{B}_0$, $\tilde{B}_*$ are the parameter estimates produced by the logistic regression procedure and based on the original random variable $\mathbf{X}$ and $\tilde{G}_0$, $\tilde{G}_*$ are the corresponding estimates based on the random variable $\mathbf{Y}$, then it can be shown that

$$\tilde{B}_0 = \tilde{G}_0 + (\mathbf{a}' \otimes I_J)\mathrm{vec}\tilde{G}_* \quad \text{and} \quad \mathrm{vec}\tilde{B}_* = (A' \otimes I_J)\mathrm{vec}\tilde{G}_*.$$

It follows that

$$\begin{aligned}\mathrm{Var}(\tilde{B}_0) &= \mathrm{Var}(\tilde{G}_0) + (\mathbf{a}' \otimes I_J)\mathrm{Var}(\mathrm{vec}\tilde{G}_*)(\mathbf{a} \otimes I_J) \\ &\quad + (\mathbf{a}' \otimes I_J)\mathrm{Cov}(\mathrm{vec}\tilde{G}_*, \tilde{G}_0) + \mathrm{Cov}(\tilde{G}_0, \mathrm{vec}\tilde{G}_*)(\mathbf{a} \otimes I_J), \\ \mathrm{Var}(\mathrm{vec}\tilde{B}_*) &= (A' \otimes I_J)\mathrm{Var}(\mathrm{vec}\tilde{G}_*)(A \otimes I_J)\end{aligned}$$

and

$$\begin{aligned}\mathrm{Cov}(\tilde{B}_0, \mathrm{vec}\tilde{B}_*) &= (\mathbf{a}' \otimes I_J)\mathrm{Var}(\mathrm{vec}\tilde{G}_*)(A \otimes I_J) \\ &\quad + \mathrm{Cov}(\tilde{G}_0, \mathrm{vec}\tilde{G}_*)(A \otimes I_J).\end{aligned}$$

Lemma 3. We obtain the standard situation in which $\mathbf{Y} \sim N_p(\mathbf{v}_j, I_p)$ with probability π_j, by the linear transformation $\mathbf{Y} = A\mathbf{X} + \mathbf{a}$.

The transformation takes the population means $\boldsymbol{\mu}_0, \ldots, \boldsymbol{\mu}_j$ in the metric Σ^{-1} in p-space into the corresponding population means $\mathbf{v}_0, \ldots, \mathbf{v}_J$ in the metric I. This is equivalent to a change of coordinates from an affine co-ordinate system in $X_1, \ldots, X_p$ to an orthogonal co-ordinate system in $Y_1, \ldots, Y_p$ involving:

(i) translation of the origin to the point $\boldsymbol{\mu}_0$,

(ii) change of axes from an oblique orientation (i.e., the axes X_s and X_t are *not* perpendicular) to the usual cartesian orientation (the axes $Y_s Y_t$ *are* perpendicular), and

(iii) rotation of axes around the origin so that $\mathbf{v}_1$ is on the Y_1 axis, $\mathbf{v}_2$ is in the $Y_1 \times Y_2$ plane, $\mathbf{v}_3$ is in the $Y_1 \times Y_2 \times Y_3$ plane, and similarly.

We then have $\mathbf{v}_0 = \mathbf{0}$ and $\mathbf{v}_j = V\mathbf{i}_j$ $(j = 1, 2, \ldots, J)$, the jth column of the $p \times J$ matrix V. V is such that $\mathbf{v}_1, \mathbf{v}_2, \ldots, \mathbf{v}_m$ comprise a set of linearly independent vectors and $\mathbf{v}_{m+1}, \ldots, \mathbf{v}_J$ are each a linear combination of $\mathbf{v}_1, \ldots, \mathbf{v}_m$. Here m, the dimension of the space spanned by $\mathbf{v}_1, \ldots, \mathbf{v}_J$, is less than or equal to the minimum of J and p.

The matrix V may be expressed in partitioned form as:

$$V = \left[\frac{V_1 \mid V_2}{0_{(p-m)\times J}}\right], \text{ with } \quad V_1 = \{v_{ij}\}, \quad V_2 = \{v_{ik}\},$$

where

$$v_{ij} = \begin{cases} \Delta_j \sin\theta_{i-1,j} \prod_{\ell=i}^{j} \cos\theta_{\ell j} & i \le j \\ 0 & i > j, \ (i, j = 1, 2, \ldots, m), \end{cases}$$

$$v_{ik} = \begin{cases} \Delta_k \sin\theta_{i-1,k} \ \prod_{\ell=i}^{m-1} \cos\theta_{\ell k} & 1 \le i \le m-1 \\ \Delta_k \sin\theta_{i-1,k} & i = m, \ (k = m+1, \ldots, J), \end{cases}$$

and

$$\Delta_j = \left[(\boldsymbol{\mu}_j - \boldsymbol{\mu}_0)'\Sigma^{-1}(\boldsymbol{\mu}_j - \boldsymbol{\mu}_0)\right]^{\frac{1}{2}},$$
$$\cos\theta_{\ell j} = (\boldsymbol{\mu}_\ell - \boldsymbol{\mu}_0)'\Sigma^{-1}(\boldsymbol{\mu}_j - \boldsymbol{\mu}_0)/\Delta_\ell\Delta_j,$$

and

$$\sin\theta_{\ell j} = (1 - \cos^2\theta_{\ell j})^{\frac{1}{2}}.$$

It follows that $\mathbf{v}_j'\mathbf{v}_j = \Delta_j^2$ $(j = 1, 2, \ldots, J)$. The transformation to Y is specified by $A_{p\times p}$ and $\mathbf{a}_{1\times p}$:

$$A = \left[\frac{A_A}{A_B}\right] = \left[\frac{(V_1')^{-1}(\Sigma^{-1}U_A)'}{(\Sigma^{-1}U_B)'}\right],$$

$$\mathbf{a} = \left[\frac{\mathbf{a}_A}{\mathbf{a}_B}\right] = \left[\frac{A_A\boldsymbol{\mu}_0}{A_B\boldsymbol{\mu}_0}\right],$$

where $U_A = (\boldsymbol{\mu}_1|\ldots|\boldsymbol{\mu}_m) - \boldsymbol{\mu}_0\mathbf{1}_m'$, and U_B is constructed such that

$$U_A'\Sigma^{-1}U_B = 0_{m\times(p-m)} \quad \text{and} \quad U_B'\Sigma^{-1}U_B = I_{p-m}.$$

Note that $A'A = \Sigma^{-1}$ implies that $A_A'A_A + A_B'A_B = \Sigma^{-1}$.

Theorem 2. The logistic parameter estimates, $\tilde{B}_0$ and $\tilde{B}_*$, produced by the logistic regression procedure have the asymptotic distribution:

$$L[\sqrt{n}\,\text{vec}(\tilde{B} - B)] \longrightarrow N(\mathbf{0}, \Sigma_2(B)),$$

where

$$\frac{1}{n}\Sigma_2(B) = \begin{bmatrix} \text{Var}(\tilde{B}_0) & \text{Cov}(\tilde{B}_0, \text{vec}\tilde{B}_*) \\ \text{Cov}(\text{vec}\tilde{B}_*, \tilde{B}_0) & \text{Var}(\text{vec}\tilde{B}_*) \end{bmatrix},$$

$$\text{Var}(\tilde{B}_0) = \frac{1}{n}\left\{\left(\mathbf{a}_A' \otimes I_J - \Omega_{00}^{-1}B_2\right)E_3^{-1}(\mathbf{a}_A \otimes I_J - B_2'\Omega_{00}^{-1}) + (1 + \mathbf{a}_B'\mathbf{a}_B)\Omega_{00}^{-1}\right\},$$

$$\text{Cov}(\tilde{B}_0, \text{vec}\tilde{B}_*) = \frac{1}{n}\left\{\left(\mathbf{a}_A' \otimes I_J - \Omega_{00}^{-1}B_2\right)E_3^{-1}(A_A \otimes I_J) + \left(\mathbf{a}_B'A_B \otimes \Omega_{00}^{-1}\right)\right\},$$

$$\text{Var}(\text{vec}\tilde{B}_*) = \frac{1}{n}\left\{(A_A' \otimes I_J)E_3^{-1}(A_A \otimes I_J) + A_B'A_B \otimes \Omega_{00}^{-1}\right\},$$

with

$$E_3 = B_3 - B_2'\Omega_{00}^{-1}B_2,$$
$$B_2 = (\Omega_{01}|\Omega_{02}|\ldots|\Omega_{0m})$$

and

$$B_3 = \{\Omega_{st}\} \quad (s,t = 1,2,\ldots,m),$$

where the Ω_{st} are $J \times J$ submatrices of the Fisher information matrix evaluated in the standard situation.

Corollary 2.1. The large sample variances and covariances of the logistic parameter estimates, $\tilde{\beta}_{sj}$ ($s = 1,2,\ldots,p$; $j = 1,2,\ldots,J$), produced by the logistic regression procedure are:

$$\mathrm{Var}(\tilde{\beta}_{sj}) = \frac{1}{n}\left\{(\mathbf{a}_s' \otimes \mathbf{i}_j')E_3^{-1}(\mathbf{a}_s \otimes \mathbf{i}_j) + (\sigma^{ss} - \mathbf{a}_s'\mathbf{a}_s)(\mathbf{i}_j'\Omega_{00}^{-1}\mathbf{i}_j)\right\},$$
$$\mathrm{Cov}(\tilde{\beta}_{sj}, \tilde{\beta}_{sk}) = \frac{1}{n}\left\{(\mathbf{a}_s' \otimes \mathbf{i}_j')E_3^{-1}(\mathbf{a}_s \otimes \mathbf{i}_k) + (\sigma^{ss} - \mathbf{a}_s'\mathbf{a}_s)(\mathbf{i}_j'\Omega_{00}^{-1}\mathbf{i}_k)\right\},$$
$$\mathrm{Cov}(\tilde{\beta}_{sj}, \tilde{\beta}_{tj}) = \frac{1}{n}\left\{(\mathbf{a}_s' \otimes \mathbf{i}_j')E_3^{-1}(\mathbf{a}_t \otimes \mathbf{i}_j) + (\sigma^{st} - \mathbf{a}_s'\mathbf{a}_t)(\mathbf{i}_j'\Omega_{00}^{-1}\mathbf{i}_j)\right\},$$

where $\mathbf{a}_s = A_A\mathbf{e}_s$ and $\Sigma^{-1} = \{\sigma^{st}\}$; $s,t = 1,2,\ldots,p$.

Proof of Theorem 2. In contravention of the usual maximum likelihood theory assumptions, the multinomial random variables Z_i are not identically distributed. Rather, the distribution of each z_i is parameterized by $\theta_j(\mathbf{x}_i)$ ($j = 1,2,\ldots,J$) which depend on $\mathbf{x}_i$. However, the distributions are related in that the logistic parameters β_{0j} and $\boldsymbol{\beta}_j$ ($j = 1,2,\ldots,J$) are shared.

This situation, involving related, or associated, populations was considered by Bradley and Gart (1962). They give conditions under which the usual properties of consistency and asymptotic normality hold for maximum likelihood estimators from associated populations. Hauck (1981) adopted assumptions more restrictive than those of Bradley and Gart in order to convert the associated populations situation into the identically distributed situation. Usual maximum likelihood theory can then be applied.

Of interest here is the case in which the number of possible values of $\mathbf{x}$ is infinite. We therefore assume:

(i) the pairs $(z_i, \mathbf{x}_i)$ are independent and identically distributed with density $g(z,\mathbf{x}) = f(z|\mathbf{x})h(\mathbf{x})$, where

$$f(z|\mathbf{x}) = \left[\prod_{j=0}^{J} \theta_j(\mathbf{x})^{\delta_j}\right];$$

(ii) $h(\mathbf{x})$, the marginal density of $\mathbf{x}$, does not depend on any of the logistic parameters and is given by

$$h(\mathbf{x}) = \sum_{j=0}^{J} \pi_j \frac{|\Sigma|^{-\frac{1}{2}}}{(2\pi)^{p/2}} \exp\left\{-\frac{1}{2}(\mathbf{x}-\boldsymbol{\mu}_j)'\Sigma^{-1}(\mathbf{x}-\boldsymbol{\mu}_j)\right\};$$

(iii) $g(z, \mathbf{x})$ satisfies all the conditions required for consistency and asymptotic normality of maximum likelihood estimation (see Cox and Hinkley, 1974, Chapter 9).

Following Hauck's derivation, the log likelihood is then

$$\begin{aligned}\ell(B,\boldsymbol{\gamma}) &= \sum_{i=1}^{n} \ln\{g(z_i,\mathbf{x}_i)\} \\ &= \sum_{i=1}^{n}\{\ln f(z_i|\mathbf{x}_i) + \ln h(\mathbf{x}_i)\},\end{aligned}$$

where $\boldsymbol{\gamma}$ denotes the parameters of $h(\mathbf{x})$. By assumption (ii), we have:

$$\frac{\partial \ell(B,\boldsymbol{\gamma})}{\partial \mathrm{vec} B} = \frac{\partial}{\partial \mathrm{vec} B}\left\{\sum_{i=1}^{n} \ln f(z_i|\mathbf{x}_i)\right\},$$

$$\frac{\partial^2 \ell(B,\boldsymbol{\gamma})}{\partial \mathrm{vec} B\, \partial \boldsymbol{\gamma}'} = 0,$$

and

$$\frac{\partial^2 \ell(B,\boldsymbol{\gamma})}{\partial \mathrm{vec} B\, \partial \mathrm{vec}' B} = \sum_{i=1}^{n} \frac{\partial^2}{\partial \mathrm{vec} B\, \partial \mathrm{vec}' B}\{\ln f(z_i|\mathbf{x}_i)\}.$$

Thus, vec $\tilde{B}$ is asymptotically independent of $\boldsymbol{\gamma}$ with distribution such that

$$L[\sqrt{n}(\mathrm{vec}\tilde{B} - \mathrm{vec}B)] \longrightarrow N_{j(p+1)}(\mathbf{0}, \Sigma_2(B)),$$

where

$$\Sigma_2^{-1} = \frac{1}{n} E_{g(z,\mathbf{x})}\left\{\frac{-\partial^2 \ell(B,\boldsymbol{\gamma})}{\partial \mathrm{vec} B\, \partial \mathrm{vec}' B}\right\} = E_{h(\mathbf{x})}\left\{\frac{-\partial^2 \ln f(z|\mathbf{x})}{\partial \mathrm{vec} B\, \partial \mathrm{vec}' B}\right\}.$$

This follows from

$$\begin{aligned}E_{z,\mathbf{x}}\left(\sum_{i=1}^{n} w(\mathbf{x}_i)\right) &= E_{\mathbf{x}}\left[\sum_{i=1}^{n} E_{z|\mathbf{x}}(w(\mathbf{x}_i))\right] \\ &= E_{\mathbf{x}}\left[\sum_{i=1}^{n} w(\mathbf{x})\right] \\ &= n\, E_{\mathbf{x}}[w(\mathbf{x})],\end{aligned}$$

where $w(\mathbf{x})$ is a function of $\mathbf{x}$ and $\boldsymbol{\theta}$ only.

Let

$$\Omega = E_{\mathbf{x}}\left\{\begin{pmatrix}1\\ \mathbf{x}\end{pmatrix}(1\ \mathbf{x}') \otimes P(\mathbf{x})\right\} = \int \left[\begin{pmatrix}1\\ \mathbf{x}\end{pmatrix}(1\ \mathbf{x}') \otimes P(\mathbf{x})h(\mathbf{x})\right] d\mathbf{x},$$

where $P(\mathbf{x}) = P_D(\mathbf{x}) - P_V(\mathbf{x}) = \operatorname{diag}[\theta_1(\mathbf{x}), \ldots, \theta_J(\mathbf{x})] - \boldsymbol{\theta}'\boldsymbol{\theta}$ with $\boldsymbol{\theta} = (\theta_1(\mathbf{x}), \ldots, \theta_J(\mathbf{x}))$. Furthermore, $\Omega = \{\Omega_{st}\}$ $(s, t = 0, 1, 2, \ldots, p)$, where each Ω_{st} is a $J \times J$ matrix:

$$\Omega_{st} = \int x_s x_t P(\mathbf{x})h(\mathbf{x})d\mathbf{x} \quad \text{with} \quad x_0 = 1.$$

In the standard situation, as described in Lemma 3, we have $\Sigma_y = I_p, \mathbf{v}_0 = \mathbf{0}$, $\beta_{0k} = \log(\pi_k/\pi_0) - \frac{1}{2}\Delta_k^2$ and $\boldsymbol{\beta}_k = \mathbf{v}_k$, where $\Delta_k^2 = \mathbf{v}_k'\mathbf{v}_k = \boldsymbol{\mu}_k'\Sigma^{-1}\boldsymbol{\mu}_k$ and $V = (\mathbf{v}_1|\mathbf{v}_2|\ldots|\mathbf{v}_J)$ is a $p \times J$ matrix with entries $v_{sj} = 0$ whenever $s > j$ or $s > m$. Usually m will be equal to the minimum of J and p, but if the population means $\mathbf{v}_0, \ldots, \mathbf{v}_J$ lie in an r-dimensional subspace of p, then $m = r$.

Thus,

$$\theta_j(\mathbf{y}) = \exp\{\ln(\pi_j/\pi_0) - \frac{1}{2}\Delta_j^2 + \mathbf{v}_j'\mathbf{y}\} \cdot \left[1 + \sum_{k=1}^{J} \exp\left\{(\ln(\pi_k/\pi_0)) - \frac{1}{2}\Delta_k^2 + \mathbf{v}_k'\mathbf{y}\right\}\right]^{-1}$$

and

$$\theta_j(\mathbf{y})h(\mathbf{y}) = \pi_j \ \exp\left\{-\frac{1}{2}\Delta_j^2 + \mathbf{v}_j'\mathbf{y}\right\} g(\mathbf{y}),$$

where $g(\mathbf{y}) = (2\pi)^{-p/2}\exp\{-\frac{1}{2}\mathbf{y}'\mathbf{y}\}$.

Consider $\int y_s y_t P_D(\mathbf{y})h(\mathbf{y})d\mathbf{y}$, which is a $J \times J$ diagonal matrix with diagonal terms:

$$\int y_s y_t \theta_j(\mathbf{y})h(\mathbf{y})d\mathbf{y} = \pi_j(2\pi)^{-p/2}\int(\mathbf{y}'C_{st}\mathbf{y} + \mathbf{y}'\mathbf{c}_{st} + c_{st})\exp\{-\frac{1}{2}\mathbf{y}'\mathbf{y} + \mathbf{y}'\mathbf{v}_j - \frac{1}{2}\Delta_j^2\}d\mathbf{y},$$

where

$$C_{st} = \begin{cases} \mathbf{e}_s\mathbf{e}_t' & s,\ t \geq 1 \\ 0_{p\times p} & \text{otherwise,} \end{cases}$$

$$\mathbf{c}_{st} = \begin{cases} \mathbf{e}_s & s \geq 1,\ t = 0 \\ \mathbf{e}_t & t \geq 1,\ s = 0 \\ \mathbf{0} & \text{otherwise,} \end{cases}$$

$$c_{st} = \begin{cases} 1 & s = t = 0 \\ 0 & \text{otherwise.} \end{cases}$$

Then, by Graybill (1969, Theorem 10.5.1), we obtain

$$\pi_j[\text{tr}(C_{st}) + \mathbf{v}_j'C_{st}\mathbf{v}_j + \mathbf{v}_j'\mathbf{c}_{st} + c_{st}].$$

However,

$$\text{tr}(C_{st}) = \begin{cases} 1 & s = t \geq 1 \\ 0 & \text{otherwise} \end{cases}$$

and

$$\mathbf{v}_j'\mathbf{e}_s = \begin{cases} v_{sj} & 1 \leq s \leq m \\ 0 & s > m. \end{cases}$$

Therefore,

$$\int y_s y_t P_D(\mathbf{y})h(\mathbf{y})d\mathbf{y} = \begin{cases} R & s = t = 0 \\ RV_sV_t & 0 \leq s \text{ and } t \leq m,\quad s \neq t \\ R + RV_s^2 & 0 \leq s \text{ and } t \leq m,\quad s = t \\ 0_{J\times J} & m < s \text{ or } t,\quad s \neq t \\ R & m < s,\quad s = t \end{cases}$$

where $R = \text{diag}[\pi_1, \ldots, \pi_J]$, $V_0 = I_J$, $V_S = \text{diag}[v_{s1}, v_{s2}, \ldots, v_{sJ}]$.

Secondly, consider $\int y_s y_t P_V(\mathbf{y})h(\mathbf{y})d\mathbf{y}$, which is a $J \times J$ symmetric matrix with terms

$$\int y_s y_t \theta_j(\mathbf{y})\theta_k(\mathbf{y})h(\mathbf{y})d\mathbf{y} \quad (j, k = 1, \ldots, J)$$

$$= \pi_j\pi_k \exp\{-\frac{1}{2}(\Delta_j^2 + \Delta_k^2)\} \int y_s y_t \exp\{\mathbf{y}'(\mathbf{v}_j + \mathbf{v}_k)\} f(\mathbf{y})g(\mathbf{y})d\mathbf{y},$$

where $f(\mathbf{y}) = [\pi_0 + \sum_{i=1}^{j} \pi_i \exp\{\mathbf{y}'\mathbf{v}_i - \frac{1}{2}\Delta_i^2\}]^{-1}$.

Because of the structure of V, $f(\mathbf{y})$ will be a function of $y_1, y_2, \ldots, y_m$ (m = dimension of subspace), and $\exp\{\mathbf{y}'(\mathbf{v}_j + \mathbf{v}_k)\}$ will be a function of

$y_1, y_2, \ldots, y_n$, $(n = \min\{m, \max(j,k)\})$. Thus,

$$\int y_s y_t \exp\{\mathbf{y}'(\mathbf{v}_j + \mathbf{v}_k)\} f(\mathbf{y}) g(\mathbf{y}) d\mathbf{y}$$
$$= \int_{y_1} \cdots \int_{y_m} \exp\{\mathbf{y}'(\mathbf{v}_j + \mathbf{v}_k)\} f(\mathbf{y}) (2\pi)^{-p/2} \cdot \exp\left\{-\frac{1}{2}\sum_{r=1}^{m} y_r^2\right\} H(\mathbf{y}) dy_1, \ldots dy_m$$

and

$$H(\mathbf{y}) = \int_{y_{m+1}} \cdots \int_{y_p} y_s y_t \exp\left\{-\frac{1}{2}\sum_{r=m+1}^{p} y_r^2\right\} dy_{m+1} \ldots dy_p$$

$$= \begin{cases} y_s y_t (2\pi)^{(p-m)/2} & 0 \le s \text{ and } t \le m \\ 0 & s > m \text{ or } t > m,\ s \ne t \\ (2\pi)^{(p-m)/2} & s = t > m \\ 1 & m = p. \end{cases}$$

Therefore,

$$\int y_s y_t P_V(\mathbf{y}) h(\mathbf{y}) d\mathbf{y} = \begin{cases} RDA_{st}D'R' & 0 \le s \text{ and } t \le m \\ 0_{J\times J} & m < s \text{ or } t,\ s \ne t \\ RDA_{00}D'R' & m < s = t \end{cases}$$

where $D = \text{diag}[\exp(-\Delta_1^2/2), \ldots, \exp(-\Delta_J^2/2)]$ and A_{st} is a $J \times J$ matrix with entries $A_{st}(j,k) = \int y_s y_t \exp\{\mathbf{y}'(\mathbf{v}_j + \mathbf{v}_k)\} f(y) Q(y) dy_1 \ldots dy_m$ and now $\mathbf{y}' = (y_1, y_2, \ldots, y_m)$, $y_0 = 1$, and $Q(\mathbf{y}) = (2\pi)^{-m/2} \exp\{-\frac{1}{2}\mathbf{y}'\mathbf{y}\}$.

Combining the two integral expressions, we obtain the following expression for Ω_{st}:

$$\Omega_{st} = \Omega_{ts} = \begin{cases} R - RDA_{00}D'R' & s = t = 0 \\ RV_sV_t - RDA_{st}D'R' & 0 \le s \text{ and } t \le m;\ s \ne t \\ R + RV_s^2 - RDA_{ss}D'R' & 0 \le s \le m,\ s = t \\ 0_{J\times J} & m < s \text{ or } t,\ s \ne t \\ R - RDA_{00}D'R' & m < s,\ s = t. \end{cases}$$

Therefore,

$$\Omega = \left[\begin{array}{c|c} B_1 & 0_{mJ\times(p-m)J} \\ \hline 0_{(p-m)J\times mJ} & I_{p-m} \otimes \Omega_{00} \end{array}\right]$$

and inverting,

$$\Omega^{-1} = \left[\begin{array}{c|c} B_1^{-1} & 0 \\ \hline 0 & I_{p-m} \otimes \Omega_{00}^{-1} \end{array}\right],$$

where

$$B_1 = \begin{bmatrix} \Omega_{00} & B_2 \\ B_2' & B_3 \end{bmatrix} = \left[\begin{array}{c|c} \Omega_{00}^{-1} + \Omega_{00}^{-1} B_2 E_3^{-1} B_2' \Omega_{00}^{-1} & \Omega_{00}^{-1} B_2 E_3^{-1} \\ \hline -E_3^{-1} B_2' \Omega_{00}^{-1} & E_3^{-1} \end{array}\right]^{-1},$$

$$B_2 = (\Omega_{01}|\ldots|\Omega_{0m}), \quad B_3 = \{\Omega_{st}\} \quad s,t = 1,2,\ldots,m,$$

and $E_3 = [B_3 - B_2'\Omega_{00}^{-1}B_2]$ by Graybill (1969, Theorem 8.2.1).

Then,

$$\mathrm{Var}(\tilde{G}_0) = \frac{1}{n}\left(\Omega_{00}^{-1} + \Omega_{00}^{-1} B_2 E_3^{-1} B_2' \Omega_{00}^{-1}\right),$$

$$\mathrm{Cov}(\tilde{G}_0, \mathrm{vec}\tilde{G}_*) = -\frac{1}{n}\left(\Omega_{00}^{-1} B_2 E_3^{-1} | 0_{J\times(p-m)J}\right),$$

and

$$\mathrm{Var}(\mathrm{vec}\tilde{G}_*) = \frac{1}{n}\left[\begin{array}{c|c} E_3^{-1} & 0 \\ \hline 0 & I_{p-m} \otimes \Omega_{00}^{-1} \end{array}\right],$$

where $\tilde{G}_0$ and $\tilde{G}_*$ are the estimates obtained in the standard situation. Applying Lemma 2 and partitioning A and $\mathbf{a}$ as in Lemma 3, we obtain the expressions given in Theorem 2. Corollary 2.1 follows from substitution of $\Sigma^{-1} - A_A'A_A = A_B'A_B$.

5. RELATIVE EFFICIENCY

Theorem 3. Following Kendall and Stuart (1979, §17.29), we define large sample relative efficiency of estimation by the ratio of $\mathrm{Var}(\hat{\beta}_{sj})$ to $\mathrm{Var}(\tilde{\beta}_{sj})$.

Therefore, based on the results of Theorems 1 and 2:

$$\mathrm{RE} = \frac{\sigma^{ss}\left(\pi_0^{-1} + \pi_j^{-1} + \Delta_j^2\right) + \beta_{sj}^2}{\left[(\mathbf{a}_s' \otimes \mathbf{i}_j')E_3^{-1}(\mathbf{a}_s \otimes \mathbf{i}_j) + (\sigma^{ss} - \mathbf{a}_s'\mathbf{a}_s)(\mathbf{i}_j'\Omega_{00}^{-1}\mathbf{i}_j)\right]},$$

where $\mathbf{a}_s = (V_1')^{-1}B_A'\mathbf{e}_s$, $B_A = (\boldsymbol{\beta}_1|\ldots|\boldsymbol{\beta}_m)$, $\mathbf{e}_s'B_A = (\beta_{s1},\ldots,\beta_{sm})$, and hence $\mathbf{a}_s'\mathbf{a}_s = \mathbf{e}_s'B_A(B_A'\Sigma B_A)^{-1}B_A'\mathbf{e}_s$. We can rewrite $\Delta_j^2 = (\boldsymbol{\mu}_j -$

$\boldsymbol{\mu}_0)'\Sigma^{-1}(\boldsymbol{\mu}_j - \boldsymbol{\mu}_0)$ as $\Delta_j^2 = \boldsymbol{\beta}_j'\Sigma\boldsymbol{\beta}_j$. V_1 is as defined in Lemma 3, E_3 and Ω_{00} are as defined in Theorem 2. Furthermore, m is the dimension of the subspace spanned by the vectors $\boldsymbol{\beta}_1, \boldsymbol{\beta}_2, \ldots, \boldsymbol{\beta}_J$. When $m = p$, $\mathbf{a}_s'\mathbf{a}_s = \sigma^{ss}$ and the denominator of RE reduces to the first term.

Proof. The result follows from Corollaries 1.1 and 2.1 by making the substitution $(\boldsymbol{\mu}_j - \boldsymbol{\mu}_0) = \Sigma\boldsymbol{\beta}_j$.

Corollary 3.1. For the case considered by Efron (1975), $J = 1$ and p arbitrary, we have:

$$\mathrm{RE} = \frac{\sigma^{ss}\left(\pi_0^{-1} + \pi_1^{-1} + \Delta^2\right) + \beta_s^2}{\left[a_s^2 E_3^{-1} + (\sigma^{ss} - a_s^2)\Omega_{00}^{-1}\right]}$$

with $a_s^2 = \beta_s^2/\Delta^2$ and $\Delta^2 = \boldsymbol{\beta}'\Sigma\boldsymbol{\beta}$. When $p = 1$, $a_s^2 = \sigma^{ss}$.

According to Kendall and Stuart (1979, §25.13), the relative efficiency evaluated at $\beta_s = 0$ is precisely equal to the asymptotic relative efficiency (ARE). Therefore,

$$\begin{aligned}\mathrm{ARE} &= \Omega_{00}\left(\pi_0^{-1} + \pi_1^{-1} + \Delta^2\right) \\ &= (1 + \pi_0\pi_1\Delta^2)\Omega_{00}/\pi_0\pi_1.\end{aligned}$$

It can be shown, from the definition of Ω_{00} given in the proof of Theorem 2, that when $J = 1$:

$$\Omega_{00} = \frac{\pi_0\pi_1}{(2\pi)^{1/2}}\exp\{-\Delta^2/8\}\int\frac{\exp\{-z^2/2\}}{\left[\pi_0\exp\{-\Delta z/2\} + \pi_1\exp\{\Delta z/2\}\right]}dz.$$

Substituting for Ω_{00} in the expression for ARE produces the asymptotic relative efficiency of logistic regression to normal discrimination for estimating the angle of the discriminant boundary derived by Efron (1975).

A FORTRAN program was developed to evaluate the relative estimating efficiency of the logistic regression estimate of β_{sj} for two or three response groups and two explanatory variables. The program was run on the CDC Cyber 170 computer system, using International Mathematical and Statistical Libraries, Edition 8, subroutines. The subroutine DMLIN, a Gaussian numerical integration method, evaluates the entries in the J by J symmetric Ω_{st} matrices. These matrices are then manipulated using theory for partitioned matrices (Graybill, 1969). Matrix inversion is done with subroutine LINV2P, which invokes iterative improvement until machine accuracy is reached.

We consider two configurations of the explanatory variables, assuming them to be standardized and uncorrelated (that is, the variance-covariance matrix equal to the identity matrix). In the first case, X_1, but not X_2, is associated with a response in group 1 and in group 2. This corresponds to $\boldsymbol{\beta}_1' = (\beta_{11}, 0)$ and $\boldsymbol{\beta}_2' = (\beta_{12}, 0)$. In the second case, X_1 is associated with a response in group 1 but not in group 2, while X_2 is associated with a response in group 2 but not in group 1. This corresponds to $\boldsymbol{\beta}_1' = (\beta_{11}, 0)$ and $\boldsymbol{\beta}_2' = (0, \beta_{22})$. Without loss of generality, we evaluate RE for β_{11}, the log odds ratio that compares group 1 to the reference population with respect to X_1, the first explanatory variable.

Figure 1a illustrates the behaviour of RE as a function of β_1 and β_2 for the first case. The population response probabilities are assumed to be equal, so that π_j is equal to one-third for all three groups. When β_{11} is equal to zero, RE is equal to 1.00 regardless of the value of β_{12}. When β_{12} is equal to zero, RE declines monotonically as β_{11} increases, dropping below 0.80 when β_{11} is 2.0. However, for non-zero values of β_{12}, the rate of decline in RE depends on β_{12} and is not necessarily monotonic. RE decline is slowest for parameter combinations such that β_{12} is equal to one-half of β_{11}.

When the population response probabilities are not equal in the first case, the effect of β_{12} on the rate of decline in RE with β_{11} is much weaker. Figure 1b displays RE for π_1 and π_2 both equal to one-tenth. RE will only be greater than 0.80 provided β_{11} is less than 1.5.

Figure 2 indicates that RE is consistently lower in the second case, described above, than it is in the first case. The rate of decline in RE with β_{11} increases gradually as β_{22} increases for both equal and unequal π_j's. There is virtually no effect due to β_{22} for unequal π_j's.

6. SUMMARY AND CONCLUSIONS

When there are three response groups involved and the response probabilities are equal, RE at β_{11} equal to 3.0 ranges from 46 to 88 percent in the first case and from 38 to 46 percent in the second case, for the group 2 parameters (either β_{12} or β_{22} respectively) considered. Similarly, when the reference group response probability is large and β_{11} is equal to 3.0, RE ranges from 33 to 56 percent in the first case and is between 32 and 33 percent in the second. Generally for three response groups, the results in the first case tend to be higher than those reported by Efron (1975) for two response groups, while those for the second case are lower.

Although logistic regression is always less efficient than normal discrimination, the former compares more favourably to the latter for log odds ratios close to zero, for observations equally distributed among the response groups and for explanatory variables related to more than one response. Further investigation is required of the effects of increasing the number of explana-

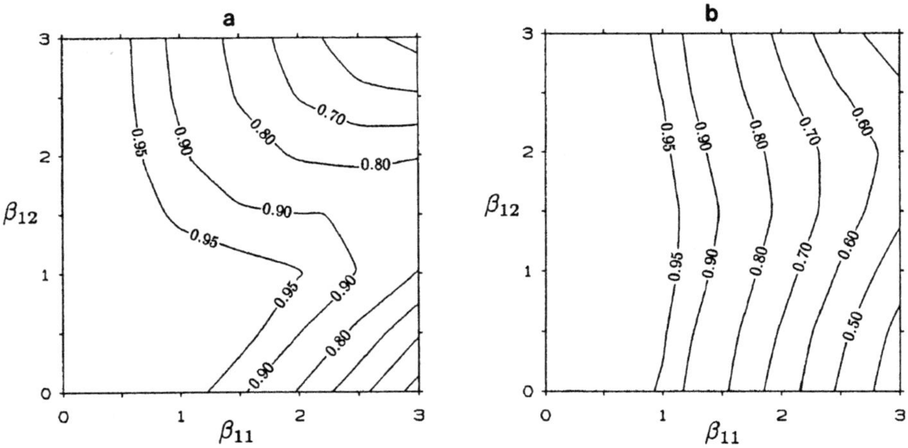

Figure 1. *Relative Estimating Efficiency. Contours of constant RE are plotted for combinations of the log odds ratio parameters* $\boldsymbol{\beta}_1' = (\beta_{11}, 0)$ *and* $\boldsymbol{\beta}_2' = (\beta_{12}, 0)$. *The point specified by a* β_{11}, β_{12} *combination corresponds to a value of RE between the closest contour line values. (a) The population response probabilities are equal. (b) The reference population response probability is large.*

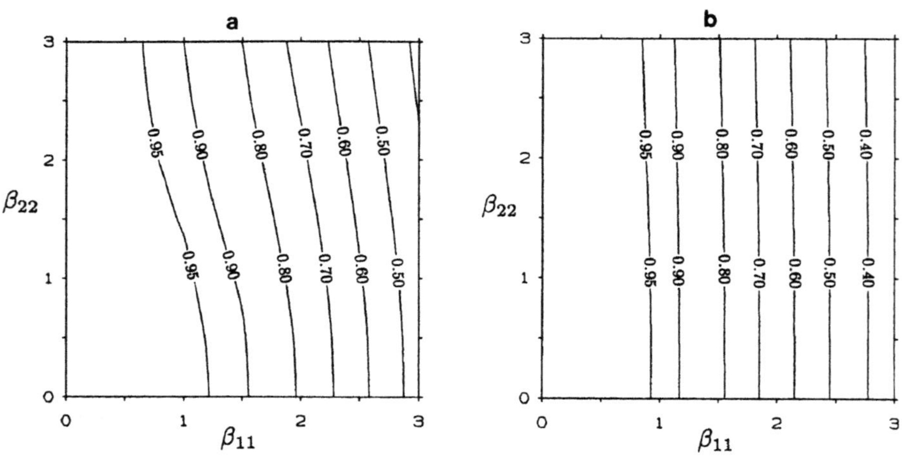

Figure 2. *Relative Estimating Efficiency. Contours of constant RE are plotted for combinations of the log odds ratio parameters* $\boldsymbol{\beta}_1' = (\beta_{11}, 0)$ *and* $\boldsymbol{\beta}_2' = (0, \beta_{22})$. *The point specified by a* β_{11}, β_{22} *combination corresponds to a value of RE between the two closest contour line values. (a) The population response probabilities are equal. (b) The reference population response probability is large.*

tory variables and of trends with the number of response categories. In particular, the discovery of the ameliorating influence of log odds ratios for a secondary category on relative efficiency needs to be confirmed in more than three response groups.

REFERENCES

Bishop, Y. M. M., S. E. Fienberg, and W. Holland (1975). *Discrete Multivariate Analysis: Theory and Practice.* Cambridge, MA: MIT Press.

Bradley, R. A., and J. J. Gart (1962), "The asymptotic properties of ML estimators when sampling from associated populations." *Biometrika* **49**, 205–214.

Bull, S. B. (1983). "Theoretical properties of multinomial logistic regression." Ph.D. Thesis, The University of Western Ontario.

Cornfield, J. (1962), "Joint dependence of risk of coronary heart disease on serum cholesterol and systolic blood pressure: A discriminant function analysis". *Federation Proceedings* **21**, Supplement 11, Part 2, 58–61.

Cox, D. R. (1970). *The Analysis of Binary Data.* London: Methuen.

Cox, D. R., and D. V. Hinkley (1974). *Theoretical Statistics.* London: Chapman and Hall.

Dwyer, P. S., and M. S. MacPhail (1948), "Symbolic matrix derivatives". *Annals of Mathematical Statistics* **19**, 517–534.

Efron, B. (1975), "The efficiency of logistic regression compared to normal discriminant analysis". *Journal of the American Statistical Association* **70**, 892–898.

Graybill, F. A. (1969). *Introduction to Matrices with Applications in Statistics.* Belmont, California: Wadsworth.

Greenhouse, S. W. (1982), "Jerome Cornfield's contributions to epidemiology". *Biometrics* **38**, supplement, 33–45.

Hauck, W. W. (1981). *Logit Analysis,* Preliminary draft. To be published by Wiley, New York.

Henderson, H. V., and S. R. Searle (1979), "Vec and vech operators for matrices, with some uses in Jacobians and multivariate statistics". *Canadian Journal of Statistics* **7**, 65–81.

Kendall, M., and A. Stuart (1979). *The Advanced Theory of Statistics,* Volume 2, Fourth edition. New York: Macmillan.

Lachenbruch, P. A. (1975). *Discriminant Analysis.* New York: Hafner Press.

MacRae, E. C. (1974), "Matrix derivatives with an application to an adaptive linear decision problem". *Annals of Statistics* **2**, 337–346.

Magnus, J. R., and H. Neudecker (1979), "The commutation matrix: some properties and applications". *Annals of Statistics* **7**, 381–394.

Mantel, N. (1966), "Models for complex contingency tables and polychotomous dosage response curves". *Biometrics* **22**, 83–95.

McCulloch, C. E. (1982), "Symmetric matrix derivatives with applications". *Journal of the American Statistical Association* **77**, 679–682.

John Koval [1] and Allan Donner [2]

ESTIMATION UNDER THE CORRELATED LOGISTIC MODEL

ABSTRACT

Rosner (1983, 1984) proposed a model for correlated binary outcomes in the presence of covariates based on the beta-binomial distribution. Although calculation of maximum likelihood estimates under this model is costly, alternative estimators based on the usual logistic model, with or without dummy varibles, are asymptotically biased (in general) and yield standard errors that are incorrect. An adjustment for the standard error of the usual logistic estimator is suggested for the case of one dichotomous independent variable. Also a familiar conditional approach is shown to have low asymptotic relative efficiency.

1. INTRODUCTION

Regression analysis provides a means of determining the effect of explanatory variables on a continuous outcome measure; for example, the effect of age and sex on blood pressure. The usual method of estimation in regression analysis is ordinary least squares. The regression model may be refined to allow for correlation among the outcome measures; for example, when the blood pressure scores come from the members of the same family, the method of estimation becomes generalized least squares.

Interest has been shown in the effects of the presence of intraclass correlation on ordinary least squares estimation. A large number of references has been given by Scott and Holt (1982). These authors have shown that

[1] Department of Statistics and Actuarial Science, University of Waterloo, Waterloo, Ontario N2L 3G1

[2] Department of Epidemiology and Biostatistics, The University of Western Ontario, London, Ontario N6A 5C1

I. B. MacNeill and G. J. Umphrey (eds.), Biostatistics, 199–224.

the ordinary least squares estimator is unbiased but is not as efficient as the generalized least squares estimator, although the loss of efficiency is small. However, the estimate of the covariance matrix of the least squares estimator is also incorrect, and may underestimate or overestimate the true value, depending on both the usual intracluster correlation and on the sample intracluster correlation of the explanatory variables (also see Campbell, 1979; Donner, 1984).

Correlated binary outcomes occur in many family studies, for example, in the investigation of hypertension, in other cross-sectional studies, in studies with repeated measurements on an individual (e.g., measurements of physical ability at different levels of physical exertion), and in complex surveys (e.g., Rao and Scott, 1981). When the outcome measure is discrete, in particular binary, the usual model for assessing the effects of explanatory variables is the logistic regression model (Berkson, 1955; Cox, 1966; Day and Kerridge, 1967). Until recently no extensions of this model have been available that would allow for the modelling of intracluster correlation.

Williams (1982) generalized the beta-binomial model by introducing covariates measured at the cluster level and requiring only a specific relationship between the mean and variance of the cluster-level response and the covariates. Brier (1980) generalized the beta-binomial model by allowing more than two levels of response, and, more importantly, covariates measured at the unit (as opposed to the cluster) level; however, the covariates are only discrete (in the example in his paper, the covariate and the response variable are actually ordered but this is ignored in the methodology).

Pierce and Sands (1975) introduced a correlated logistic-normal model which allowed the scores of units within a cluster to be correlated, but they only considered the case in which scores are measured on the cluster level. The generalization of this correlated logistic-normal model by Laird and Ware (1981) and Stiratelli *et al.* (1984), and the correlated logistic model by Rosner (1983, 1984) are models that allow both for a mixture of continuous and discrete explanatory variables, and for correlation between the dependent binary variables.

The correlated logistic model resembles more closely the usual logistic model than does its counterpart, the correlated logistic-normal model. Although somewhat underdeveloped at present, the correlated logistic model offers scope for expansion to multinomial outcomes and to more complex error structures.

In this paper, the work of Scott and Holt is extended to discrete outcomes by using Rosner's correlated logistic model to represent the true probabilisitic mechanism, and the first and second order properties of the estimators of "regression" parameters obtained under several incorrect but commonly used methods are examined.

When there are only two observations per cluster and the explanatory variable is itself discrete, the following results are obtained:

1. Several methods yield estimators that are asymptotically biased, except when the regression parameter is zero, that is, when there is no effect of the covariate.
2. When the regression parameter is zero, the true variances of two commonly used estimators are, in general, larger than that of the maximum likelihood estimator, but under some circumstances one of these estimators has a smaller variance than that of the maximum likelihood estimator (without contradicting the usual theory of inference).
3. The estimated variance of the usual logistic estimator is incorrect and, when the regression parameter is zero, may overestimate or underestimate the true value depending on the value of both the usual intracluster correlation and the intracluster correlation of the distribution of the independent variables.
4. In order to test the hypothesis that the parameter is zero, the usual logistic method may be used with a simple correction for the estimated variance of the estimator of the regression parameter. This generalizes a result of Brier (1980).
5. A conditional estimator based on the Breslow-Day methodology for case-control studies is shown to have low efficiency.

2. ROSNER'S MODEL

Let Y_1 and Y_2 denote the outcome (or dependent) variables, and $\mathbf{X}_1$ and $\mathbf{X}_2$ the corresponding vectors of explanatory (or independent) variables (or covariates). Further let $P(y_1, y_2 \mid \mathbf{x}_1, \mathbf{x}_2)$ denote the conditional probability $\Pr(Y_1 = y_1, Y_2 = y_2 \mid \mathbf{X}_1 = \mathbf{x}_1, \mathbf{X}_2 = \mathbf{x}_2)$. Rosner (1983, 1984) proposed the model

$$P(0,0 \mid \mathbf{x}_1, \mathbf{x}_2) = 1/d,$$
$$P(1,0 \mid \mathbf{x}_1, \mathbf{x}_2) = \exp(\alpha_1 + \boldsymbol{\beta}'\mathbf{x}_1)/d,$$
$$P(0,1 \mid \mathbf{x}_1, \mathbf{x}_2) = \exp(\alpha_1 + \boldsymbol{\beta}'\mathbf{x}_2)/d,$$

and

$$P(1,1 \mid \mathbf{x}_1, \mathbf{x}_2) = \exp[\alpha_2 + \boldsymbol{\beta}'(\mathbf{x}_1 + \mathbf{x}_2)],$$

where d is the normalizing constant, namely,

$$d = 1 + \exp(\alpha_1 + \boldsymbol{\beta}'\mathbf{x}_1) + \exp(\alpha_1 + \boldsymbol{\beta}'\mathbf{x}_2) + \exp[\alpha_2 + \boldsymbol{\beta}'(\mathbf{x}_1 + \mathbf{x}_2)].$$

When both $\mathbf{x}_1$ and $\mathbf{x}_2$ are null vectors, the distribution of the sum $(Y_1 + Y_2)$ is that of a beta-binomial random variable with parameters 2,

a and b, such that

$$\alpha_1 = \ln[a/(b+1)]$$

and

$$\alpha_2 = \ln\{a(a+1)/[b(b+1)]\}.$$

In addition Y_1 and Y_2 have the correlation calculated by Crowder (1979), namely,

$$\mathrm{corr}(Y_1, Y_2 \mid 0, 0) = 1/(a+b+1).$$

However, this simple correlation does not exist for all values of $\mathbf{x}_1$ and $\mathbf{x}_2$; in general, it may be shown that:

$$\begin{aligned}\mathrm{corr}(Y_1, Y_2 \mid \mathbf{x}_1, \mathbf{x}_2) = &\left\{[\exp(\alpha_2) - \exp(2\alpha_1)][\exp(\boldsymbol{\beta}'\{\mathbf{x}_1 + \mathbf{x}_2\})]^{1/2}\right\} / \\ &\Big(\big\{[\exp\alpha_1 + \exp(\alpha_2 + \boldsymbol{\beta}'\mathbf{x}_1)][\exp\alpha_1 + \exp(\alpha_2 + \boldsymbol{\beta}'\mathbf{x}_2)] \\ &\times [1 + \exp(\alpha_1 + \boldsymbol{\beta}'\mathbf{x}_1)][1 + \exp(\alpha_1 + \boldsymbol{\beta}'\mathbf{x}_2)]\big\}^{1/2}\Big). \quad (1)\end{aligned}$$

A more general way of writing Rosner's model is in terms of the extended exponential family of Dempster (1971); that is,

$$\begin{aligned}\ln P(y_1, y_2 \mid \mathbf{x}_1, \mathbf{x}_2) = &\alpha_1(y_1 + y_2) + \boldsymbol{\beta}'(\mathbf{x}_1 y_1 + \mathbf{x}_2 y_2) \\ &+ (\alpha_2 - 2\alpha_1) y_1 y_2 - \ln d,\end{aligned}$$

where d is defined above, and

1. the condition $\alpha_2 > 2\alpha_1$ yields the beta-binomial when $\mathbf{x}_1$ and $\mathbf{x}_2$ are both null,
2. the condition $\alpha_2 = 2\alpha_1$ yields the usual (or uncorrelated) logistic model.

Because of the complexity of the model, explicit solutions to the maximum likelihood equations are usually not attainable, although in the case of a single binary covariate some simplification is possible. Alternative methods for estimation are provided by:

1. the usual logistic model, that is, ignoring the effect of clustering,
2. the usual logistic model with dummy variables to handle the effect of clustering, that is, with variables such that each takes on a value of 1 in only one cluster and 0 elsewhere,
3. the conditional approach of Breslow and Day (1980),
4. the usual regression model with binary dependent variable treated as a continuous variable,

5. the beta-binomial model and its extensions as proposed by Williams (1982),
6. the correlated logistic-normal model of Laird and Ware (1981), and
7. the correlated probit model of Ochi and Prentice (1984).

The usual regression model is known to be inappropriate in the estimation of the parameters of the usual logistic model, and hence will fare even worse in estimating parameters of the correlated logistic model. Williams' models allow the explanatory variables to be only cluster-specific, that is, they take on the same value for every member of the same cluster. The correlated logistic-normal model and the correlated probit model are quite different from Rosner's model, and no easy comparisons are possible.

The remainder of the paper discusses parameter estimation using the first three alternatives listed above. In particular, we examine the (asymptotic) bias and variance of the estimators of the coefficient of a single binary explanatory variable when all clusters contain exactly two units.

3. ASYMPTOTIC BIAS

Under the unconditional model, the maximum likelihood estimator of all parameters of the model is consistent (and hence asymptotically unbiased). This is also true for the estimator of the regression coefficient under the conditional model. However, it is not true, in general, for the two other estimators.

3.1 Usual Logistic Model

Consider a sample of m independent clusters, each containing 2 units. The usual logistic model ignores the intracluster correlation and provides an explicit estimator of β, namely, the usual log odds ratio

$$\ln[n_{00}n_{11}/(n_{01}n_{10})],$$

where n_{ij} is the observed number of units that have $x = i$ and $y = j$, for $i = 0, 1$ and $j = 0, 1$, and may be considered as the observed value of a random variable N_{ij} which is the sum of m independent random variables R_{ij}, where R_{ij} can assume values 0, 1 or 2, according to the following probability distribution function:

$$\begin{aligned}
\Pr(R_{ij}=0) &= \Pr(X=i \text{ does not occur in same unit as } Y=j),\\
\Pr(R_{ij}=1) &= \Pr(\text{one of } X_1 \text{ and } X_2 \text{ is } i\\
&\qquad\qquad \text{and the corresponding } Y \text{ has value } j)\\
&\quad + \Pr(\text{both } X_1 \text{ and } X_2 \text{ are } i\\
&\qquad\qquad \text{and only one of } Y_1 \text{ and } Y_2 \text{ is } j),\\
\Pr(R_{ij}=2) &= \Pr(\text{both } X_1 \text{ and } X_2 \text{ are } i \text{ and both } Y_1 \text{ and } Y_2 \text{ are } j).
\end{aligned}$$

Since each random variable N_{ij} is the sum of m independent R_{ij}'s,

$$E(N_{ij}/m) = E(R_{ij}),$$

so that as $m \to \infty$,

$$\frac{n_{00}n_{11}}{n_{01}n_{10}} \xrightarrow{\text{P}} \frac{E(R_{00})E(R_{11})}{E(R_{01})E(R_{10})}. \tag{2}$$

It may be shown (Koval and Donner, 1984) that each expectation in (2), that is, $E(R_{ij})$, $i=1,2$, $j=1,2$, may be written as a simple function of the marginal distribution of $\mathbf{X}$ and the conditional distribution of $\mathbf{Y}$. Hence the right-hand side of (2) becomes:

$$\frac{p_{00}[P(0,0\mid 0,0)+P(1,0\mid 0,0)]+p_{10}[P(0,0\mid 1,0)+P(1,0\mid 1,0)]}{p_{00}[P(0,1\mid 0,0)+P(1,1\mid 0,0)]+p_{10}[P(0,1\mid 1,0)+P(1,1\mid 1,0)]}$$
$$\times\frac{p_{11}[P(1,0\mid 1,1)+P(1,1\mid 1,1)]+p_{10}[P(1,0\mid 1,0)+P(1,1\mid 1,0)]}{p_{11}[P(0,0\mid 1,1)+P(1,0\mid 1,1)]+p_{10}[P(1,0\mid 1,0)+P(0,0\mid 1,0)]}$$

which, when we assume Rosner's correlated logistic model for the conditional distribution of $\mathbf{Y}$, becomes

$$e^{\beta}\frac{[p_{00}d_{10}(1+e^{\alpha_1})+p_{10}d_{00}(1+e^{\alpha_1+\beta})]}{[p_{10}d_{11}(1+e^{\alpha_1})+p_{11}d_{10}(1+e^{\alpha_1+\beta})]}
\times\frac{[p_{11}d_{10}(e^{\alpha_1}+e^{\alpha_2+\beta})+p_{10}d_{11}(e^{\alpha_1}+e^{\alpha_2})]}{[p_{10}d_{00}(e^{\alpha_1}+e^{\alpha_2+\beta})+p_{00}d_{10}(e^{\alpha_1}+e^{\alpha_2})]}, \tag{3}$$

where

$$\begin{aligned}
d_{00} &= 1+2e^{\alpha_1}+e^{\alpha_2},\\
d_{01} = d_{10} &= 1+e^{\alpha_1}+e^{\alpha_1+\beta}+e^{\alpha_2+\beta},
\end{aligned}$$

and

$$d_{11} = 1 + 2e^{\alpha_1+\beta} + e^{\alpha_2+\beta}.$$

The expression (3) is equal to e^β when

1. $\beta = 0$, that is, when there is no effect of covariate on the dependent variable, or
2. $\alpha_2 = 2\alpha_1$, that is, when there is independence of units within the cluster.

Hence the usual logistic estimator of β yields a consistent estimator of β if one of the above conditions hold; however, these are not necessary conditions.

In order to investigate the size of the asymptotic bias of the usual logistic estimator of β, and to determine other conditions under which this estimator is consistent, we introduce a symmetric correlated binary distribution for $\mathbf{X}$. Assume that the marginal distributions of X_1 and X_2 are identical to write:

$$\begin{aligned} p_1 &= \Pr(X_i = 1), \qquad i = 1,2, \\ p_0 &= \Pr(X_i = 0) = 1 - p_1, \qquad i = 1,2, \\ p_{jk} &= \Pr(X_1 = j, X_2 = k), \qquad j,k = 0,1. \end{aligned}$$

The joint probabilities may then be written in terms of ρ_x in the following way:

$$\begin{aligned} p_{11} &= p_1(p_1 + \rho_x p_0), \\ p_{00} &= p_0(p_0 + \rho_x p_1), \\ p_{01} &= p_{10} = p_0 p_1(1 - \rho_x). \end{aligned}$$

This binary distribution handles several special cases. Using this joint distribution of $\mathbf{X}$ we can write the second part of (3), known as the *inflation factor*, as

$$\frac{[d_{10}(p_0 + \rho_x p_1)(1 + e^{\alpha_1}) + d_{00}p_1(1 - \rho_x)(1 + e^{\alpha_1+\beta})]}{[d_{11}p_0(1 - \rho_x)(1 + e^{\alpha_1}) + d_{10}(p_1 + \rho_x p_1)(1 + e^{\alpha_1+\beta})]} \times \frac{[d_{10}(p_1 + \rho_x p_0)(e^{\alpha_1} + e^{\alpha_2+\beta}) + d_{11}p_0(1 - \rho_x)(e^{\alpha_1} + e^{\alpha_2})]}{[d_{00}p_1(1 - \rho_x)(e^{\alpha_1} + e^{\alpha_2+\beta}) + d_{10}(p_0 + \rho_x p_1)(e^{\alpha_1} + e^{\alpha_2})]}.$$

Table 1 gives values of this inflation factor for various values of ρ_x , $\rho_{y|x}$ and β, where $\rho_{y|x}$ is defined as $(a + b + 1)^{-1}$. When $a, b \to \infty$, such that $a/b \to$ a constant, then $\rho_{y|x} \to 0$. In particular, we assume that $a = b$ so that $p_0 = p_1 = 1/2$. From Table 1 it may be seen that

1. The estimator $\ln[(n_{00}n_{11})/(n_{01}n_{10})]$ is consistent for β only if
   (a) $\beta = 0$,
   (b) $\rho_{y|x} = 0$ (independence), or

Table 1. *Inflation Factor for the Usual Estimator of the Odds Ratio*

		Values of β (followed by values of $\exp(\beta)$)			
ρ_x	$\rho_{y\|x}$	0.000 (1.0)	0.693 (2.0)	1.609 (5.0)	2.303 (10.0)
−1.00	0.050	1.000	0.967	0.935	0.921
	0.200	1.000	0.875	0.765	0.719
	0.500	1.000	0.714	0.500	0.419
	0.800	1.000	0.579	0.304	0.209
−0.50	0.050	1.000	0.983	0.967	0.959
	0.200	1.000	0.935	0.871	0.842
	0.500	1.000	0.843	0.692	0.625
	0.800	1.000	0.756	0.529	0.432
0.00	0.050	1.000	1.000	1.000	0.999
	0.200	1.000	0.999	0.994	0.990
	0.500	1.000	0.995	0.966	0.941
	0.800	1.000	0.989	0.918	0.860
0.10	0.050	1.000	1.003	1.006	1.008
	0.200	1.000	1.013	1.021	1.023
	0.500	1.000	1.029	1.035	1.024
	0.800	1.000	1.044	1.028	0.990
0.50	0.050	1.000	1.017	1.034	1.041
	0.200	1.000	1.068	1.138	1.170
	0.500	1.000	1.179	1.371	1.458
	0.800	1.000	1.301	1.653	1.816
1.00	0.050	1.000	1.034	1.069	1.085
	0.200	1.000	1.143	1.307	1.391
	0.500	1.000	1.400	1.998	2.382
	0.800	1.000	1.726	3.281	4.778

(c) a very special combination of values of ρ_x, $\rho_{y|x}$ and β exists.

2. This estimator is asymptotically negatively biased when ρ_x is negative or zero or has small positive values.
3. This estimator is asymptotically positively biased for positive values of ρ_x, in particular, for values greater than 0.10.

3.2 Dummy Variables Model

The logistic model for a single observation may be written as:

$$f(y \mid \mathbf{x}) = y[\exp(\gamma + \boldsymbol{\beta}'\mathbf{x})]/[1 + \exp(\gamma + \boldsymbol{\beta}'\mathbf{x})].$$

Hence the log likelihood is

$$l(\gamma, \boldsymbol{\beta}) = y(\gamma + \boldsymbol{\beta}'\mathbf{x}) - \ln[1 + \exp(\gamma + \boldsymbol{\beta}'\mathbf{x})].$$

In the dummy variables approach the logistic model is written with a different γ parameter for each cluster, so that the log likelihood for the jth unit in the ith cluster is

$$l_{ij}(\gamma_i, \boldsymbol{\beta}) = y(\gamma_i + \boldsymbol{\beta}'\mathbf{x}_{ij}) - \ln[1 + \exp(\gamma_i + \boldsymbol{\beta}'\mathbf{x}_{ij})],$$
$$j = 1, \ldots, n_i;\ i = 1, \ldots, m.$$

The maximum likelihood equations for this model consist of two sets of equations:

$$\sum_{j=1}^{n_i} y_{ij} = \sum_{j=1}^{n_i} \hat{f}_{ij}, \qquad i = 1, \ldots, m, \tag{4}$$

where

$$\hat{f}_{ij} = \exp(\hat{\gamma}_i + \hat{\boldsymbol{\beta}}'\mathbf{x}_{ij})/[1 + \exp(\hat{\gamma}_i + \hat{\boldsymbol{\beta}}'\mathbf{x}_{ij})]$$

and

$$\sum_{i=1}^{m}\sum_{j=1}^{n_i} y_{ij}\mathbf{x}_{ij} = \sum_{i=1}^{m}\sum_{j=1}^{n_i} \hat{f}_{ij}\mathbf{x}_{ij}. \tag{5}$$

Consider those clusters with all Y values the same. For those clusters with all Y's equal to 1, equation (4) becomes

$$n_i = \sum_{j=1}^{n_i} \hat{f}_{ij},$$

which implies that $\hat{f}_{ij}$ is 1 for every member of cluster i and hence that

$$\hat{\gamma}_i + \hat{\boldsymbol{\beta}}'\mathbf{x}_{ij} \to \infty.$$

Since this holds for all values of $\mathbf{x}_{ij}$, and we assume the values of $\hat{\beta}_1, \hat{\beta}_2, \ldots, \hat{\beta}_p$ are finite, it follows that

$$\hat{\gamma}_i \rightarrow \infty. \tag{6}$$

Similarly when all Y's in a cluster are 0, then

$$\hat{\gamma}_i \rightarrow -\infty. \tag{7}$$

Hence if all the Y's in a cluster are the same, we cannot use that cluster for the estimation of $\boldsymbol{\beta}$. We say that the cluster contains no information about the parameter $\boldsymbol{\beta}$, and may be discarded from the estimation procedure. In particular, in any numerical minimization, when either (6) or (7) or both occur, the solution to the minimization problem is considered unbounded, and no numerical value is returned for any of the parameter estimates.

In the rest of this section we consider there to be only a single binary covariate, X, which takes values of 0 and 1, and only two units per cluster.

3.2.1 Cluster-Specific Covariates. In the case of a cluster-specific covariate, X_i has the same value x_i for every unit in the cluster; hence $\hat{f}_{ij}$ is the same for every unit in the cluster. Let us denote the cluster-specific $\hat{f}_{ij}$ by $\hat{f}_i$. From the preceding section we need only consider those clusters with exactly one Y equal to 0 and the other Y equal to 1. For these clusters, equation (4) implies that $\hat{f}_i = 1/2$, so that $x_i = 0 \Rightarrow \hat{\gamma}_i = 0$, but there is no information about β. Similarly, $x_i = 1 \Rightarrow \hat{\gamma}_i + \hat{\beta} = 0$, that is, $\hat{\gamma}_i = -\hat{\beta}$. However, substitution into the second set of the maximum likelihood equations (5) does not produce a solution for $\hat{\beta}$. Hence cluster-specific covariates will not yield a maximum likelihood estimator of β. Moreover, if the covariates are unit-specific, but in a specific cluster, the observed x's are the same for all units, that cluster provides no information for the estimation of β and may be discarded from the estimation procedure.

3.2.2 Unit-Specific Covariates. By the arguments given in the preceding two sections we need only consider the clusters with dissimilar X values. Let us say that for cluster i we have $x_{i1} = 1$ and $x_{i2} = 0$. Equation (4) evaluated for this cluster yields

$$1 = \hat{f}_{i1} + \hat{f}_{i2},$$

where $\hat{f}_{i1}$ and $\hat{f}_{i2}$ are the estimated probabilities, so that the equation may be written as

$$1 = \frac{\exp(\hat{\gamma}_i + \hat{\beta})}{1 + \exp(\hat{\gamma}_i + \hat{\beta})} + \frac{\exp(\hat{\gamma}_i)}{1 + \exp(\hat{\gamma}_i)},$$

which becomes $1 = \exp(2\hat{\gamma}_i + \hat{\beta})$. This may be rewritten as

$$\hat{\gamma}_i = -\hat{\beta}/2, \tag{8}$$

so that $\hat{f}_{i1} = \exp(\hat{\beta}/2)/[1 + \exp(\hat{\beta}/2)] = \hat{f}$, say, for all these clusters, and $\hat{f}_{i2} = 1 - \hat{f}$. For the other clusters, that is those with $x_{i1} = 0$ and $x_{i2} = 1$, we get $\hat{f}_{i1} = 1 - \hat{f}$ and $\hat{f}_{i2} = \hat{f}$.

Hence equation (4) evaluated over the clusters with both dissimilar x's and dissimilar y's yields

$$r_{11} = r_1 \hat{f}, \tag{9}$$

where r_1 is the total number of clusters with dissimilar x's and dissimilar y's and r_{11} is the number of these clusters within which each unit has the same x and y value. Solving (9) for $\hat{f}$ and then solving (8) for $\hat{\beta}$, we get

$$\exp(\hat{\beta}) = (r_{11}/r_{10})^2,$$

where r_{10} is the number of clusters in which each unit has an x value different from its y value.

In a manner similar to that used for the usual logistic model in Section 3.1, it may be shown that, conditional upon the total r_1 $(= r_{11} + r_{10})$,

$$\frac{r_{11}}{r_{10}} \xrightarrow{\text{P}} \frac{\Pr(Y_1 = 1, Y_2 = 0 \mid X_1 = 1, X_2 = 0)}{\Pr(Y_1 = 0, Y_2 = 1 \mid X_1 = 1, X_2 = 0)},$$

which, under Rosner's model, is

$$\frac{\exp(\alpha_1 + \beta)}{\exp(\alpha_1)} = \exp(\beta).$$

Hence the dummy variables approach produces an estimator of $\exp(\beta)$ with an inflation factor equal to the value being estimated.

3.3 *The Conditional Model*

Breslow *et al.* (1978) proposed a conditional logistic model for use in the analysis of case-control studies. In particular, for each cluster, they condition upon the total number of occurrences, that is, they condition upon all possible outcomes with the same number of Y's equal to 1 and 0. If we consider the case of only two units per cluster and condition upon the existence of only (0,1) or (1,0) outcomes for (Y_1, Y_2), we get

$$\Pr(1,0 \mid \mathbf{x}_1, \mathbf{x}_2) = \exp(\boldsymbol{\beta}'\mathbf{x}_1)/d,$$
$$\Pr(0,1 \mid \mathbf{x}_1, \mathbf{x}_2) = \exp(\boldsymbol{\beta}'\mathbf{x}_2)/d,$$

where

$$d = \exp(\boldsymbol{\beta}'\mathbf{x}_1) + \exp(\boldsymbol{\beta}'\mathbf{x}_2).$$

This is the same function as obtained by Breslow and Day (1980, p. 248). According to Breslow and Day (p. 250), this conditional logistic model, in the case of a single binary explanatory variable, has the maximum likelihood estimator $\ln(r_{11}/r_{10})$.

4. ASYMPTOTIC VARIANCE

Althc gh the asymptotic variance can be calculated for all parameter values, the initial case of interest is that with a single binary covariate where the true value of β is 0, that is, when the covariate has no effect on the dependent variable. In this case, the correlation between Y_1 and Y_2 is the same regardless of the value of X_1 and X_2, and the following four estimators are consistent:

1. the maximum likelihood estimator under Rosner's (unconditional) model, $\hat{\beta}_R$,
2. the estimator under the usual logistic model, $\hat{\beta}_u$,
3. the estimator under the usual logistic model with dummy variables, $\hat{\beta}_d$, and
4. the maximum likelihood estimator under the conditional (Breslow-Day) logistic model, $\hat{\beta}_c$.

We now compute the asymptotic variance of these estimators.

4.1 Unconditional Maximum Likelihood Estimator

For a single observation, (y_1, y_2), with covariate values (x_1, x_2), the log likelihood function may be written as:

$$l(\alpha_1, \alpha_2, \beta) = \alpha_1(y_1 + y_2) + \beta(x_1 y_1 + x_2 y_2) + (\alpha_2 - 2\alpha_1)y_1 y_2 - \ln d,$$

where

$$d = 1 + \exp(\alpha_1 + \beta x_1) + \exp(\alpha_1 + \beta x_2) + \exp(\alpha_2 + \beta\{x_1 + x_2\}).$$

By defining the following probabilities:

$$f_{jk} = \Pr(y_1 = j, y_2 = k \mid \mathbf{X} = \mathbf{x}),$$
$$f_{1+} = f_{10} + f_{11},$$
$$f_{+1} = f_{01} + f_{11},$$

one can calculate the second derivatives of the likelihood function with respect to α_1, α_2, and β to obtain the upper triangular part of the matrix containing second derivatives as:

$$\begin{pmatrix} (f_{10}+f_{01})(1-f_{10}-f_{01}) & -f_{11}(f_{10}+f_{01}) & \begin{array}{c} x_1(f_{10}f_{00}-f_{01}f_{11}) \\ +x_2(f_{01}f_{00}-f_{10}f_{11}) \end{array} \\ & f_{11}(1-f_{11}) & f_{11}(x_1 f_{0+} + x_2 f_{+0}) \\ & & \begin{array}{c} x_1^2 f_{1+}f_{0+} + x_2^2 f_{+1}f_{+0} \\ +2x_1x_2(f_{11}-f_{1+}f_{+1}) \end{array} \end{pmatrix}$$

In the case when β is 0, taking expectation of the above matrix with respect to the distribution of $\mathbf{X}$, and writing f_{00} as h_0, f_{11} as h_2, and $(f_{01}+f_{10})$ as h_1, where

$$h_i = \Pr(\text{exactly } i\ Y'\text{s are } 1),$$

we obtain the following matrix:

$$\begin{pmatrix} h_1(1-h_1) & -h_1h_2 & \mu h_1(h_0-h_2) \\ -h_1h_2 & h_2(1-h_2) & 2\mu h_2 f_0 \\ \mu h_1(h_0-h_2) & 2\mu h_2 f_0 & \begin{array}{c} 2E(X^2)f_1f_0 \\ +2E(X_1X_2)[h_2-f_1^2] \end{array} \end{pmatrix}.$$

We call this matrix I, denoting the Fisher information matrix. Let I_{ij} be the (i,j)th element of I, and I^{ij} the corresponding element of I^{-1}. We wish to obtain I^{33} which, when divided by m, is an expression for the asymptotic variance of $\hat{\beta}_R$, the maximum likelihood estimator under Rosner's correlated logistic model. Adapting the partitioning method while inverting the matrix I, I^{33} may be obtained as:

$$I^{33} = 2\sigma_x^2[f_0f_1 + \rho_x(h_2 - f_1^2)].$$

However,

$$\rho_{y/x} = (h_2 - f_1^2)/(f_0f_1)$$

and hence

$$I^{33} = 2\sigma_x^2 f_0 f_1(1 + \rho_{y|x}\rho_x).$$

This expression may be described as the information about β in a single cluster (when β is 0), and its inverse divided by m is the variance of $\hat{\beta}_R$ in a sample of m clusters, so that:

$$\text{Var}(\hat{\beta}_R) = [2m\sigma_x^2 f_0 f_1(1 + \rho_x\rho_{y|x})]^{-1}.$$

4.2 Usual Logistic Estimator

If an incorrect model, namely the usual (or uncorrelated) logistic model, is used, estimates of β and α are obtained. When the true value of β is 0, this estimator of β, denoted by $\hat{\beta}_u$, is consistent, but the estimator of α, denoted by $\hat{\alpha}_u$, is, in general, not consistent. Regardless, we can calculate the asymptotic variance of $\hat{\beta}_u$ under Rosner's model, and compare this with variance of the true maximum likelihood estimator, $\hat{\beta}_R$. The methodology in this section was suggested by Kulperger (1979). An explanation is given in Appendix 1. Let α_0 and β_0 denote the values of the parameters to which the estimators, $\hat{\alpha}_u$ and $\hat{\beta}_u$, approach asymptotically. When there is a single binary covariate, a closed form expression can be obtained for $\hat{\alpha}_u$, namely,

$$\hat{\alpha}_u = \ln(n_{10}/n_{00}),$$

where n_{i0} is the number of Y's equal to i when the corresponding X is 0. Using the methodology of Section 3.1, it can be shown that

$$\exp(\hat{\alpha}_u) = (n_{10}/n_{00}) \xrightarrow{\text{P}} E(R_{10})/E(R_{00}),$$

where the notation was defined in Section 3.1. Assuming Rosner's model to be correct, we can write this expression as

$$\frac{d_{10}p_{00}(e^{\alpha_1} + e^{\alpha_2}) + d_{00}p_{10}(e^{\alpha_1} + e^{\alpha_2+\beta})}{d_{10}p_{00}(1 + e^{\alpha_1}) + d_{00}p_{10}(1 + e^{\alpha_1+\beta})}. \tag{11}$$

When β is 0, d_{00} is equal to d_{10}, and (11) becomes

$$\frac{e^{\alpha_1} + e^{\alpha_2}}{1 + e^{\alpha_1}}; \tag{12}$$

hence α_0 is equal to the logarithm of this expression.

The equations for obtaining the maximum likelihood estimators, also known as the estimating equations, are the following:

$$\sum_{i=1}^{m}(y_{i1} + y_{i2} - f_{i1} - f_{i2}) = 0,$$

$$\sum_{i=1}^{m} x_{i1}(y_{i1} - f_{i1}) + x_{i2}(y_{i2} - f_{i2}) = 0.$$

Let us denote the left-hand side of these equations by $\mathbf{S}_m(\alpha, \beta)$. We require the expected value of $\mathbf{S}_m(\alpha_0, \beta_0)$ with respect to Rosner's model, that is

$E(\mathbf{S}_m(\alpha_0, \beta_0))$. Since the clusters are independent, $E(\mathbf{S}_m(\alpha_0, \beta_0))$ can be written as the vector

$$D = \begin{pmatrix} mE(Y_1 + Y_2 - F_1 - F_2) \\ \\ mE[X_1(Y_1 - F_1) + X_2(Y_2 - F_2)] \end{pmatrix}$$

When β is 0, under Rosner's model

$$E(Y_1) = \frac{e^{\alpha_1} + e^{\alpha_2}}{1 + 2e^{\alpha_1} + e^{\alpha_2}}.$$

Each value of F_1, namely f_1, is of the form

$$\exp(\alpha + \beta x)/[1 + \exp(\alpha + \beta x)].$$

However, at the value $(\alpha_0, \beta_0 = 0)$, this expression is

$$e^{\alpha_0}/(1 + e^{\alpha_0}),$$

which is a constant. Substituting from (12), we have

$$E(F_1) = \frac{e^{\alpha_1} + e^{\alpha_2}}{1 + 2e^{\alpha_1} + e^{\alpha_2}} = E(Y_1),$$

so that D_1 is 0. For D_2 we evaluate $E[X(Y - F)]$ by taking expectation with respect to the conditional distribution of Y given X and then with respect to the marginal distribution of X, so that

$$E[X(Y - F)] = E_X[XE_{Y|X}(Y - F)],$$

but it was shown above that $E_{Y|X}(Y - F)$ is 0 so that $E[X(Y - F)] = 0$. Thus $\mathbf{S}_m(\alpha_0, \beta_0)$ has expectation $\mathbf{0}$. Moreover it is a sum of m independent and identically distributed terms, each of which has mean $\mathbf{0}$ and variance V (to be calculated). Hence by the multivariate central limit theorem,

$$\frac{1}{\sqrt{m}}\mathbf{S}_m(\alpha, \beta) \xrightarrow{\text{D}} N_2(\mathbf{0}, V),$$

where N_2 is the bivariate normal distribution.

The calculation of the elements of V is fairly straightforward. We obtain

$$\begin{aligned} v_{11} &= \text{Var}(Y_1 + Y_2 - F_1 - F_2) \\ &= 2f_0 f_1(1 + \rho_{y|x}), \end{aligned}$$

by the definition of $\rho_{y|x}$. To evaluate the other entries in V we use the rule

$$\mathrm{Var}_{XY} = E_X \mathrm{Var}_{Y|X} + \mathrm{Var}_X E_{Y|X}.$$

In this case $E_{Y|X}$ of both elements of $\mathbf{S}_m(\alpha_0, \beta_0)$ is 0; moreover, the distribution of $\mathbf{Y}$ is not influenced by $\mathbf{x}$ (when β_0 is 0), so that $\mathrm{Var}_{Y|X}$ is actually Var_Y; hence

$$\begin{aligned} v_{22} &= E_X \mathrm{Var}_Y [X_1(Y_1 - F_1) + X_2(Y_2 - F_2)] \\ &= 2 f_0 f_1 [E(X^2) + \rho_{y|x} E(X_1 X_2)] \end{aligned}$$

and finally

$$v_{12} = 2 f_0 f_1 \mu_X (1 + \rho_{y|x}).$$

Now we consider the asymptotic behaviour of the first derivative of $\mathbf{S}_m(\alpha, \beta)/m$ as m approaches infinity. Let $\boldsymbol{\theta}$ denote the vector of parameters, namely α and β, and $\hat{\boldsymbol{\theta}}$ denote the value of $\boldsymbol{\theta}$ at $\hat{\alpha}_u$ and $\hat{\beta}_u$, and $\boldsymbol{\theta}_0$ denote the value at α_0 and β_0. Then,

$$-\frac{1}{m} \frac{\partial S_m(\boldsymbol{\theta})}{\partial \boldsymbol{\theta}}$$

is a matrix U_m of the following form:

$$\frac{1}{m} \sum_{i=1}^{m} \begin{pmatrix} f_{i1} g_{i1} + f_{i2} g_{i2} & x_{i1} f_{i1} g_{i1} + x_{i2} f_{i2} g_{i2} \\ x_{i1} f_{i1} g_{i1} + x_{i2} f_{i2} g_{i2} & x_{i1}^2 f_{i1} g_{i1} + x_{i2}^2 f_{i2} g_{i2} \end{pmatrix},$$

where

$$g_{ij} = 1 - f_{ij}, \qquad i = 1, \ldots, m, \;\; j = 1, 2.$$

It follows that as $m \to \infty$, $f_{ij} g_{ij} \xrightarrow{\mathrm{P}} f_1 f_0$, and

$$-\frac{1}{m} \frac{\partial S_m(\boldsymbol{\theta})}{\partial \boldsymbol{\theta}} \xrightarrow{\mathrm{P}} 2 f_0 f_1 \begin{pmatrix} 1 & \mu \\ \mu & E(X^2) \end{pmatrix}.$$

Let us denote the latter matrix by U. As stated in Appendix 1, the asymptotic distribution of

$$\sqrt{m} \left[\begin{pmatrix} \hat{\alpha}_u \\ \hat{\beta}_u \end{pmatrix} - \begin{pmatrix} \alpha_0 \\ \rho_0 \end{pmatrix} \right]$$

is $N_2(\mathbf{0}, U^{-1} V U^{-1})$. For the asymptotic variance of $\hat{\beta}_u$, we need only evaluate the entry in position $(2, 2)$ of the matrix $U^{-1} V U^{-1}$, which can be shown to be

$$(1 + \rho_{y|x} \rho_x)/(2 \sigma_x^2 f_0 f_1).$$

This is the contribution to the variance of $\hat{\beta}_u$ by a single cluster. The variance of $\hat{\beta}_u$ in a sample of m clusters is

$$(1 + \rho_{y|x}\rho_x)/(2m\sigma_x^2 f_0 f_1).$$

4.3 Conditional Maximum Likelihood Estimator

With a variation of an argument used by Farewell (1979) we show that the contribution to the asymptotic variance of $\hat{\beta}_c$ (when β is 0), by a cluster with one Y equal to 0 and the other Y equal to 1, is

$$2/[\sigma_x^2(1 - \rho_x)].$$

The log likelihood of the conditional logistic model may be written

$$l(\beta) = \beta x_1 - \ln d,$$

where

$$d = \exp(\beta x_1) + \exp(\beta x_2).$$

Differentiating $l(\beta)$ twice, we obtain

$$\frac{\partial^2 l}{\partial \beta^2} = -x_1^2 f_1 g_1 - x_2^2 f_2 g_2 + 2x_1 x_2 f_1 f_2,$$

where

$$g_i = 1 - f_i, \quad i = 1, 2,$$

and f_i is the conditional probability

$$P(Y_i = 1 \mid X_i = x_i), \quad i = 1, 2.$$

Now when β is 0, $f_i = g_i = \frac{1}{2}$, $i = 1, 2$, and we simply obtain

$$E\left(-\frac{\partial^2 l}{\partial \beta^2}\right) = \sigma_x^2(1 - \rho_x)/2. \tag{13}$$

This is the information in a single cluster in which there are dissimilar values of Y_1 and Y_2. Clusters yielding similar values are discarded (see Section 3.3). In a sample of m clusters, on average, only $2f_{01}$ of the clusters have dissimilar values of Y_1 and Y_2, and have the information described in the right-hand side of (13). The remaining $m(1 - 2f_{01})$ clusters have similar values of Y, and hence contain no information about β. The variance of a sample of m clusters containing both types of clusters is $[m\sigma_x^2 f_{01}(1 - \rho_x)]^{-1}$.

4.4 *Dummy Variables Estimator*

In Section 3.2, it was shown that $\hat{\beta}_d$, the dummy variables estimator of β, can be written as $2\ln(r_{11}/r_{10})$, that is, as twice $\hat{\beta}_c$, the conditional estimator. Hence the asymptotic variance of $\hat{\beta}_d$ is four times that of $\hat{\beta}_c$, that is, the contribution to the information about β from a single cluster with one Y equal to 0 and the other equal to 1 is $\sigma_x^2(1-\rho_x)/8$, and the large sample variance of $\hat{\beta}_d$ is $4/[m\sigma_x^2 f_{01}(1-\rho_x)]$.

5. COMPARISONS AND CORRECTIONS

5.1 *Asymptotic Relative Efficiency*

To examine the asymptotic relative efficiency of various estimators when β is 0, we consider the ratio of the asymptotic variance of the unconditional maximum likelihood estimator to the asymptotic variance of the other estimators. Let $\text{re}(\hat{\beta}_u)$ denote the relative efficiency of $\hat{\beta}_u$, the usual or uncorrelated logistic estimator. By the formulae developed in Section 4,

$$\text{re}(\hat{\beta}_u) = (1+\rho_{y|x}\rho_x)^{-2}.$$

Figure 1 shows plots of $\text{re}(\hat{\beta}_u)$ against $\rho_{y|x}$ for several different non-negative values of ρ_x and Figure 2 shows plots for several different non-positive values of ρ_x. In general, $\rho_{y|x}$ is constrained to be non-negative in Rosner's model, but ρ_x may be either negative or positive. When ρ_x is positive, $\text{re}(\hat{\beta}_u)$ is less than 1, but when ρ_x is negative, $\text{re}(\hat{\beta}_u)$ is greater than 1. In this case $\hat{\beta}_R$, the unconditional maximum likelihood estimator, is not fully efficient. This does not violate the general principle of the optimality of the maximum likelihood estimator for a single parameter, because we are dealing with a multiparameter problem.

In most family studies both $\rho_{y|x}$ and ρ_x are positive and small, in particular, they are less than 0.5, so that in these types of studies, the minimum value of $\text{re}(\hat{\beta}_u)$ is 0.64. However, when both $\rho_{y|x}$ and ρ_x are quite small, say 0.1, $\text{re}(\hat{\beta}_u)$ is 0.99, and little efficiency is lost by estimating β with $\hat{\beta}_u$. Moreover, for $\rho_x = 0.1$, $\text{re}(\hat{\beta}_u)$ is always above 0.90.

But ρ_x can be negative, for example, in designed studies where ρ_x is equal to or close to -1. In this case, $\text{re}(\hat{\beta}_u) \to \infty$. However, for $\rho_x = -0.1$, $\text{re}(\hat{\beta}_u)$ never exceeds 1.21.

The asymptotic relative efficiency of $\hat{\beta}_c$, denoted by $\text{re}(\hat{\beta}_c)$, is

$$[f_{01}(1-\rho_x)]/[2f_0f_1(1+\rho_{y|x}\rho_x)]. \qquad (14)$$

It can be shown (see Appendix 2) that $\text{re}(\hat{\beta}_c)$ is always less than or equal to 1, with equality coming only when ρ_x is -1, that is, when both x values

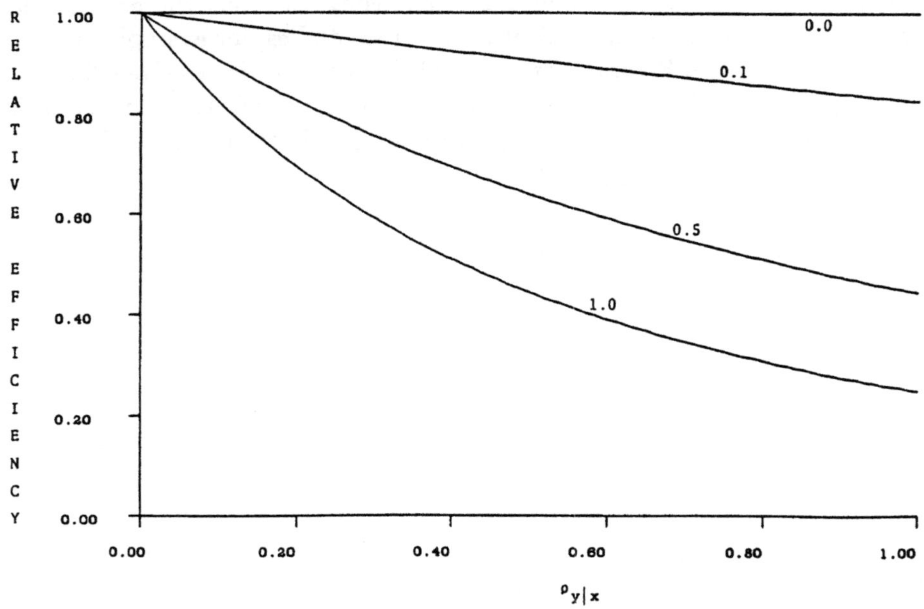

Figure 1. *Asymptotic relative efficiency of $\hat{\beta}_u$ for non-negative values of ρ_x.*

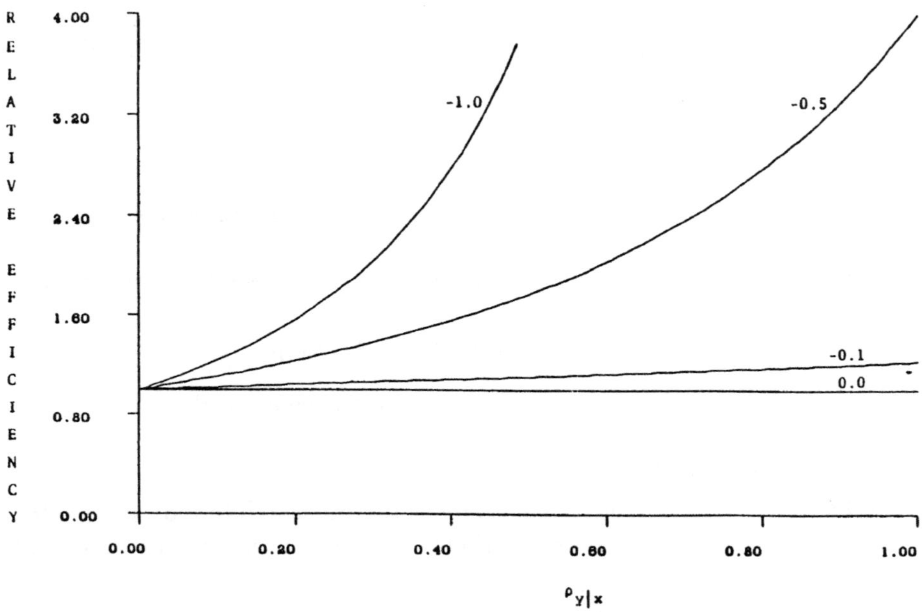

Figure 2. *Asymptotic relative efficiency of $\hat{\beta}_u$ for non-positive values of ρ_x.*

are dissimilar, one being 0 and the other being 1. Thus the conditional estimator of β is only fully efficient in designed studies, for example in case-control studies. See Figure 3 for a plot of the asymptotic relative efficiency of $\hat{\beta}_c$ against $\rho_{y|x}$ for several values of ρ_x. For $\rho_x = 0.0$, re($\hat{\beta}_c$) is at most 0.50 and falls to zero for increasing values of $\rho_{y|x}$. Even for $\rho_x = -0.50$, the maximum value of re($\hat{\beta}_c$) is 0.75.

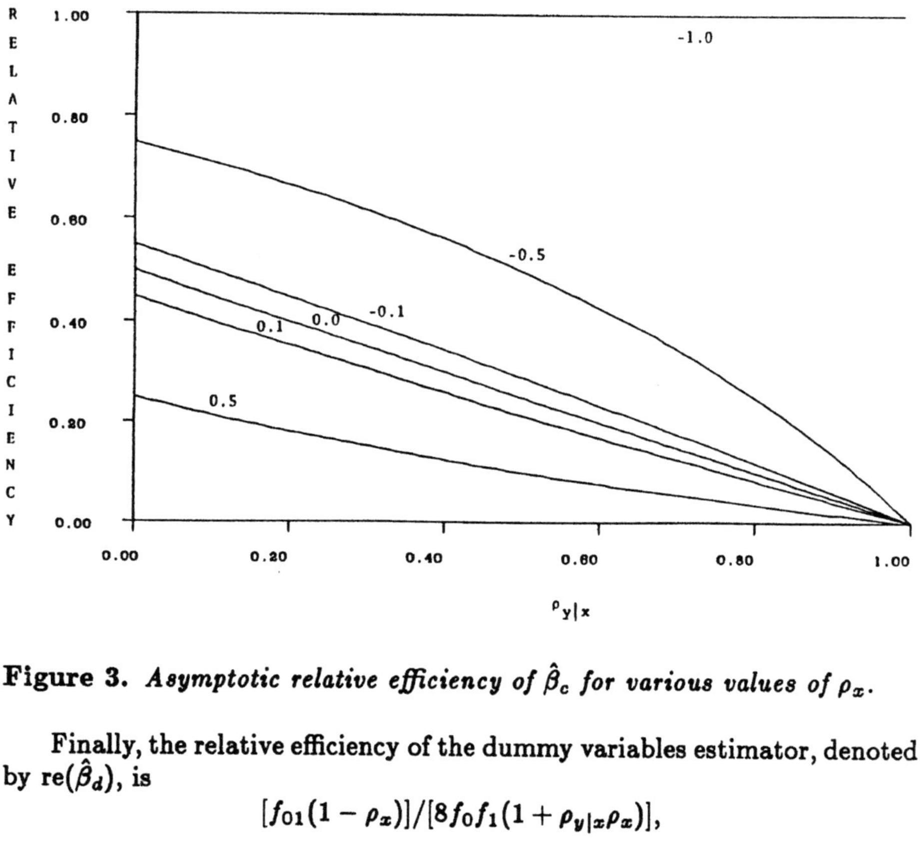

Figure 3. *Asymptotic relative efficiency of $\hat{\beta}_c$ for various values of ρ_x.*

Finally, the relative efficiency of the dummy variables estimator, denoted by re($\hat{\beta}_d$), is

$$[f_{01}(1-\rho_x)]/[8f_0f_1(1+\rho_{y|x}\rho_x)],$$

that is, one-quarter of the asymptotic relative efficiency of $\hat{\beta}_c$.

5.2 *Second-Order Properties*

When the wrong model, namely the usual or uncorrelated logistic model, is used, the estimated variance-covariance matrix is based on the inverse of the matrix

$$\frac{1}{N}\sum_{i=1}^{N}\hat{f}_i(1-\hat{f}_i)\begin{pmatrix}1 & x_i\\ x_i & x_i^2\end{pmatrix}, \tag{15}$$

where

$$N = \sum_{i=1}^{m} n_i.$$

As indicated in Section 4.2, as $m \to \infty$, this matrix approaches

$$2f_0f_1 \begin{pmatrix} 1 & \mu \\ \mu & E(X^2) \end{pmatrix}.$$

By using the inverse of the matrix in (15), we are consistently estimating

$$(2\sigma_x^2 f_0 f_1)^{-1},$$

whereas we should be estimating $\mathrm{Var}(\hat{\beta}_u)$, which, by the results of Section 4.2, is

$$(1 + \rho_{y|x}\rho_x)/(2\sigma_x^2 f_0 f_1).$$

This discrepancy corresponds exactly to that noticed for the continuous case by Scott and Holt (1982, p. 850). The factor $(1 + \rho_{y|x}\rho_x)$ is called the *misspecification factor* of the variance estimator. Thus for the purpose of testing the hypothesis $\beta = 0$, one can use the output of a standard computer program to get $\hat{\beta}_u$ and an estimate of $\mathrm{Var}(\hat{\beta}_u)$, and then adjust by multiplying the estimate of $\mathrm{var}(\hat{\beta}_u)$ by $(1 + \hat{\rho}_{y|x}\hat{\rho}_x)$, where $\hat{\rho}_x$ and $\hat{\rho}_{y|x}$ are any consistent estimates of the intraclass correlation between the x's and between the y's; for example, the usual pairwise correlation may be used. Conversely, at least in this case, $(1 + \rho_{y|x}\rho_x)$ may be estimated by $\hat{C}$, where $\hat{C}$ was defined by Brier (1980, p. 594). The approach in this paper gives a more general interpretation to Brier's parameter C because $C - 1$, at least under the null hypothesis, is the product of the intraclass correlations of the dependent and independent variables.

6. DISCUSSION

When Rosner's correlated logistic model provides the correct probabilistic mechanism for pairs of correlated binary outcomes in the presence of a single binary correlated covariate, three commonly used estimators of the regression coefficient have been shown, in general, to be lacking consistency and efficiency.

1. The Breslow-Day conditional estimator, although consistent, may have very low asymptotic relative efficiency.
2. The usual logistic estimator is consistent when the true regression coefficient is zero, and in some other situations, and, when the regresion

coefficient is zero, this estimator may be more efficient than the unconditional maximum likelihood estimator.

3. The usual logistic model with dummy variables produces an estimator that is consistent only when the true regression coefficient is zero, and, in this case, has very low asymptotic relative efficiency.

It would be of interest to see whether these conclusions hold for multiple covariates, continuous covariates, and mixtures of discrete and continuous covariates, and more than two units per cluster.

This last point may be investigated using a version of the model in this paper expanded in a way suggested by Rosner (1983), that is, a model based on the beta-binomial with several units per cluster. However, of more interest is a model that would allow more complex correlation structure in logistic models, for example a model that allows second-order correlation different from Rosner's (and hence the beta-binomial), or a model which allows unequal correlation between the units within a cluster. Such models would permit the modelling of quite complex patterns of correlated binary outcomes in the presence of covariates.

APPENDIX 1: ASYMPTOTIC DISTRIBUTION UNDER AN INCORRECT MODEL

The asymptotic distribution of an estimator, $\hat{\boldsymbol{\theta}}$, in this case, the maximum likelihood estimator, can be obtained by expanding the estimating equations, $\mathbf{S}_m(\hat{\boldsymbol{\theta}})$, in this case, the maximum likelihood equations, in a multivariate Taylor series expansion about the correct value, $\boldsymbol{\theta}$, of the parameter (see Cox and Hinkley, 1974, p. 294–302). Hence

$$\mathbf{S}_m(\hat{\boldsymbol{\theta}}) = \mathbf{S}_m(\boldsymbol{\theta}) + S'_m(\boldsymbol{\theta})(\hat{\boldsymbol{\theta}} - \boldsymbol{\theta}), \tag{A1.1}$$

where $S'_m(.)$ is the matrix of first derivatives of $\mathbf{S}_m(\boldsymbol{\theta})$ with respect to $\boldsymbol{\theta}$. We are ignoring terms of $o_p(m^{-1/2})$. Since

$$\mathbf{S}_m(\hat{\boldsymbol{\theta}}) = \mathbf{0},$$

(A1.1) becomes

$$(\hat{\boldsymbol{\theta}} - \boldsymbol{\theta}) = -[S'_m(\boldsymbol{\theta})]^{-1}\mathbf{S}_m(\boldsymbol{\theta})$$

or

$$\sqrt{m}(\hat{\boldsymbol{\theta}} - \boldsymbol{\theta}) = m[S'_m(\boldsymbol{\theta})]^{-1}[-\mathbf{S}_m(\boldsymbol{\theta})/\sqrt{m}].$$

In the usual case where we have the correct model, by the multivariate central limit theorem,

$$\frac{1}{\sqrt{m}}\mathbf{S}_m(\boldsymbol{\theta}) \xrightarrow{\text{D}} N_2(\mathbf{0}, I),$$

where I is the Fisher Information matrix. Moreover,

$$-\frac{1}{m}S'_m(\boldsymbol{\theta}) \xrightarrow{\text{P}} I$$

so that

$$\sqrt{m}(\hat{\boldsymbol{\theta}} - \boldsymbol{\theta}) \xrightarrow{\text{D}} N_2(0, I^{-1}).$$

However, in the case of an incorrectly specified probability distribution,

$$\hat{\boldsymbol{\theta}} \xrightarrow{\text{P}} \boldsymbol{\theta}_0,$$

where $\boldsymbol{\theta}_0$ is usually not $\boldsymbol{\theta}$, the value under the correct model, so that we must examine the behaviour of

$$\sqrt{m}(\hat{\boldsymbol{\theta}} - \boldsymbol{\theta}_0) = m[S'_m(\boldsymbol{\theta}_0)]^{-1}[\mathbf{S}_m(\boldsymbol{\theta}_0)/\sqrt{m}] \tag{A1.2}$$

under the correct model.

If $E[\mathbf{S}_m(\hat{\boldsymbol{\theta}}_0)] = \mathbf{0}$ and $\text{Var}[\mathbf{S}_m(\boldsymbol{\theta}_0)] = V$, then, by the multivariate central limit theorem,

$$\frac{1}{\sqrt{m}}\mathbf{S}_m(\boldsymbol{\theta}_0) \xrightarrow{\text{D}} N_2(0, V),$$

(where V is not necessarily the Fisher Information matrix for either the correct or the incorrect model). Now

$$-\frac{1}{m}S'_m(\boldsymbol{\theta}_0) \xrightarrow{\text{P}} U$$

(where U is also not necessarily the Fisher Information matrix), so that

$$\sqrt{m}(\hat{\boldsymbol{\theta}} - \boldsymbol{\theta}_0) \xrightarrow{\text{D}} N_2(0, U^{-1}VU^{-1}),$$

and the asymptotic variance of $\sqrt{m}\hat{\theta}$ is given by $U^{-1}VU^{-1}$.

APPENDIX 2: ASYMPTOTIC RELATIVE EFFICIENCY OF THE CONDITIONAL ESTIMATOR

The asymptotic relative efficiency of $\hat{\beta}_c$, as compared to $\hat{\beta}_R$, is given by (14) in Section 5.1 as

$$[f_{01}(1 - \rho_x)]/[2f_0f_1(1 + \rho_{y|x}\rho_x)]. \tag{A2.1}$$

When β is 0,

$$f_{01} = e^{\alpha_1}/d, \; f_0 = (1 + e^{\alpha_1})/d, \; f_1 = (e^{\alpha_1} + e^{2\alpha_1})/d,$$

where

$$d = 1 + 2e^{\alpha_1} + e^{\alpha_2}.$$

Moreover, from (1) in Section 2, when β is 0,

$$\rho_{y|x} = (e^{\alpha_2} - e^{2\alpha_1})/[(1 + e^{\alpha_1})(e^{\alpha_1} + e^{\alpha_2})].$$

Substituting into (A2.1), we have

$$\begin{aligned} \mathrm{re}(\hat{\beta}_c) &= \frac{de^{\alpha_1}(1 - \rho_x)}{2[(1 + e^{\alpha_1})(e^{\alpha_1} + e^{\alpha_2}) + \rho_x(e^{\alpha_2} - e^{2\alpha_1})]} \\ &= \frac{(e^{\alpha_1} + 2e^{2\alpha_1} + e^{\alpha_2 + \alpha_1})(1 - \rho_x)}{2[e^{\alpha_1} + e^{2\alpha_1}(1 - \rho_x) + e^{\alpha_2 + \alpha_1} + e^{\alpha_2}(1 + \rho_x)]} \end{aligned} \qquad \text{(A2.2)}$$

When ρ_x is -1, that is, when the two x values are always different from each other, then we have $\mathrm{re}(\hat{\beta}_c) = 1$. For other values of ρ_x , we compare the two estimators by subtracting the numerator of (A2.2) from its denominator to get

$$\begin{aligned} & e^{\alpha_1}(1 + \rho_x) + e^{\alpha_2 + \alpha_1}(1 + \rho_x) + 2e^{\alpha_2}(1 + \rho_x) \\ &= (1 + \rho_x)(e^{\alpha_1} + e^{\alpha_2 + \alpha_1} + 2e^{\alpha_2}), \end{aligned}$$

which is non-negative for all values of the parameters, and is 0 only when ρ_x is -1. Hence the conditional estimator, when β is 0, is never more efficient than the maximum likelihood estimator, and is as efficient as that estimator only when ρ_x is -1.

REFERENCES

Berkson, J. (1955), "Maximum likelihood and minimum χ^2 estimates of the logistic function". *Journal of the American Statistical Association* **50**, 130–162.

Breslow, N. E., and N. E. Day (1980), *Statistical Methods in Cancer Research, Volume 1—The Analysis of Case-Control Studies*. Lyon: International Agency for Research on Cancer..

Breslow, N. E., N. E. Day, K. T. Halvorsen, R.L. Prentice, and C. Sabal (1978), "Estimation of multiple relative risk functions in matched case-control studies." *American Journal of Epidemiology* **108**, 299–307.

Brier, S. S. (1980), "Analysis of contingency tables under cluster sampling." *Biometrika* **67**, 591–596.

Brillinger, D. R. (1975), "Statistical inference for stationary point processes." In *Stochastic Processes and Related Topics*, ed. M. L. Puri, pp. 1–96. New York: Academic Press.

Campbell, C. (1979), "Properties of ordinary and weighted least squares estimates in two stage samples." *Proceedings of the Social Statistics Section, American Statistical Association,* pp. 800–805.

Cox, D. R. (1966), "Some procedures associated with the logistic qualitative response curve." In *Research Papers in Statistics: Festschrift for J. Neyman*, ed. F. N. David, pp. 55–71. New York: Wiley and Sons.

Cox, D. R., and D. V. Hinkley (1974), *Mathematical Statistics*. London: Chapman and Hall.

Crowder, M. J. (1979), "Inference about the intraclass correlation coefficient in the beta-binomial ANOVA for proportions." *Journal of the Royal Statistical Society,* Series B **41**, 230–234.

Day, N. E., and D. F. Kerridge (1967), "A general maximum likelihood discriminant." *Biometrics* **23**, 313–323.

Dempster, A. P. (1979), "An overview of multivariate logistic models." *Journal of Multivariate Analysis* **1**, 316–346.

Donner, A. P. (1984), "Linear regression analysis with repeated measurements." *Journal of Chronic Diseases* **37**, 441–488.

Farewell, V. T. (1979), "Some results on the estimation of logistic models based on retrospective data." *Biometrika* **66**, 27–32.

Laird, N., and T. Ware (1981), "Random effects models for serial observations with dichotomous response." Annual meeting, American Statistical Association, Detroit, Michigan.

Koval, J. J., and A. P. Donner (1984), "The effects of clustering on parameter estimation in logistic regression." Annual meeting, American Statistical Association, Philadelphia, Pennsylvania.

Kulperger, R. (1979), "Parametric estimation for a simple branching diffusion process." *Journal of Multivariate Analysis* **9**, 101–105.

Ochi, Y., and R. L. Prentice (1984), "Likelihood inference in a correlated probit regression model." *Biometrika* **71**, 531–543.

Pierce, D. A., and B. R. Sands (1975), "Extra-Bernoulli variation in binary data". Technical Report No. 46, Department of Statistics, Oregon State University.

Rao, J. N. K., and A. J. Scott (1981), "The analysis of categorical data from complex surveys: chi-squared tests for goodness of fit and independence in two-way tables." *Journal of the American Statistical Association* **76**, 221–230.

Rosner, B. (1983), "Multivariate methods in opthamology." Annual meeting, American Statistical Association, Toronto, Ontario.

Rosner, B. (1984), "Multivariate methods in opthamology with application to other paired-data situations." *Biometrics* **40**, 1025–1035.

Scott, A. J., and D. Holt (1982), "The effect of two-stage sampling on ordinary least squares methods." *Journal of the American Statistical Association* **77**, 848–854.

Stiratelli, R., N. Laird, and J. Ware (1984), "Random effects models for serial observations with binary response." *Biometrics* **40**, 961–972.
Williams, D. A. (1982), "Extra-binomial variation in logistic linear models." *Applied Statistics* **31**, 144–148.

Sadanori Konishi [1] and A. K. Gupta [2]

INFERENCES ABOUT INTERCLASS AND INTRACLASS CORRELATIONS FROM FAMILIAL DATA

ABSTRACT

A brief review of estimation and tests of hypotheses concerning interclass and intraclass correlations from familial data is given. A modified likelihood ratio test is proposed for testing the equality of intraclass correlations in two multivariate normal populations. The asymptotic null distribution of the proposed test statistic is obtained. A test procedure based on Fisher's z-transformation is also discussed.

1. INTRODUCTION

Estimation of the degree of familial resemblance concerning a certain characteristic, such as blood pressure, stature, body weight, lung capacity or I.Q., is of great importance in the biometrical study of inheritance. Inferences are made on the basis of a sample of N families represented by $\mathbf{X}_\alpha = (X_{0\alpha}, X_{1\alpha}, X_{2\alpha}, \ldots, X_{k_\alpha,\alpha})'$, $\alpha = 1, 2, \ldots, N$, each of which consists of the mother's score $X_{0\alpha}$, in general the parent's score and his or her k_α siblings' scores $X_{1\alpha}, \ldots, X_{k_\alpha,\alpha}$. This implies that each family may have different numbers of siblings. We assume that there is no difference among siblings with regard to the attribute under consideration.

A model based on this familial data is constructed by assuming that $\mathbf{X}_\alpha$ follows a $(k_\alpha + 1)$-variate normal distribution with mean vector

$$\boldsymbol{\mu}_\alpha = (\mu_m, \mu_s, \ldots, \mu_s)'$$

[1] Institute of Statistical Mathematics, 4-6-7, Minami-Azabu, Minato-Ku, Tokyo 106, Japan

[2] Department of Mathematics and Statistics, Bowling Green State University, Bowling Green, Ohio 43403, U.S.A.

I. B. MacNeill and G. J. Umphrey (eds.), Biostatistics, 225–233.

and covariance matrix

$$\Sigma_\alpha = \begin{bmatrix} \sigma_m^2 & \boldsymbol{\sigma}_{ms}' \\ \boldsymbol{\sigma}_{ms} & \Sigma_{ss} \end{bmatrix}, \tag{1.1}$$

where $\boldsymbol{\sigma}_{ms} = (\rho_{ms}\sigma_m\sigma_s, \rho_{ms}\sigma_m\sigma_s, \ldots, \rho_{ms}\sigma_m\sigma_s)'$ is a k_α-dimensional vector and

$$\Sigma_{ss} = \sigma_s^2[(1-\rho_{ss})I_\alpha + \rho_{ss}\mathbf{e}_\alpha\mathbf{e}_\alpha']$$

with I_α the identity matrix of order k_α and $\mathbf{e}_\alpha = (1,1,\ldots,1)'$ a k_α-dimensional vector.

In the analysis of familial data there are two types of correlations: (i) a parent-child correlation and (ii) a sib-sib correlation. The degree of resemblance between a parent and siblings is measured by the interclass correlation, ρ_{ms}, and the degree of resemblance among siblings by the intraclass correlation, ρ_{ss}. Various kinds of estimators of the interclass and intraclass correlations have been proposed, some of which were obtained on a more or less intuitive basis while others are based on the method of maximum likelihood. A brief survey of estimation and tests of hypotheses that are relevant to interclass and intraclass correlations is given in Section 2.

Another object of the present paper is to consider a test of hypothesis concerning intraclass correlations. In Section 3 a modified likelihood ratio test statistic is proposed for testing the equality of intraclass correlations in two multivariate normal populations. An improvement to a test procedure based on Fisher's z-transformation is also discussed. Large sample theory is applied to obtain some results on distributions of the test statistics.

2. CORRELATIONS FROM FAMILIAL DATA

2.1 Interclass Correlations

One approach to estimation is to use the method of maximum likelihood. Unfortunately, an explicit solution of likelihood equations does not exist in the case where the number of siblings varies among families. So several algorithms have been proposed for finding maximum likelihood estimates of the interclass correlation ρ_{ms}.

An algorithm proposed by Rosner (1979) involves iterative maximization of an implicit function of the parameters ρ_{ms} and ρ_{ss}, and seems to be difficult to implement. A simpler algorithm was proposed by Mak and Ng (1981), using a linear model approach discussed by Kempthorne and Tandon (1953). Srivastava (1984) gave an alternative algorithm that requires solving only one equation, in which observations are transformed to simplify the covariance structure in (1.1). For this type of transformation, see Olkin and Pratt (1958) and Anderson (1963).

The disadvantage of the maximum likelihood estimates is that they cannot be expressed in closed form. Hence, several estimators have been proposed that use methods analogous to that of computing an ordinary product-moment correlation. These are called the pairwise, sib-mean and random-sib estimators, and were discussed in detail by Rosner *et al.* (1977). In addition to these traditional estimators, Rosner *et al.* (1977) proposed the so-called ensemble estimator obtained by calculating an expected value for the random-sib estimator over all possible choices of random sibs from each family.

These four estimators have been compared in terms of mean squared error, using Monte Carlo simulation (Rosner *et al.*, 1977) and applying large sample theory (Konishi, 1982). This work showed that the pairwise and ensemble estimators are far superior to the sib-mean and random-sib estimators, and that the pairwise estimator is more effective than the ensemble estimator at low values ($\leq .3$) of ρ_{ss}. Comparison of the estimators was extended to include the maximum likelihood estimator by Rosner (1979); the maximum likelihood estimator was shown to be roughly equivalent to the pairwise estimator. It is known (Rosner, 1979) that the pairwise estimator reduces to the maximum likelihood estimator in the case of an equal number of siblings per family. Recently Srivastava (1984) gave two alternative sets of estimators for ρ_{ms}, adding a slight modification in the process of deriving the maximum likelihood estimators.

Asymptotic distributions of the pairwise and ensemble estimators were derived by Konishi (1982). He also showed that the sib-mean estimator is not a consistent estimator for ρ_{ms}, and proposed a modified sib-mean estimator which is consistent. The asymptotic variance of the maximum likelihood estimator was derived by Rosner (1982).

Regarding the interclass correlation ρ_{ms}, it is of interest to test the hypothesis $H_0 : \rho_{ms} = 0$ versus $H_1 : \rho_{ms} > 0$. The traditional test procedures—the classical pairwise, conservative pairwise and sib-mean tests—are based on the t-distribution with appropriate degrees of freedom. These are analogous to the usual procedure for a correlation coefficient in a bivariate normal sample (for details, see Rosner *et al.*, 1979). However, the problem of assessing the aggregate degrees of freedom over N families of varying sibship sizes k_α still remains. Rosner *et al.* (1979) suggested improvements on the t-approximations of the classical and conservative test statistics and gave the adjusted pairwise test. Konishi (1982, 1985b) proposed alternative test procedures based on asymptotic distributions of the estimators. These compared favorably with the previous test procedures based on the studentized version of the pairwise estimator (see also Donner and Bull, 1984).

It may also be of interest to test the hypothesis $H_0 : \rho_{ms} = \rho_0$ versus

$H_1 : \rho_{ms} \neq \rho_0$ where ρ_0 is a specified constant. For this testing problem, Konishi (1985b) proposed test procedures based on large sample theory to obtain asymptotic distributions of the pairwise and ensemble estimators.

Confidence intervals for interclass correlations can be constructed using the close relationship between tests of hypotheses and confidence sets (see Konishi, 1985b).

2.2 Intraclass Correlations

In the model for the familial data discussed in Section 1, we consider only the scores of siblings, so that $\mathbf{X}_\alpha = (X_{1\alpha}, X_{2\alpha}, \ldots, X_{k_\alpha,\alpha})'$, $\alpha = 1, 2, \ldots, N$ is assumed to have a k_α-variate normal distribution with mean vector $\boldsymbol{\mu}_\alpha = (\mu_s, \mu_s, \ldots, \mu_s)'$ and covariance matrix Σ_{ss} with homogeneous variances and covariances.

The maximum likelihood estimate of the intraclass correlation ρ_{ss} also cannot be expressed in a closed form. Algorithms for finding maximum likelihood estimates have been given by Donner and Koval (1980) and Smith (1980a,b). Donner and Koval (1980) compared the maximum likelihood estimator with previous estimators—the analysis of variance and pairwise intraclass correlations—using Monte Carlo simulation (also see Smith, 1957; Snedecor and Cochran, 1967). They suggested that if the value of ρ_{ss} is expected to be small or of moderate size ($< .5$), then the analysis of variance intraclass correlation is effective and if ρ_{ss} is thought to be large or if no prior knowledge about ρ_{ss} exists, then the maximum likelihood estimator is effective. It is known (Elston, 1975) that if the number of siblings is the same in each family, then the pairwise intraclass correlation yields the maximum likelihood estimates for ρ_{ss}.

Asymptotic distribution theory can be applied to test the hypotheses or to construct confidence intervals concerning the intraclass correlations. The results are messy, however, when the families have unequal numbers of siblings.

3. A TEST FOR EQUALITY OF INTRACLASS CORRELATIONS

We consider a test for the equality of intraclass correlations in two multivariate normal populations. Let $\mathbf{X}_1^{(\alpha)}, \mathbf{X}_2^{(\alpha)}, \ldots, \mathbf{X}_{N_\alpha}^{(\alpha)}$ be a random sample of size N_α from a p_α-variate normal distribution with mean vector $\boldsymbol{\mu}_\alpha = (\mu_\alpha, \mu_\alpha, \ldots, \mu_\alpha)'$ and covariance matrix Σ_α $(\alpha = 1, 2)$. Assume that Σ_α has homogeneous variances and covariances, so that

$$\Sigma_\alpha = \sigma_\alpha^2[(1 - \rho_\alpha)I_\alpha + \rho_\alpha \mathbf{e}_\alpha \mathbf{e'}_\alpha] \quad \text{for} \quad \alpha = 1, 2.$$

On the basis of the observations, we wish to test the hypothesis

$$H_0 : \rho_1 = \rho_2 \; (= \rho)$$

where ρ is unspecified. For related work see Hahn (1975) and Gupta and Nagar (1986) and the references given there.

3.1 Modified Likelihood Ratio Test

One possible test for the equality of two intraclass correlations is the likelihood ratio test. Donner and Bull (1983) discussed the likelihood ratio test, which involves, however, an iterative maximization of the likelihood function.

It is known (e.g., see Elston, 1975) that under the alternative hypothesis H_1 the maximum likelihood estimates of σ_α^2 and ρ_α for $\alpha = 1, 2$ are, respectively, given by

$$s_\alpha^2 = \frac{1}{p_\alpha} \sum_{i=1}^{p_\alpha} s_{ii}^{(\alpha)} \quad \text{and} \quad r_\alpha = \frac{\sum_{i \neq j}^{p_\alpha} s_{ij}^{(\alpha)} / \{p_\alpha(p_\alpha - 1)\}}{\sum_{i=1}^{p_\alpha} s_{ii}^{(\alpha)} / p_\alpha}, \tag{3.1}$$

where

$$S^{(\alpha)} = \left(s_{ij}^{(\alpha)}\right) = \frac{1}{N_\alpha} \sum_{\ell=1}^{N_\alpha} \left(X_\ell^{(\alpha)} - \overline{\mathbf{X}}_\alpha\right) \left(\mathbf{X}_\ell^{(\alpha)} - \overline{\mathbf{X}}_\alpha\right)' \tag{3.2}$$

for $\overline{\mathbf{X}}_\alpha = (\overline{X}_\alpha, \overline{X}_\alpha, \ldots, \overline{X}_\alpha)'$ with $\overline{X}_\alpha = \frac{1}{N_\alpha p_\alpha} \sum_{i=1}^{p_\alpha} \sum_{j=1}^{N_\alpha} X_{ij}^{(\alpha)}$.

Under the null hypothesis H_0 we must maximize the likelihood function with respect to ρ, σ_1^2 and σ_2^2. It has been shown (Donner and Bull, 1983) that the estimate of the common intraclass correlation ρ is given as the root of a cubic equation and must be solved numerically. Hence we propose a modified likelihood ratio test statistic as follows:

$$\begin{aligned} -2 \log \Lambda &= -2 \log L(\hat{\omega}) + 2 \log L(\hat{\Omega}) \\ &= \sum_{\alpha=1}^{2} N_\alpha \left\{ p_\alpha \log \frac{\hat{\sigma}_\alpha^2}{s_\alpha^2} + (p_\alpha - 1) \log \frac{1 - r}{1 - r_\alpha} \right. \\ &\quad \left. + \log \frac{1 + (p_\alpha - 1) r}{1 + (p_\alpha - 1) r_\alpha} \right\}, \end{aligned} \tag{3.3}$$

where

$$r = \sum_{\alpha=1}^{2} N_\alpha p_\alpha (p_\alpha - 1) r_\alpha \Big/ \left\{ \sum_{\alpha=1}^{2} N_\alpha p_\alpha (p_\alpha - 1) \right\} \tag{3.4}$$

and

$$\hat{\sigma}_\alpha^2 = s_\alpha^2 \left[\frac{(p_\alpha - 1) r (r_\alpha - r)}{(1 - r)\{1 + (p_\alpha - 1) r\}} \right] \quad \text{for } \alpha = 1, 2.$$

Since the proposed test statistic does not, in general, have an asymptotic χ^2-distribution under the null hypothesis, we apply large sample theory to derive the asymptotic distribution of the test statistic (3.3).

Let $N = N_1 + N_2$ and $f_\alpha = N_\alpha/N$ for $\alpha = 1, 2$. It can be shown that $s_{ij}^{(\alpha)}/\sigma_\alpha^2$ and $s_{ii}^{(\alpha)}/\sigma_\alpha^2$ given by (3.2) converge to ρ_α and 1, respectively, as N tends to infinity. So we set

$$\frac{s_{ij}^{(\alpha)}}{\sigma_\alpha^2} = \rho_\alpha + \frac{1}{\sqrt{N}} v_{ij}^{(\alpha)} \quad \text{for} \quad i \neq j = 1, 2, \ldots, p_\alpha$$

and

$$\frac{s_{ii}^{(\alpha)}}{\sigma_\alpha^2} = 1 + \frac{1}{\sqrt{N}} v_{ii}^{(\alpha)} \quad \text{for} \quad i = 1, 2, \ldots, p_\alpha.$$

Then we can expand r_α and r in a power series of order $1/\sqrt{N}$ as

$$\begin{aligned} r_\alpha &= \rho_\alpha + \frac{1}{\sqrt{N}} v_1^{(\alpha)} + \frac{1}{N} v_2^{(\alpha)} + O_p(N^{-3/2}), \\ r &= \sum_{\alpha=1}^{2} a_\alpha \rho_\alpha + \frac{1}{\sqrt{N}} \sum_{\alpha=1}^{2} a_\alpha v_1^{(\alpha)} + \frac{1}{N} \sum_{\alpha=1}^{2} a_\alpha v_2^{(\alpha)} + O_p(N^{-3/2}), \end{aligned} \tag{3.5}$$

where $a_\alpha = f_\alpha p_\alpha(p_\alpha - 1)/\{\sum_{\alpha=1}^{2} f_\alpha p_\alpha(p_\alpha - 1)\}$ and

$$v_1^{(\alpha)} = \frac{1}{p_\alpha(p_\alpha - 1)} \sum_{i \neq j}^{p_\alpha} v_{ij}^{(\alpha)} - \frac{\rho_\alpha}{p_\alpha} \sum_{i=1}^{p_\alpha} v_{ii}^{(\alpha)},$$

$$v_2^{(\alpha)} = \frac{\rho_\alpha}{p_\alpha^2} \left(\sum_{i=1}^{p_\alpha} v_{ii}^{(\alpha)} \right)^2 - \frac{1}{p_\alpha^2(p_\alpha - 1)} \sum_{i=1}^{p_\alpha} v_{ii}^{(\alpha)} \sum_{j \neq k}^{p_\alpha} v_{jk}^{(\alpha)}.$$

Substituting (3.5) in the modified likelihood ratio test statistic given by (3.3) and expanding the result in a Taylor's series, we have, under the null hypothesis,

$$-2 \log \Lambda = \frac{1}{2(1-\rho)^2} \sum_{\alpha \neq \beta}^{2} \frac{f_\alpha p_\alpha (p_\alpha - 1) a_\beta^2}{\{1 + (p_\alpha - 1)\rho\}^2} \left(v_1^{(1)} - v_1^{(2)} \right)^2.$$

The variate $v_1^{(\alpha)}$ is a linear function of $v_{ij}^{(\alpha)}$ $(i, j = 1, 2, \ldots, p_\alpha)$ which are asymptotically jointly normally distributed with zero means and covariances

$$E[v_{ii}^{(\alpha)^2}] = 2, \qquad E[v_{ij}^{(\alpha)^2}] = 1 + \rho^2 \ (i \neq j),$$

$$E[v_{ij}^{(\alpha)} v_{ii}^{(\alpha)}] = 2\rho \quad (i \neq j), \ E[v_{ij}^{(\alpha)} v_{ik}^{(\alpha)}] = \rho(1+\rho) \ (i \neq j \neq k),$$

$$E[v_{ij}^{(\alpha)} v_{kr}^{(\alpha)}] = 2\rho^2 \quad (i, j) \neq (k, r).$$

Hence it can be shown that under H_0 the limiting distribution of $v_1^{(1)} - v_1^{(2)}$, noting that $v_1^{(1)}$ and $v_1^{(2)}$ are mutually independent, is normal with mean 0 and variance

$$\psi^2 = 2\sum_{\alpha=1}^{2} \frac{1}{p_\alpha(p_\alpha - 1)}(1-\rho)^2\{1+(p_\alpha - 1)\rho\}^2. \tag{3.6}$$

Then we have the following result.

Theorem 3.1. *Under the null hypothesis H_0, the asymptotic distribution of the modified likelihood ratio test statistic $-2\log\Lambda$ as $N = N_1+N_2 \to +\infty$ is*

$$\lim Le(-2\log\Lambda) = c\chi_1^2,$$

where χ_1^2 is a chi-squared variate with one degree of freedom and the coefficient c is given by

$$c = \frac{\psi^2}{2(1-\rho)^2} \sum_{\alpha\neq\beta}^{2} \frac{f_\alpha p_\alpha(p_\alpha - 1)a_\beta^2}{\{1+(p_\alpha - 1)\rho\}^2}$$

with ψ^2 given by (3.6) and a_β by (3.5).

In practice we have to estimate the unknown parameter ρ included in the coefficient c by (3.4) or by some other procedure.

It is of interest to note that in the special case when $p_1 = p_2$, the coefficient c reduces to $1 - f_1^2 - f_2^2$ which is independent of the unknown parameter. Then we have the following result.

Corollary 3.1. *In the case when $p_1 = p_2$, the asymptotic null distribution of $-2\log\Lambda$ as $N \to +\infty$ is*

$$\lim Le(-2\log\Lambda) = (1 - f_1^2 - f_2^2)\chi_1^2,$$

where $f_\alpha = N_\alpha/(N_1+N_2)$ for $\alpha = 1, 2$.

3.2 Fisher's z-transformation

In the particular case when $p_1 = p_2 = p$, say, we can use Fisher's z-transformation

$$Z_p(r_\alpha) = \left(\frac{p-1}{2p}\right)^{1/2} \log\frac{1+(p-1)r_\alpha}{1-r_\alpha}, \tag{3.7}$$

where r_α is the maximum likelihood estimator of ρ_α defined by (3.1). It is known (Fisher, 1958, Chapter 7) that this transformation stabilizes the variance for any value of dimension p, and that the asymptotic distribution of $Z_p(r_\alpha)$ is normal with

$$\text{mean}: \ Z_p(\rho_\alpha) \text{ and variance}: \ \frac{1}{N_\alpha - 2}. \tag{3.8}$$

However this normal approximation becomes poorer for the values of p greater than 2, since for $p \geq 3$ the variance stabilization and normalization cannot be achieved simultaneously by the same transformation (3.7), unlike the case of the correlation coefficient in a bivariate normal sample (Konishi, 1981, 1985a).

It has been shown (Konishi, 1985a) that $Z_p(r_\alpha)$ is asymptotically normally distributed with

$$\text{mean}: \ Z_p(\rho_\alpha) + \frac{1}{N_\alpha} \cdot \frac{(7-5p)}{\{18p(p-1)\}^{1/2}}, \quad \text{and variance}: \ \frac{1}{N_\alpha}.$$

This normal approximation is far superior to that of (3.8) and the second term of the mean does not depend on the unknown parameter. Therefore, for testing the equality of two intraclass correlations, we suggest using the statistic

$$Z_p(r_1) - Z_p(r_2)$$

which, under the null hypothesis H_0, is asymptotically normally distributed with

$$\text{mean}: \ \frac{(7-5p)}{\{18p(p-1)\}^{1/2}} \left(\frac{1}{N_1} - \frac{1}{N_2} \right) \text{ and variance}: \ \frac{1}{N_1} + \frac{1}{N_2}.$$

REFERENCES

Anderson, T. W. (1963), "Asymptotic theory for principal component analysis". *Annals of Mathematical Statistics* **34**, 122–148.

Donner, A., and S. Bull (1983), "Inferences concerning a common intraclass correlation coefficient". *Biometrics* **39**, 771–775.

Donner, A., and S. Bull (1984), "A comparison of significance-testing procedures for parent-child correlations computed from family data". *Applied Statistics* **33**, 278–284.

Donner, A., and J. J. Koval (1980), "The estimation of intraclass correlation in the analysis of family data". *Biometrics* **36**, 19–25.

Elston, R. C. (1975), "On the correlation between correlations". *Biometrika* **62**, 133–140.

Fisher, R. A. (1958), *Statistical Methods for Research Workers*. Edinburgh: Oliver and Boyd.

Gupta, A. K., and D. K. Nagar (1986), "Testing equality of covariance matrices under intraclass correlation structure". *Tamkang Journal of Mathematics* **17**. (To appear.)

Hahn, C. P. (1975), "Testing the equality of covariance matrices under intraclass correlation models". *Annals of the Institute of Statistical Mathematics* **27**, 349–356.

Kempthorne, O., and O. B. Tandon (1953), "The estimation of heritability by regression of offspring on parent". *Biometrics* **9**, 90–100.

Konishi, S. (1981), "Normalizing transformations of some statistics in multivariate analysis". *Biometrika* **68**, 647–651.

Konishi, S. (1982) "Asymptotic properties of estimators of interclass correlation from familial data". *Annals of the Institute of Statistical Mathematics* **34**, 505–515.

Konishi, S. (1985a), "Normalizing and variance stabilizing transformations for intraclass correlations". *Annals of the Institute of Statistical Mathematics* **37**, 87–94.

Konishi, S. (1985b), "Testing hypotheses about interclass correlations from familial data". *Biometrics* **41**, 167–176.

Mak, T. K., and K. W. Ng (1981), "Analysis of familial data: linear-model approach". *Biometrika* **68**, 457–461.

Olkin, I., and J. W. Pratt (1958), "Unbiased estimation of certain correlation coefficients". *Annals of Mathematical Statistics* **29**, 201–211.

Rosner, B. (1979), "Maximum likelihood estimation of interclass correlations". *Biometrika* **66**, 533–538.

Rosner, B. (1982), "On the estimation and testing of interclass correlations: the general case of multiple replicates for each variable". *American Journal of Epidemiology* **116**, 722–730.

Rosner, B., A. Donner, and C. H. Hennekens (1977), "Estimation of interclass correlation from familial data". *Applied Statistics* **26**, 179–187.

Rosner, B., A. Donner, and C. H. Hennekens (1979), "Significance testing of interclass correlations from familial data". *Biometrics* **35**, 461–471.

Smith, C. A. B. (1957), "On the estimation of intraclass correlation". *Annals of Human Genetics* **21**, 363–373.

Smith, C. A. B. (1980a), "Estimating genetic correlations". *Annals of Human Genetics* **43**, 265–284.

Smith, C. A. B. (1980b), "Further remarks on estimating genetic correlations". *Annals of Human Genetics* **44**, 95–105.

Snedecor, G., and W. G. Cochran (1967), *Statistical Methods*, 6th edition. Ames, Iowa: Iowa State University Press.

Srivastava, M. S. (1984), "Estimation of interclass correlations in familial data". *Biometrika* **71**, 177–185.

R. Viveros [1] and D. A. Sprott [2]

MAXIMUM LIKELIHOOD ESTIMATION IN QUANTAL RESPONSE BIOASSAY

ABSTRACT

Approximate confidence intervals for the LD50 and LD90 doses and for the relative potency of one drug with respect to another in Bioassay are usually obtained using the Fieller method. A procedure is presented in this paper based on transformations of the parameters of interest, designed to facilitate normal, t, or log F approximations to the observed maximized or profile likelihood function. Comparisons with the Fieller method and applications to specific data sets are discussed. The procedure proposed illustrates methods and principles of statistical estimation that have a wide applicability.

1. INTRODUCTION

1.1 Logistic Model and Parameters of Interest

A model frequently used in Bioassay is the linear logistic binomial model in which observations of a random count Y are considered for which the distribution

$$f\,(y; b_1, b_2) = \binom{k}{y} p^y (1-p)^{k-y}, \; y = 0, 1, \ldots, k, \tag{1.1}$$

is assumed, where $p = p(x)$ is given by

$$\log[p/(1-p)] = b_1 + b_2 x. \tag{1.2}$$

[1] Escuela de Ciencias Fis. Mat., Universidad Autónoma de Puebla, Av. San Claudio y 14 Sur, 72320 Puebla, Puebla, Mexico.

[2] Department of Statistics and Actuarial Science, University of Waterloo, Waterloo, Ontario N2L 3G1

I. B. MacNeill and G. J. Umphrey (eds.), Biostatistics, 235–250.

The random variable Y is the number of successes observed when an amount x of a drug, usually measured by the log concentration, is administered to k individuals.

Two parameters of interest are the LD100p_0 dose $(0 < p_0 < 1)$ and the relative potency. The LD100p_0 dose is the value $x = \theta$ such that $p(\theta) = p_0$. For two drugs such that

$$\log\left[p_1/\left(1-p_1\right)\right] = b_1 + b_2 x, \qquad \log\left[p_2/\left(1-p_2\right)\right] = b_3 + b_2 x, \tag{1.3}$$

the relative potency of the second compared to the first is defined by $p_2(x) = p_1(x+\gamma)$. From (1.2) and (1.3)

$$\theta = \left(t_0 - b_1\right)/b_2, \qquad \gamma = \left(b_3 - b_1\right)/b_2, \tag{1.4}$$

where $t_0 = \log\left[p_0/\left(1-p_0\right)\right]$. Two specific LD's are of principal interest: the LD50 and the LD90 obtained when $p_0 = 0.5,\ 0.9\,(t_0 = 0, 2.1972)$, respectively.

Assumption (1.3) is equivalent to assuming that (log) dose x of the second drug produces the same effect as (log) dose $x+\gamma$ of the first drug, for all x. In Bioassay experiments the first drug is called the standard preparation while the second drug is the test preparation. In terms of the original doses the above assumption is expressed by stating that the test preparation behaves like a dilution/concentration of the standard preparation with dilution factor $\exp\{\gamma\}$. Although this quantity is often called the relative potency of the second drug with respect to the first, in this paper γ will be referred to as the relative potency.

The joint likelihood $L(b_1, b_2; \mathbf{y})$ of (b_1, b_2) based on the numbers of responses $\mathbf{y} = (y_1, \ldots, y_A)$ observed on A groups of sizes $\mathbf{k} = (k_1, \ldots, k_A)$ to which the respective doses $\mathbf{x} = (x_1, \ldots, x_A)$ were administered, is given by

$$\log L\left(b_1, b_2; \mathbf{y}\right) = \sum y_i \log p_i + \sum \left(k_i - y_i\right) \log\left(1 - p_i\right) \tag{1.5}$$

where

$$\log\left[p_i/\left(1-p_i\right)\right] = b_1 + b_2 x_i. \tag{1.6}$$

1.2 Fieller Method

Both the LD100p_0 dose and the relative potency involve ratios of regression parameters. The usual procedure for estimating ratios is based on the Fieller pivotal, discussed in some detail by Sprott and Viveros (1985). Its use in Bioassay is discussed and exemplified by Finney (1971, 1978). Here, however, the Fieller procedure is approximate, since none of the assumptions required when applied to the normal model is satisfied.

Noting that the LD50 dose θ is defined by $b_1 + \theta b_2 = 0$, and that the maximum likelihood (ML) estimate $(\hat{b}_1, \hat{b}_2)$ has an asymptotic bivariate normal distribution about (b_1, b_2) with estimated covariance matrix $V = I^{-1}$, it follows that

$$Z = (\theta \hat{b}_2 + \hat{b}_1) / \left(\theta^2 v_{22} + 2\theta v_{12} + v_{11}\right)^{\frac{1}{2}} \tag{1.7}$$

has an asymptotic standard normal distribution. Here, I is the observed information matrix with components

$$I_{ij} = -\partial^2 \log L\left(b_1, b_2; \mathbf{y}\right) / \partial \hat{b}_i \partial \hat{b}_j, (i, j = 1, 2). \tag{1.8}$$

Taking Z as a standard normal variate, setting Z^2 equal to a specified value and solving the resulting quadratic equation yields approximate confidence intervals for θ. This is the Fieller procedure for estimating the confidence limits for the LD50 dose. Note that the denominator of (1.7) is a random variable, whereas the exact use of the Fieller pivotal requires it be a constant.

1.3 Maximum Likelihood

The purpose here is to examine the direct application of maximum likelihood estimation to make inferences about θ, γ, valid and accurate in small samples. This will be done by the application of the methods developed by Sprott and Viveros (1984) and discussed in more detail for a related problem by Sprott and Viveros (1985).

The analysis will be based on the maximized or profile likelihood for the parameter of interest. Suppose the observations $\mathbf{y} = (y_1, y_2, \ldots, y_k) \in S$, S being the sample space, produce the joint likelihood $L(\theta; \mathbf{y})$ of the vector parameter $\theta = (\theta_1, \theta_2)$, where θ_1 and θ_2 may also be vectors. The associated maximized or profile relative likelihood of θ_1 is defined as

$$R_M\left(\theta_1; \mathbf{y}\right) = L(\theta_1, \tilde{\theta}_2; \mathbf{y}) / L(\hat{\theta}_1, \hat{\theta}_2; \mathbf{y}), \tag{1.9}$$

where $\tilde{\theta}_2 = \tilde{\theta}_2\left(\theta_1\right)$ is the maximum likelihood estimate of θ_2 for specified θ_1. For more details, see Edwards (1972, p. 118) and Sprott (1980).

2. ESTIMATION OF THE LD50 DOSE

2.1 Maximized Likelihood and Transformations

The asymptotic bivariate normality of $(\hat{b}_1, \hat{b}_2)$ would imply an asymptotic log maximized relative likelihood function $\log R_M(\theta; \mathbf{y}) = r_M(\theta; \mathbf{y}) =$

$-z^2/2$ for the LD50 dose θ, which belongs to the family of likelihoods discussed by Sprott and Viveros (1985), except that the denominator of z here depends on random variables. The skewness in $R_M(\theta;\mathbf{y})$ can be eliminated by the use of

$$\phi = \tan^{-1}\left[(\theta v_{22} + v_{12}) / \left(v_{11}v_{22} - v_{12}^2\right)^{\frac{1}{2}}\right] \tag{2.1}$$

for which

$$F_i(\hat{\phi}) = \left[\partial^i \log R_M(\phi;\mathbf{y})/\partial\hat{\phi}^i\right]/I_\phi^{i/2} \tag{2.2}$$

is zero for all $\mathbf{y} \in S$ and $i = 3, 5, \ldots$, where $I_\phi = -\partial^2 \log R_M(\phi;\mathbf{y})/\partial\hat{\phi}^2$ is the observed Fisher information of ϕ based on $R_M(\phi;\mathbf{y})$. Thus $R_M(\phi;\mathbf{y})$ will be a symmetric function of ϕ with respect to $\hat{\phi}$.

In terms of ϕ, the Fieller pivotal (1.7) is

$$Z = I_\phi^{\frac{1}{2}} \sin(\phi - \hat{\phi}) \quad \text{where} \quad I_\phi = \hat{b}_2^2/v_{22}\cos^2(\hat{\phi}). \tag{2.3}$$

If I_ϕ is sufficiently large, $z \doteq U_\phi = I_\phi^{\frac{1}{2}}(\phi - \hat{\phi})$ in the region of non-negligible likelihood $|z| \le 3$ and so ϕ has an approximate normal likelihood centered at $\hat{\phi}$ and scaled by $I_\phi^{-\frac{1}{2}}$. By this is meant that $R_M(\phi;\mathbf{y}) \doteq \exp\{-u_\phi^2/2\} = \exp\{-I_\phi(\phi - \hat{\phi})^2/2\}$.

Since, however, in finite samples $(\hat{b}_1, \hat{b}_2)$ does not have a normal distribution, $R_M(\phi;\mathbf{y})$ is only an approximation to the maximized relative likelihood of ϕ and $F_i(\hat{\phi})$, $i = 3, 5, 7, \ldots$ will not be exactly zero in general. Thus the observed $R_M(\phi;\mathbf{y})$ arising directly from (1.5) will have to be examined. This can be done by writing the joint likelihood of (b_1, b_2) as a function of (θ, b_2), yielding a likelihood $L(\theta, b_2;\mathbf{y})$ proportional to (1.5) with (1.6) becoming

$$\log\left[p_i/(1 - p_i)\right] = b_2(x_i - \theta). \tag{2.4}$$

Then the corresponding $R_M(\theta;\mathbf{y})$ is

$$R_M(\theta;\mathbf{y}) = L(\theta, \tilde{b}_2;\mathbf{y})/L(\hat{\theta}, \hat{b}_2;\mathbf{y}), \tag{2.5}$$

where $\tilde{b}_2 = \tilde{b}_2(\theta)$ is the ML estimate of b_2 for specified θ.

The standardized third and fourth derivatives $F_3(\hat{\phi})$ and $F_4(\hat{\phi})$, which are the coefficients associated with the powers 3 and 4 of the Taylor series of $r_M(\phi;\mathbf{y})$ around $\hat{\phi}$, can be calculated numerically for any given sample $\mathbf{y} \in S$ using the results of Sprott (1980). These quantities are the basis of the approximations discussed below.

As mentioned previously, both ϕ of (2.1) and the denominator of the Fieller pivotal (1.7) arising from model (1.5) are random variables. Although this is probably not a serious drawback, strictly speaking the logic of the procedure requires them both to be fixed constants, as they were when applied to the normal parent distribution by Sprott and Viveros (1985). To avoid this in the case of (2.1), the transformation

$$\Psi = \tan^{-1}\left[(w_{22}\theta + w_{12}) / \left(w_{11}w_{22} - w_{12}^2\right)^{\frac{1}{2}}\right] \tag{2.6}$$

can be examined, where $W = (X'X)^{-1}$, X being the $A \times 2$ matrix of 1's in the first column and the x_i's in the second.

2.2 Inferences Using Normal Approximations

It can be shown that if $F_3(\hat{\phi})$ and $F_4(\hat{\phi})$ are small for every $\mathbf{y} \in S$ then the Taylor expansion about $\hat{\phi}$ of $r_M(\phi; \mathbf{y})$ coincides up to the fourth power with that of the logarithm of the normal likelihood centered at $\hat{\theta}$ and scaled by $I_\phi^{\frac{1}{2}}$ (see Viveros, 1985). In other words, $u_\phi = I_\phi^{\frac{1}{2}}(\phi - \hat{\phi})$ has an approximate standard normal likelihood for every $\mathbf{y} \in S$. Treating U_ϕ as an approximate standard normal variate (Welch and Peers, 1963; Sprott, 1973), the information in the data about θ can be summarized in the form of a complete set of approximate confidence intervals:

$$\phi = \hat{\phi} \pm u / I_\phi^{\frac{1}{2}}, \quad \theta = c_1 \tan\left[\hat{\phi} \pm u / I_\phi^{\frac{1}{2}}\right] + c_2, \tag{2.7}$$

where from (2.1) $c_1 = \left(v_{11}v_{22} - v_{12}^2\right)^{\frac{1}{2}} / v_{22'}$ $c_2 = -v_{12}/v_{22'}$ and u is the value of a $N(0,1)$ variate required to produce the desired confidence level. This is equivalent to linearising the Fieller pivotal as in Section 2.1.

Example 1. Finney (1978, p. 365) gives the numerical data $A = 4, \mathbf{x} = (1.3863, 1.7918, 2.1972, 2.6027), \mathbf{k} = (12, 24, 24, 10), y = (1, 8, 15, 8)$. In this case $\hat{b}_1 = -6.1775$, $\hat{b}_2 = 3.0124$, so that $\hat{\theta} = 2.0507$, and the calculation of V and W leads to

$$\phi = \tan^{-1}(3.0959\,\theta - 6.2109), \quad \Psi = \tan^{-1}(2.2059\,\theta - 4.3997),$$

from which

$$\hat{\phi} = 0.1358, \quad \hat{\Psi} = 0.1225, \quad I_\phi = 13.0072, \quad I_\Psi = 25.1578.$$

The quantities F_3 and F_4 are

	$\hat{\theta}$	$\hat{\phi}$	$\hat{\Psi}$
F_3	0.1971	−0.0301	0.0507
F_4	0.5818	0.0047	0.3034

Since $F_3(\hat{\phi})$ and $F_4(\hat{\phi})$ are small, $U_\phi = I_\phi^{\frac{1}{2}}(\phi - \hat{\phi})$ can be taken as an approximate standard normal variate in the proposed procedure. The summary of the data is, using (2.7),

$$\phi = 0.1358 \pm 0.27727\, u, \quad \theta = 0.3230 \tan(0.1358 \pm 0.27727\, u) + 2.0062,$$

where u is the required value of a standard normal variate.

The approximate 95% confidence interval obtained by setting $u = 1.96$ is $\phi = (-0.4076,\ 0.6792)$, $\theta = (1.866,\ 2.267)$. Inserting these values of ϕ into the Fieller pivotal (2.3) gives $z = \pm 1.86$, with a resulting confidence coefficient of 0.937. However, this also is only approximate (see Section 5).

2.3 Inferences Using t Approximations

If $F_4(\hat{\phi})$ is not so small then a t distribution may be used for inferences about ϕ. Suppose that for every $\mathbf{y} \in S$, $F_3(\hat{\phi})$ is small and $0 \le F_4(\hat{\phi}) < 6$, and let

$$T = q^{\frac{1}{2}} U_\phi, \qquad q = n/(n+1), \qquad n = \left[6/F_4(\hat{\phi})\right] - 1. \tag{2.8}$$

Sprott (1982) showed that the Taylor series of $r_M(\phi; \mathbf{y})$ around $\hat{\phi}$ agrees up to the fourth power with that of the $t_{(n)}$ likelihood centered at $\hat{\phi}$ and scaled by $[qI_\phi]^{-\frac{1}{2}}$, that is, $R_M(\phi;\mathbf{y}) \doteq \left\{1 + t^2/n\right\}^{-(n+1)/2}$, where $t = q^{\frac{1}{2}} u_\phi = [qI_\phi]^{\frac{1}{2}}(\phi - \hat{\phi})$ of (2.8) (see Viveros, 1985, for details). Equivalently, the quantity t has an approximate standard $t_{(n)}$ likelihood for every $\mathbf{y} \in S$. The inferences about ϕ (and hence about θ) can be obtained by treating T as an approximate standard $t_{(n)}$ variate.

Example 2. In Example 1, $F_3(\hat{\Psi}) = 0.0507$ but $F_4(\hat{\Psi}) = 0.3034$ is not very small. From (2.8) and the numerical results presented in Example 1, $n \doteq 19$ and the resulting summary of the data is

$$\Psi = 0.1225 \pm 0.2045 t_{(19)}, \quad \theta = 0.4533 \tan\left[0.1225 \pm 0.2045 t_{(19)}\right] + 1.9945.$$

The approximate 95% confidence interval obtained by setting $t_{(19)} = 2.093$ is $\Psi = (-0.3056,\ 0.5506)$, $\theta = (1.851,\ 2.273)$, which is essentially the same as that obtained in Example 1.

2.4 Inferences Using Log F Approximations

Cases for which $F_3(\hat{\phi})$ is not so small are indicative of residual skewness in the likelihood of ϕ. These may be handled by use of the skew log F distribution. Suppose $F_3^2(\hat{\phi}) + F_4(\hat{\phi}) \geq 0$, and let

$$\begin{array}{rclrcl} H & = & q^{\frac{1}{2}} U_\phi, & q & = & 2/m + 2/n, \\ m & = & 4/c\left[c - F_3(\hat{\phi})\right], & n & = & 4/c\left[c + F_3(\hat{\phi})\right], \end{array} \tag{2.9}$$

where $c = [3F_3^2(\hat{\phi}) + 2F_4(\hat{\phi})]^{\frac{1}{2}}$. Viveros (1985) showed that the Taylor series of $r_M(\phi; \mathbf{y})$ about $\hat{\phi}$ agrees up to the fourth power with that of the logarithm of the log $F_{(m,\ n)}$ likelihood centered at $\hat{\phi}$ and scaled by $[qI_\phi]^{-\frac{1}{2}}$, that is, $R_M(\hat{\phi}; \mathbf{y}) \doteq \exp(mh/2)\ \{[1 + m/n] \,/\, [1 + (m/n)\exp(h)]\}^{(m+n)/2}$ where $h = q^{\frac{1}{2}} u_\phi = [qI_\phi]^{\frac{1}{2}}(\phi - \hat{\phi})$ as in (2.9). Equivalently, the quantity h has an approximate standard log $F_{(m,\ n)}$ likelihood for every $\mathbf{y} \in S$. This suggests treating H as a standard log $F_{(m,\ n)}$ variate in obtaining approximate inferences about ϕ.

The form taken by the inferences will be

$$\begin{aligned} &\phi = \hat{\phi} + [qI_\phi]^{-\frac{1}{2}}\,(h_1, h_2)\,, \\ &\phi = c_1 \tan\left\{\hat{\phi} + [qI_\phi]^{-\frac{1}{2}}\,(h_1, h_2)\right\} + c_2, \end{aligned} \tag{2.10}$$

where c_1 and c_2 are as in (2.7), and h_1 and h_2 usually selected according to the criterion of equal probability tails or of highest probability density for the log F distribution. The latter will be preferable since it automatically assures the approximate likelihood property of the resulting confidence intervals, that is, values of θ inside the interval will have likelihoods greater than or equal to those of values outside the interval. These intervals will be referred to as likelihood confidence intervals. Notice that the use of a symmetric model, like normal or t, and the form of statement adopted in the previous sections gives rise to approximate likelihood confidence intervals.

The use of the log F approximation will be illustrated with the estimation of the LD90 dose in Section 3.

2.5 Remarks

As seen in the examples presented, the confidence coefficients assigned by the Fieller method to the intervals obtained by the procedure proposed here tend to be smaller than the stated ones. However, the Fieller method is also an approximation and, in fact, none of the assumptions required by that procedure is satisfied in this case. Moreover, Jarret *et al.* (1984) show that the Fieller method tends to produce wide confidence intervals in the sense that the resulting *exact* coverage frequency is larger than the stated one. These facts together with the conformity of the intervals obtained in reflecting the likelihood arising from the exact binomial logistic model is a good indication that the likelihood-based procedure proposed here may be more accurate. A more thorough discussion is given in Section 5.

3. ESTIMATION OF THE LD90 DOSE

The analysis in Section 2 can easily be extended to the estimation of the LD90 dose. The approximate Fieller pivotal (1.7) becomes

$$Z = \left(\hat{b}_2\theta + \hat{b}_1 - t_0\right) / \left(\theta^2 v_{22} + 2\theta v_{12} + v_{11}\right)^{\frac{1}{2}}, \tag{3.1}$$

where θ now denotes the LD90 dose and $t_0 = 2.1972$; the transformations (2.1) and (2.6) remain the same, where V and W are as defined in Section 2. As before, the maximum likelihood estimate $\hat{\phi}$ as well as the observed Fisher information I_ϕ and the standardized coefficients $F_3(\hat{\phi})$ and $F_4(\hat{\phi})$ can be calculated numerically; the normal, t or log F approximations can be used in the same way as in Section 2. The corresponding calculations can be performed for parameter Ψ. Consider the following example.

Example 3. The following data arose from a medical experiment in which doses $\mathbf{d} = (2.0, 2.5, 3.0, 3.5, 4.0, 4.5, 5.0, 5.5, 6.0)$ mg/kg of the anaesthetic thiopentone were administered to 137 children allocated in $A = 9$ groups of sizes $\mathbf{k} = (15, 21, 20, 14, 12, 8, 11, 22, 14)$ producing the respective numbers of responses $\mathbf{y} = (7, 14, 15, 13, 11, 5, 10, 22, 13)$. Each of these children was given a standard (fixed) dose of the drug TDP as premedication one and a half hours earlier, and the child was classified as responding if 20 seconds after the administration of thiopentone the eyelash reflex was abolished. These data were supplied by Brian Newman and are discussed, assuming a probit model, by Duncan *et al.* (1984). For more details see their paper.

Fitting the binomial logistic model (1.1), (1.2), with $x_i = \log d_i$, the following results were obtained: $\hat{b}_1 = -1.92005$, $\hat{b}_2 = 2.78041$, so that $\hat{\theta} = 1.48082$ $[\exp(\hat{\theta}) = 4.3965]$. The calculation of V and W yields

$$\phi = \tan^{-1}(3.0956\,\theta - 3.4135), \quad \Psi = \tan^{-1}(2.8585\,\theta - 3.7977),$$

from which

$$\hat{\phi} = 0.8638, \qquad \hat{\Psi} = 0.4105, \qquad I_\phi = 35.7784, \qquad I_\Psi = 10.5663.$$

The standardized coefficients F_3 and F_4 are

	$\hat{\theta}$	$\hat{\phi}$	$\hat{\Psi}$
F_3	0.8553	−0.3188	0.0520
F_4	−0.45998	−0.0536	−0.4882

Consider the transformation ϕ. The application of (2.9) yields an approximating log $F_{(m,\ n)}$ likelihood, centered at $\hat{\phi}$ and scaled by $[qI_{\phi}]^{-\frac{1}{2}}$, to the maximized likelihood of ϕ with $m \doteq 12$ and $n \doteq 72$ df, and $q = (2/m + 2/n) = 0.1944$. A plot of $R_M(\phi; \mathbf{y})$ and of the log $F_{(m,\ n)}$, likelihood along with the normal likelihood centered at $\hat{\phi}$ and scaled by $I_{\phi}^{-\frac{1}{2}}$ is shown in Figure 1. The log F approximation appears satisfactory while some residual skewness in the maximized likelihood of ϕ makes the normal approximation less appropriate.

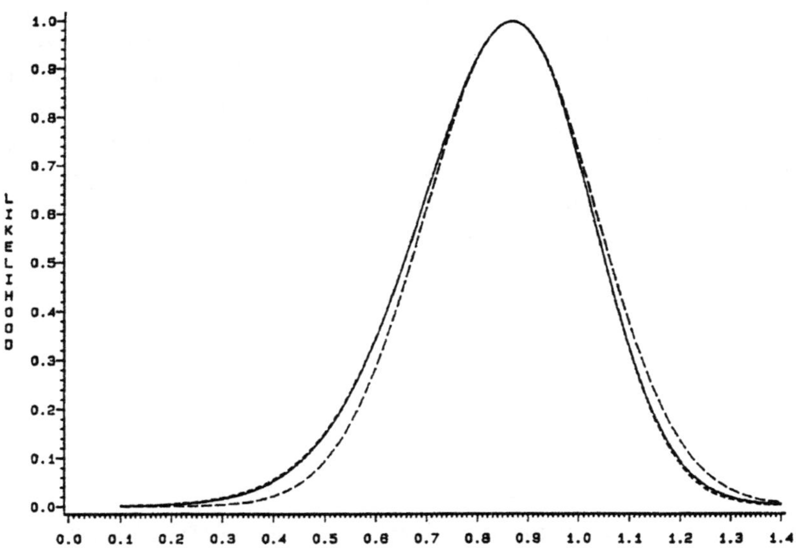

Figure 1. *Relative maximized likelihood $R_M(\phi; \mathbf{y})$ (solid line); log F likelihood (small dashes); normal likelihood (large dashes).*

The approximate inferences take the form

$$\phi = 0.8638 + 0.3791\,(h_1, h_2)\,,$$
$$\theta = 0.32304\ \tan\left[0.8638 + 0.3791\,(h_1, h_2)\right] + 1.1027,$$

where h_1 and h_2 are values of a standard log $F_{(m,\ n)}$ variate required to produce the desired confidence level. For instance, the values $(h_1, h_2) = (-0.9859,\ 0.7998)$ lead to the approximate 95% likelihood confidence interval $\phi = (0.4900,\ 1.1670)$ which in terms of θ is $\theta = (1.2750, 1.8588)$ [$\exp(\theta) = (3.5787,\ 6.4160)$] (see end of Section 2.4). It is interesting to note that the substitution of these values in the (approximate) Fieller pivotal (3.1) gives $Z = (-2.1842, 1.7861)$, yielding a confidence coefficient of 94.85% under the assumed asymptotic normality of Z. Although the confidence coefficient is very close to the stated 95%, the use of the Fieller pivotal written in terms of ϕ and its asymptotic normal distribution will produce a symmetric ranking in ϕ and hence will not reflect the skewness in the observed maximized likelihood.

4. ESTIMATION OF THE RELATIVE POTENCY

4.1 Fieller Method

The estimation of the relative potency, although more complicated, introduces nothing essentially new. Suppose the numbers of responses $\mathbf{y}_1 = (y_{11}, \ldots, y_{A1})$ are observed on A groups of respective sizes $\mathbf{k}_1 = (k_{11}, \ldots, k_{A1})$ to which doses $\mathbf{x}_1 = (x_{11}, \ldots, x_{A1})$ of the standard preparation are administered. Suppose, on the other hand, that when doses $\mathbf{x}_2 = (x_{12}, \ldots, x_{B2})$ of the test preparation are administered to B groups of sizes $\mathbf{k}_2 = (k_{12}, \ldots, k_{B2})$ the numbers of responses $\mathbf{y}_2 = (y_{12}, \ldots, y_{B2})$ are observed. The joint likelihood $L(b_1, b_2, b_3) = L(b_1, b_2, b_3; \mathbf{y}_1, \mathbf{y}_2)$ of (b_1, b_2, b_3) arising from these data is

$$\log L(b_1, b_2, b_3) = \sum \{y_{i1} \log p_{i1} + (k_{i1} - y_{i1}) \log(1 - p_{i1})\} + \sum \{y_{j2} \log p_{j2} + (k_{j2} - y_{j2}) \log(1 - p_{j2})\}, \quad (4.1)$$

where

$$\log[p_{i1}/(1 - p_{i1})] = b_1 + b_2 x_{i1},$$
$$\log[p_{j2}/(1 - p_{j2})] = b_3 + b_2 x_{j2}. \quad (4.2)$$

A common way of estimating the relative potency (Finney, 1971, 1978) is again by reference to the Fieller pivotal and its asymptotic normal distribution arising from the asymptotic distribution of $(\hat{b}_1, \hat{b}_2, \hat{b}_3)$, the ML estimate of (b_1, b_2, b_3) calculated from (4.1). As is well known, $(\hat{b}_1, \hat{b}_2, \hat{b}_3)$ has an asymptotic multivariate normal distribution about (b_1, b_2, b_3) with covariance matrix usually estimated by $V = I^{-1}$ where I is the observed Fisher information matrix with components

$$I_{ij} = -\partial^2 \log L(b_1, b_2, b_3) / \partial \hat{b}_i \partial \hat{b}_j, \quad (i, j = 1, 2, 3). \quad (4.3)$$

Noting that γ is defined by the equation $b_1+\gamma b_2-b_3=0$, then proceeding as in Section 1.2 the Fieller pivotal for γ is

$$Z=\left(\hat{b}_1+\gamma\hat{b}_2-\hat{b}_3\right)/\left(\gamma^2A_2+2\gamma A_1+A_0\right)^{\frac{1}{2}}, \tag{4.4}$$

where $A_2=v_{22}$, $A_1=v_{12}-v_{23}$ and $A_0=v_{11}-2v_{13}+v_{33}$, with asymptotic standard normal distribution.

4.2 Transformations

The transformation that symmetrizes the maximized likelihood of γ arising from the asymptotic multivariate normality of $(\hat{b}_1,\hat{b}_2,\hat{b}_3)$ is

$$\phi=\tan^{-1}\left[(A_2\gamma+A_1)/\left(A_0A_2-A_1^2\right)^{\frac{1}{2}}\right] \tag{4.5}$$

with inverse

$$\gamma=\left[\left(A_0A_2-A_1^2\right)^{\frac{1}{2}}/A_2\right]\tan\ \phi-[A_1/A_2]. \tag{4.6}$$

The Fieller pivotal in terms of ϕ is

$$Z=I_\phi^{\frac{1}{2}}\ \sin(\phi-\hat{\phi})\text{ where }I_\phi=\hat{b}_2^2/A_2\ \cos^2(\hat{\phi}). \tag{4.7}$$

As in Section 2 the functional specification of ϕ involves random quantities. Although this is not of great concern, an alternative parameter that avoids this feature is

$$\Psi=\tan^{-1}\left[(B_2\gamma+B_1)/\left(B_0B_2-B_1^2\right)^{\frac{1}{2}}\right], \tag{4.8}$$

where the B_i's are as the A_i's but calculated from the components of $W=(X'X)^{-1}$ where

$$X'=\begin{pmatrix}1 & ,\dots, & 1, & 0 & ,\dots, & 0\\ x_{11} & ,\dots, & x_{A1}, & x_{12} & ,\dots, & x_{B2}\\ 0 & ,\dots, & 0, & 1 & ,\dots, & 1\end{pmatrix}.$$

Although the parameter Ψ does not symmetrize the *asymptotic* maximized likelihood of γ, it can be as effective as ϕ in reducing the skewness of the exact maximized likelihood of γ, arising from the exact linear logistic model. This is discussed below.

4.3 Maximized Likelihood and Inferences Using Normal, t or Log F Approximations

Writing b_3 in terms of γ, b_1, b_2 as $b_3 = b_1 + \gamma b_2$ and substituting in (4.2) gives

$$\log\left[p_{j2}/\left(1-p_{j2}\right)\right] = b_1 + b_2\left(x_{j2}+\gamma\right). \tag{4.9}$$

Then the joint likelihood of (γ, b_1, b_2) is proportional to (4.1) with p_{i1} obtained from (4.2) and p_{j2} from (4.9). The relative maximized likelihood of γ is then

$$R_M(\gamma;\mathbf{y}) = L\left(\gamma,\tilde{b}_1,\tilde{b}_2\right)/L\left(\hat{\gamma},\hat{b}_1,\hat{b}_2\right), \tag{4.10}$$

where $\tilde{b}_i = \tilde{b}_i(\gamma)$ is the ML estimate of b_i for γ specified, $i = 1,\ 2$.

Although (4.5) will not eliminate the skewness in $R_M(\phi;\ \mathbf{y})$ it will reduce it considerably. The information I_ϕ based on $R_M(\phi;\ \mathbf{y})$ and the standardized derivatives $F_3(\hat{\phi})$ and $F_4(\hat{\phi})$ can be calculated numerically for every $(\mathbf{y}_1,\ \mathbf{y}_2) \in S$ as before. Then the normal, t, or log F approximations to $R_M(\phi;\ \mathbf{y})$ can be applied as in Section 2, leading to approximate inferences about ϕ which, as usual, can immediately be transformed into inferences about γ by use of (4.6).

Example 4. Table 1 presents data given by Finney (1971, p.108) arising from laboratory tests run on mice involving the drugs Morphine (standard preparation) and Phenadoxone (test preparation). These data are also discussed by Kalbfleish (1983, p. 42).

Table 1. *Data from Finney (1971)*

Standard preparation				Test preparation			
Morphine				Phenadoxone			
$\mathbf{x}_1$:	0.18	0.48	0.78	$\mathbf{x}_2$:	−0.12	0.18	0.48
$\mathbf{k}_1$:	103	120	123	$\mathbf{k}_2$:	90	80	90
$\mathbf{y}_1$:	19	53	83	$\mathbf{y}_2$:	31	54	80

The maximization of (4.1) gives $\hat{b}_1 = -2.2583$, $\hat{b}_2 = 4.0102$ and $\hat{b}_3 = -0.0359$, so that $\gamma = 0.5542$. The calculation of V and W leads to the transformations

$$\phi = \tan^{-1}(2.1714\,\gamma - 0.8933),\quad \Psi = \tan^{-1}(2.0412\,\gamma - 0.6124),$$

from which

$$\hat{\phi} = 0.3007, \quad \hat{\Psi} = 0.4786, \quad I_{\phi} = 107.8179, \quad I_{\Psi} = 167.5808.$$

The values taken by the standardized coefficients F_3 and F_4 are

	$\hat{\gamma}$	$\hat{\phi}$	$\hat{\Psi}$
F_3	0.1638	−0.0153	−0.0798
F_4	0.0625	0.0149	0.0341

from which the use of a normal approximation to $R_M(\phi;\ \mathbf{y})$ seems appropriate as indicated by the small values of $F_3(\hat{\phi})$ and $F_4(\hat{\phi})$. The form taken by the resulting inferences is

$$\phi = 0.3007 \pm 0.0963u, \quad \gamma = 0.4608\ \tan(0.3007 \pm 0.0963u) + 0.4114,$$

where u is the value of a standard normally distributed variate required to produce the desired confidence coefficient approximately. For instance, $u = 1.96$ gives the approximate 95% confidence interval $\phi = (0.1119, 0.4894)$, $\gamma = (0.46297, 0.6568)$. Inserting these values in the Fieller pivotal (4.7) gives $Z = (-1.9485, 1.9481)$, producing a confidence coefficient of 94.86% under the asymptotic standard normal distribution. The large size of the experimental groups makes the two methods agree in this case and both are probably very accurate.

The use of Ψ and a normal approximation to $R_M(\Psi;\ \mathbf{y})$ gives approximately the same results.

5. ACCURACY OF THE PROCEDURE PROPOSED

The calculation of the exact coverage frequency associated with the approximate confidence intervals for the LD50 dose obtained by the procedure proposed here, or in fact by any other procedure, is difficult when four or more doses are used in the experiment. Simulations of the respective binomial distributions, however, can be performed to check in a limited way the accuracy of the foregoing procedures.

Consider, for instance, the case $A = 4$ groups with $\mathbf{x} = (-3, -1, 1, 3)$, $\mathbf{k} = (10, 10, 10, 10)$. Take for the binomial logistic coefficients the values $b_1 = -0.5493$, $b_2 = 0.5493$, the ML estimates arising from the observations $\mathbf{y} = (1, 1, 8, 6)$ (see Kalbfleisch, 1983, p. 38), so that the LD50 dose is $\theta = 1$.

Thus the respective success probabilities for the four groups calculated from (1.6) are $\mathbf{p} = (1/10, 1/4, 1/2, 3/4)$.

The results based on 2,000 independent samples generated on the computer using the IMSL subroutine GGBN are presented in Table 2. For each sample the approximate 90%, 95% and 99% confidence intervals for θ were computed using the Fieller method, and using parameters ϕ and Ψ and the method proposed here. Then it was determined whether or not the value $\theta = 1$ lay inside the interval, and the overall percentage of times in which $\theta = 1$ lay in the interval was recorded for each of the three confidence coefficients.

Table 2

Nominal Coefficient	Fieller Method	Use of ϕ	Use of Ψ
90%	90.79%	88.94%	90.29%
95%	96.10%	94.24%	95.30%
99%	99.65%	98.95%	99.20

These results reinforce the above mentioned feature that the Fieller method tends to underestimate the precision of the inferences. On the basis of Table 2, the method presented here appears to overestimate it somewhat. However, the inferences based on parameters ϕ and Ψ show a coverage frequency globally somewhat closer to the stated one than the coverage frequency associated with the Fieller method. In fact, the application of the likelihood ratio test for the testing of

$$H_0 : \text{ coverage frequency} = 0.90, 0.95, 0.99,$$

based on $n = 2,000$ trials and the observed frequencies shown in Table 1 indicates that the observed coverage frequencies arising from the use of ϕ and Ψ do not differ significantly from 0.95 and 0.99 at the 5% level, while the Fieller coverage frequency does. The approximate p-values associated with the observed Fieller coverage frequencies calculated from the likelihood ratio test and its approximate chi-squared distribution with 1 degree of freedom are $p = 0.019$ and 0.0008 for the second and third rows respectively; the corresponding approximate p-values for the observed coverage frequencies based on ϕ and on Ψ are $p = 0.127$, and 0.534 for the second row, and $p = 0.83$ and 0.35 for the third row. The approximate p-values arising from testing the first row are $p = 0.23$, 0.12, and 0.65, so that none of the three methods produces a coverage frequency differing significantly from 0.90.

Thus, although all of the coverage frequencies may appear close to the required values, the Fieller frequencies differ significantly from these at the higher levels of .95 and .99. Further, it should be again emphasized that the intervals based on ϕ and on Ψ take account of the asymmetry of the observed likelihoods of ϕ and of Ψ while the Fieller intervals do not.

6. DISCUSSION

The analysis in this paper, like that by Sprott and Viveros (1984, 1985), is based on transformations of the parameter of interest for which normal, t or log F approximations to the observed relevant likelihood are facilitated. When normal or t approximations are appropriate, the result is a complete summary of the evidence about ϕ in the data in the simple form $\phi = \hat{\phi} \pm st$, where $\hat{\phi}$ is the maximum likelihood estimate of ϕ, s is related to its precision, and t is the value of a standard normal or standard t variate required to produce the confidence level approximately. When this is not possible, because of residual skewness in the likelihood, the use of $\phi = \hat{\phi} + s(h_1, h_2)$ is illustrated, where h_1 and h_2 are appropriate values of a standard log F variate. These statements are transformed into corresponding statements about the parameter of interest by straight algebraic substitution.

This form of statement is put forward by Barnard (1985) as the object of estimation in preference to the widespread notion of estimation as an attempt to seek single numbers which are "closest" in some sense to the true value of the parameter. Because of the conformity of the approximate pivotal used here with the relevant likelihood, the resulting statements will reflect the shape of the observed likelihood function. Thus the numerical accuracy is combined with the likelihood property in the resulting inferences.

More remains to be investigated. In particular, the determination of the exact coverage frequency produced by the estimation procedure presented is of interest, but due to the amount of programming involved it has been left for further consideration.

REFERENCES

Barnard, G. A. (1985), "A coherent view of statistical inference." Technical Report, Department of Statistics and Actuarial Science, University of Waterloo.

Duncan, B. B. A., F. Zaimi, G. B. Newman, J. G. Jenkins, and W. Aveling (1984), "Effect of premedication on the induction dose of thiopentone in children." *Anaesthesia* **39**, 426–428.

Edwards, A. W. F. (1972), *Likelihood*. Cambridge: Cambridge University Press.

Finney, D. J. (1971), *Probit Analysis*, 3rd edition. Cambridge: Cambridge University Press.

Finney, D. J. (1978), *Statistical Method in Biological Assay*. London: Griffin.

Jarret, R., I. Saunders, and R. Wong (1984), "Optimal estimation of the LD50 in Bioassay." CSIRO-DMS, Newsletter 100, 2–5.

Kalbfleisch, J. G. (1983), *Notes on Discrete Data Analysis*. Department of Statistics and Actuarial Science, University of Waterloo.

Sprott, D. A. (1973), "Normal likelihoods and their relation to large sample theory of estimation." *Biometrika* **60**, 457–465.

Sprott, D. A. (1980), "Maximum likelihood in small samples: estimation in the presence of nuisance parameters." *Biometrika* **67**, 515–523.

Sprott, D. A. (1982), "Robustness and maximum likelihood estimation." *Communications in Statistics A, Theory and Methods* **11**, 2513–2529.

Sprott, D. A., and R. Viveros (1984), "The interpretation of maximum likelihood estimation." *Canadian Journal of Statistics* **12**, 27–38.

Sprott, D. A., and R. Viveros (1985), "The estimation of ratios and related quantities." In *Contributions to Econometrics and Statistics Today. In Memoriam Gunter Menges*, ed. H. Schneeweiss and H. Strecker, pp. 242–251. Berlin: Springer-Verlag.

Viveros, R. (1985), "Estimation in small samples." Ph.D. Thesis, University of Waterloo.

Welch, B. L., and H. W. Peers (1963), "On formulae for confidence points based on integrals of weighted likelihoods." *Journal of the Royal Statistical Society*, Series B **25**, 318–329.

M. N. Walsh, J. J. Hubert and E. M. Carter [1]

ESTIMATION METHODS FOR SYMMETRIC PARABOLIC BIOASSAYS

ABSTRACT

Two univariate methods of estimation in parabolic bioassays are reviewed, and a new test for the assumption of "parallelism" is proposed. One method is extended into a multivariate situation based on a general linear model which generates a point estimator and approximate confidence limits for the associated relative potency.

1. INTRODUCTION

With linear response curves, parallel-line assays can be routinely analyzed to estimate μ, the log relative potency of a test preparation and a standard preparation (see Finney, 1978 or Hubert, 1984). Even in a multivariate situation, an explicit estimator of μ exists when the response curves are linear (Carter and Hubert, 1985).

However, occasionally significant curvature persists in a parallel-line assay, even after restricting the dosage range and making suitable transformations (Bliss, 1957; Elston, 1965). Often, a parabolic (quadratic) model will adequately fit the data over a limited dosage range (Bliss, 1957; Elston, 1965).

In this paper we concentrate only on the symmetric situation of parabolic response curves (for the general situation see Carter *et al.*, 1986). Note that these are dilution assays, implying that both parabolas are identical except that one is merely shifted horizontally with respect to the other. Therefore, the horizontal distance between the two parabolas is constant. It is this constant horizontal distance which is μ, the parameter of most interest.

First, the estimation of μ by a weighted mean in the univariate case

[1] Department of Mathematics and Statistics, University of Guelph, Guelph, Ontario N1G 2W1 (all authors)

I. B. MacNeill and G. J. Umphrey (eds.), Biostatistics, 251–258.

(Elston, 1965) will be presented and then expanded to the multivariate situation. Model testing procedures will also be suggested. Next, estimation of μ by the method of maximum likelihood in the univariate case (Williams, 1972) will be given, along with a test for the horizontal "parallelism" of the parabolas.

2. THE UNIVARIATE PARABOLIC MODEL

Let y denote the response variable and x (= log dose), the dose metameter scaled such that $\sum x_i = 0$; let α, β and γ denote the model parameters in the assumed quadratic model for the standard preparation

$$y_s = \alpha_s + \beta_s x + \gamma_s x^2 + \varepsilon_s \tag{2.1}$$

and for the test preparation

$$\begin{aligned} y_t &= \alpha_s + \beta_s(x+\mu) + \gamma_s(x+\mu)^2 + \varepsilon_t \\ &= \alpha_t + \beta_t x + \gamma_t x^2 + \varepsilon_t, \end{aligned} \tag{2.2}$$

where:

$$\begin{aligned} \gamma_t &= \gamma_s = \gamma, \\ \beta_t &= \beta_s + 2\mu\gamma, \end{aligned} \tag{2.3}$$

$$\alpha_t = \alpha_s + \mu\beta_s + \mu^2\gamma, \tag{2.4}$$

and the ε's are $N(0, \sigma^2)$ and are the error components.

The equality of γ_t and γ_s can easily be tested using regular regression techniques.

With the reparameterizations

$$\begin{gathered} \alpha = \alpha_s + \alpha_t, \quad \delta_1 = \alpha_t - \alpha_s, \quad \beta_1 = (\beta_s + \beta_t)/2, \\ \delta_2 = (\beta_t - \beta_s)/2, \quad \beta_2 = \gamma, \end{gathered}$$

the restrictions (2.3) and (2.4) can be written as

$$\delta_1 = \mu\beta_1 \quad \text{and} \quad \delta_2 = \mu\beta_2$$

and a general linear model in matrix notation will be

$$\mathbf{Y} = \mathbf{\Psi X} + \boldsymbol{\varepsilon}, \tag{2.5}$$

where:

$$\mathbf{Y} = \left[y_{s1}\ y_{s2}\ \cdots\ y_{sn}\ y_{t1}\ y_{t2}\ \cdots\ y_{tn}\right],$$

$$\mathbf{\Psi} = \left[\alpha\ \delta_1\ \beta_1\ \delta_2\ \beta_2\right],$$

$$\mathbf{X} = \begin{bmatrix} 0.5 & 0.5 & \cdots & 0.5 & 0.5 & 0.5 & \cdots & 0.5 \\ -0.5 & 0.5 & \cdots & -0.5 & 0.5 & 0.5 & \cdots & 0.5 \\ x_1 & x_2 & \cdots & x_n & x_1 & x_2 & \cdots & x_n \\ -x_1 & -x_2 & \cdots & -x_n & x_1 & x_2 & \cdots & x_n \\ x_1^2 & x_2^2 & \cdots & x_n^2 & x_1^2 & x_2^2 & \cdots & x_n^2 \end{bmatrix},$$

$\boldsymbol{\varepsilon}$ is a row vector of error terms which are $N(0, \sigma^2)$.

3. ESTIMATION BY WEIGHTED MEAN (ELSTON, 1965)

For symmetric parabolic assays, Elston (1965) suggested the following weighted mean estimator for the log relative potency parameter μ. An estimator of $\mathbf{\Psi}$ is

$$\hat{\mathbf{\Psi}} = \mathbf{YX}'(\mathbf{XX}')^{-1} = \left[\hat{\alpha}\ d_1\ b_1\ d_2\ b_2\ \right] \tag{3.1}$$

and two independent estimators of μ are

$$m_1 = d_1/b_1 \quad \text{and} \quad m_2 = d_2/b_2. \tag{3.2}$$

Taking the inverses of the asymptotic variances of these estimators, the following weights are obtained:

$$w_1 = b_1^2/\left(a_1 + m_1^2 a_2\right),$$
$$w_2 = b_2^2/\left(a_3 + m_2^2 a_4\right),$$

where the a's are elements from $\mathbf{A} = \text{diag}(a_1\ a_2\ a_3\ a_4)$ which is the lower 4×4 matrix of $(\mathbf{X}\ \mathbf{X}')^{-1}$. Therefore, the weighted mean of these two estimators is

$$m = (w_1 m_1 + w_2 m_2)/(w_1 + w_2). \tag{3.3}$$

The weights should be recalculated using m in place of m_1 and m_2 and then iteratively calculate m and the weights until stability is reached. The variance of m is estimated by

$$\hat{V}(m) = V/f(w_1 + w_2), \tag{3.4}$$

where

$$V = \mathbf{YY}' - \mathbf{YX}'(\mathbf{XX}')^{-1}\mathbf{XY}'$$

and f is the degrees of freedom for error.

An asymptotic test for the validity of the two restrictions (2.3) and (2.4) is simply a two-sample z-test for the significance of the difference between m_1 and m_2 since the asymptotic distribution of these statistics is normal.

4. A MULTIVARIATE MODEL AND MODEL TESTING

Following the procedure of Carter and Hubert (1985), the estimators m_1 and m_2 can be cast into a multivariate setting, where p characteristics are measured for each experimental unit.

As in the univariate case, the equality of the quadratic regression parameters must be tested. This is done by using Hotelling's T^2 procedure (see Srivastava and Carter, 1983).

Now the general linear model is

$$\mathbf{Y} = \mathbf{\Psi X} + \boldsymbol{\varepsilon}, \tag{4.1}$$

where:

$\mathbf{Y} = (\mathbf{y}_{s1}\ \mathbf{y}_{s2}\ \cdots\ \mathbf{y}_{sn}\ \mathbf{y}_{t1}\ \mathbf{y}_{t2}\ \cdots\ \mathbf{y}_{tn})$ where each $\mathbf{y}$ is a p vector of responses for the p characteristics,

$\mathbf{\Psi} = [\boldsymbol{\alpha}\ \boldsymbol{\delta}_1\ \boldsymbol{\beta}_1\ \boldsymbol{\delta}_2\ \boldsymbol{\beta}_2]$,

$\mathbf{X}$ is defined in (2.5),

$\boldsymbol{\varepsilon}$ is a matrix of error terms which are $N_p(\mathbf{0}, \mathbf{\Sigma})$.

To test if μ is constant over the p characteristics (i.e., $\boldsymbol{\delta}_1 = \mu\boldsymbol{\beta}_1$ and $\boldsymbol{\delta}_2 = \mu\boldsymbol{\beta}_2$), first calculate the 4×4 matrix

$$\mathbf{G} = \mathbf{T}'\hat{\mathbf{\Theta}}'(f\mathbf{S})^{-1}\hat{\mathbf{\Theta}}\mathbf{T},$$

where

$$\hat{\mathbf{\Theta}} = (\hat{\boldsymbol{\delta}}_1, \hat{\boldsymbol{\beta}}_1, \hat{\boldsymbol{\delta}}_2, \hat{\boldsymbol{\beta}}_2)$$

and

$$\mathbf{T} = \operatorname{diag}[t_1\ t_2\ t_3\ t_4] = \operatorname{diag}\left[a_1^{-1/2}\ a_2^{-1/2}\ a_3^{-1/2}\ a_4^{-1/2}\right].$$

Then from Chou and Muirhead (1979), reject the null hypothesis $H : \boldsymbol{\delta}_1 = \mu\boldsymbol{\beta}_1$ if

$$[f - (p-1)/2 + \ell_1^{-1}]\,[\ln(1 + \ell_2)] > \chi^2_{p-1;\alpha},$$

where $\ell_1 > \ell_2$ are the eigenvalues of the 2×2 upper left matrix of G. Similarly, we reject the null hypothesis $H : \boldsymbol{\delta}_2 = \mu \boldsymbol{\beta}_2$ if

$$\left[f - (p-1)/2 + \ell_3^{-1}\right]\left[\ln(1+\ell_4)\right] > \chi^2_{p-1;\alpha}, \tag{4.2}$$

where $\ell_3 > \ell_4$ are the eigenvalues of the 2×2 lower right matrix of G. If μ is found not to be constant over the p characteristics then univariate procedures need to be done for each characteristic. However, if μ is constant over the p characteristics then it can be estimated as follows.

5. MULTIVARIATE ESTIMATION BY WEIGHTED MEAN

From Kraft *et al.* (1972), two explicit estimators of μ are

$$m_1 = (t_2/t_1)\left\{g_{11} - g_{22} + \left[(g_{11} - g_{22})^2 + 4g_{12}^2\right]^{1/2}\right\}/(2g_{12})$$

and

$$m_2 = (t_4/t_3)\left\{g_{33} - g_{44} + \left[(g_{33} - g_{44})^2 + 4g_{34}^2\right]^{1/2}\right\}/(2g_{34}).$$

Using the inverses of the asymptotic variances of these two estimators the corresponding weights are

$$w_1 = fg_{22}/\left(m_1^2 + a_1/a_2\right)$$

and

$$w_2 = fg_{44}/\left(m_2^2 + a_3/a_4\right)$$

and an overall estimator of μ is given by the weighted mean

$$m = (w_1 m_1 + w_2 m_2)/(w_1 + w_2). \tag{5.1}$$

The weights and m are iteratively calculated as in Section 3 until stability is reached. An approximate standard error of m is given by

$$\mathrm{SE}(m) = (w_1 + w_2)^{-1/2},$$

and an asymptotic confidence interval for μ is

$$m \mp z_{\alpha/2}\mathrm{SE}(m),$$

where $z_{\alpha/2}$ is the $(1-\alpha/2) \times 100$ percentile of a standard normal distribution.

6. ESTIMATION BY THE METHOD OF MAXIMUM LIKELIHOOD (UNIVARIATE)

The parameter vector $\mathbf{\Psi}$ can be rewritten as follows:

$$\mathbf{\Psi} = \mathbf{\Gamma U},$$

where

$$\mathbf{\Gamma} = [\alpha\ \beta_1\ \beta_2]$$

and

$$\mathbf{U} = \begin{bmatrix} 1 & 0 & 0 & 0 & 0 \\ 0 & \mu & 1 & 0 & 0 \\ 0 & 0 & 0 & \mu & 1 \end{bmatrix}.$$

Therefore the likelihood function is

$$L = \left(2\pi\sigma^2\right)^{-N/2} \exp\left\{-\left(1/2\sigma^2\right)\left[V + (\hat{\mathbf{\Psi}} - \mathbf{\Gamma U})\mathbf{A}^{-1}(\hat{\mathbf{\Psi}} - \mathbf{\Gamma U})'\right]\right\},$$

where $N = 2n$ and $\mathbf{A} = (\mathbf{XX}')^{-1}$. If we let

$$\mathbf{Q}' = (\mathbf{A}^{-1}\ \mathbf{B}'\ \hat{\mathbf{\Gamma}}')' = [q_1\ q_2\ q_3\ q_4\ q_5],$$

where $\mathbf{B}$ is the derivative of $\mathbf{U}$ with respect to μ and is explicitly

$$\mathbf{B} = \begin{bmatrix} 0 & 0 & 0 & 0 & 0 \\ 0 & 1 & 0 & 0 & 0 \\ 0 & 0 & 0 & 1 & 0 \end{bmatrix},$$

then the maximum likelihood estimators of $\mathbf{\Gamma}$ and μ are

$$\hat{\mathbf{\Gamma}} = \hat{\mathbf{\Psi}} A^{-1}\hat{U}'(\hat{U}A^{-1}\hat{U}')^{-1}$$

and

$$m = [\hat{\mathbf{\Psi}}\mathbf{Q} - (\hat{\alpha}q_1 + \hat{\beta}_1 q_3 + \hat{\beta}_2 q_5)]/(\hat{\beta}_1 q_2 + \hat{\beta}_2 q_4),$$

respectively. These solutions are iterative requiring an initial estimate of μ, which could be $m_1 = d_1/b_1$, given in (3.1).

7. LIKELIHOOD RATIO TEST FOR "PARALLELISM"

The models being used here assume that the two parabolas are equivalent except for a horizontal shift of one with respect to the other (i.e., a dilution assay). Therefore the horizontal distance between the two parabolas is constant. This will be the definition of "parallelism". The assumption of "parallelism" can be tested using a likelihood ratio criterion.

Under the alternative hypothesis of "non-parallel" parabolas (i.e., without the restrictions (2.3) and (2.4) imposed on the model (2.2)), the maximum likelihood estimator for σ^2 is

$$\hat{\sigma}_a^2 = V/N.$$

Imposing the restrictions (2.3) and (2.4) on the model (2.2) (i.e., under the null hypothesis of "parallel" parabolas), the maximum likelihood estimator of σ^2 is

$$\hat{\sigma}_o^2 = \left[V + (\hat{\mathbf{\Psi}} - \hat{\mathbf{\Gamma}}\hat{\mathbf{U}})\mathbf{A}^{-1}(\hat{\mathbf{\Psi}} - \hat{\mathbf{\Gamma}}\hat{\mathbf{U}})'\right] / N.$$

Using these two estimators in the likelihood ratio statistic, the criterion for rejecting the null hypothesis of "parallel" parabolas is to reject the null hypothesis at a significance level of α if

$$N \ln\left[1 + V^{-1}(\hat{\mathbf{\Psi}} - \hat{\mathbf{\Gamma}}\hat{\mathbf{U}})\mathbf{A}^{-1}(\hat{\mathbf{\Psi}} - \hat{\mathbf{\Gamma}}\hat{\mathbf{U}})'\right] > \chi^2_{1;\alpha}.$$

8. REMARKS

The weighted mean method is computationally less complicated than the maximum likelihood approach and is only applicable to symmetric assays.

The method of maximum likelihood is more general in that it can be extended to apply to asymmetric assays, cubic response curves and cases involving covariables. For a multivariate version of the maximum likelihood method see Carter *et al.* (1986).

Williams (1973) and Armitage *et al.* (1974) have compared these two methods in the univariate situation. At present, we are investigating their behavior in a multivariate setting.

Computer software has been developed for this univariate and multivariate parabolic situation. This extends the existing programs developed for the quantal situation described by Carter and Hubert (1984) and the parallel-line situation described by Carter and Hubert (1985).

REFERENCES

Armitage, P., J. M. Bailey, A. Petrie, L. Annable, and M. P. Stack-Dunne (1974), "Studies in the combination of bioassay results". *Biometrics* **30**, 1–9.

Bliss, C. I. (1957), "Bioassay from a parabola". *Biometrics* **13**, 35–50.

Carter, E. M., and J. J. Hubert (1984), "A growth-curve model approach to multivariate quantal bioassay". *Biometrics* **40**, 699–706.

Carter, E. M., and J. J. Hubert (1985), "Analysis of parallel-line assays with multivariate responses". *Biometrics* **41**, 703–710.

Carter, E. M., J. J. Hubert, and M. N. Walsh (1986), "Analysis of multivariate parabolic bioassays". To be submitted.

Chou, R. J., and R. J. Muirhead (1979), "On some distributional problems in MANOVA and discriminant analysis". *Journal of Multivariate Analysis* **9**, 410–419.

Elston, R. C. (1965), "A simple method of estimating relative potency from two parabolas". *Biometrics* **21**, 140–149.

Finney, D. J. (1978), *Statistical Method in Biological Assay*, 3rd edition. London: C. Griffin.

Hubert, J. J. (1984), *Bioassay*, 2nd edition. Dubuque, Iowa: Kendall-Hunt.

Kraft, C. H., I. Olkin, and C. van Eeden (1972), "Estimation and testing for differences in magnitude or displacement in the mean vectors of two multivariate normal populations". *Annals of Mathematical Statistics* **43**, 455–467.

Srivastava, M. S., and E. M. Carter (1983), *An Introduction to Applied Multivariate Statistics*. New York: North-Holland.

Walsh, M. N. (1985), "Analysis of parabolic assays". M.Sc. thesis, University of Guelph.

Williams, D. A. (1973), "The estimation of relative potency from two parabolas in symmetric bioassays". *Biometrics* **29**, 695–700.

M. M. Shoukri [1] and P. C. Consul [2]

SOME CHANCE MECHANISMS GENERATING THE GENERALIZED POISSON PROBABILITY MODELS

1. INTRODUCTION

A new generalization of the discrete Poisson distribution was introduced by Consul and Jain (1970, 1973) and studied by Charalambides (1974), Kumar and Consul (1979), Janardan and Schaeffer (1977), Janardan *el al.* (1979), Consul and Shoukri (1984, 1985), and many others. This model contains two parameters and possesses the over-dispersion and the under-dispersion properties which make it a good descriptive model for aggregation patterns in biology, ecology and many other disciplines. Although the negative binomial distribution possesses the over-dispersion property, it is found that the parametric structure of the generalized Poisson distribution (GPD) permits a relatively simpler inferential procedure. The GPD model is given by

$$P(X=x)=\theta(\theta+\lambda x)^{x-1}e^{-\theta-\lambda x}/x! \qquad x=0,1,2,\ldots \tag{1.1}$$

and zero otherwise, where $\theta>0$, $|\lambda|\leq 1$. When λ is negative and m is the smallest positive integer for which $\theta+m\lambda<0$, we set $P(X=x)=0$ for all $x\geq m$.

It was shown by Consul and Shoukri (1985) that for $0<\theta\leq 1.5$, the maximum error due to the above truncation is less than 0.5% when the number of non-zero class probabilities is limited to three. Further error analysis was made over various subregions of the parameter space of θ and $\lambda<0$. As a general remark, it may be noted that the error due to truncation gets smaller when the number of classes is greater than four. The mean and variance exist for $\lambda<1$ and are given respectively by

$$\mu=\theta(1-\lambda)^{-1}, \qquad \sigma^2=\theta(1-\lambda)^{-3}.$$

[1] Department of Mathematics and Statistics, University of Windsor, Windsor, Ontario N9B 3P4

[2] Department of Mathematics and Statistics, The University of Calgary, Calgary, Alberta T2N 1N4

I. B. MacNeill and G. J. Umphrey (eds.), Biostatistics, 259–268.

Janardan and Schaeffer (1977), and Janardan *et al.* (1979) fitted the GPD to about 100 different sets of biological and ecological data and have found that as a descriptive model, the GPD has provided an excellent fit in every case. Therefore, it is natural that one should wonder about the underlying chance mechanisms that generate this model. In this paper we show that there are many such mechanisms.

2. THE GPD AS A LIMIT OF THE GENERALIZED NEGATIVE BINOMIAL MODEL

The generalized negative binomial distribution (GNBD) was defined and studied by Jain and Consul (1971) as a discrete probability model. The definition was subsequently modified by Consul and Gupta (1980) and the discrete probability model was given as

$$P(X=x) = \frac{n}{n+\beta x}\begin{bmatrix} n+\beta x \\ x \end{bmatrix} \alpha^x (1-\alpha)^{n+\beta x - x}, \quad x = 0, 1, 2, \ldots, \tag{2.1}$$

and zero otherwise and where $0 < \alpha < 1$ and $\beta = 0$ or $1 \leq \beta \leq \alpha^{-1}$.

The distribution was introduced as a queuing model by Mohanty (1966) and Takács (1962). Recently, Hill and Gulati (1981) showed that the probabilities given by (2.1) are those of a gambler's ruin in a random walk model associated with the game of roulette.

When n and β are very large and α is very small such that $n\alpha = \theta$ and $\beta\alpha = \lambda$, where θ and λ are finite constants, one can show, by employing Stirling's approximation formula on the combinatorial term in (2.1), that the GPD model described in (1.1) is the limiting form of the above GNBD.

3. THE GPD AS A FAMILY OF LAGRANGIAN PROBABILITY DISTRIBUTIONS

Let $g(t)$ and $f(t)$ be two functions of t, which are successively differentiable and are such that $g(1) = f(1) = 1$, $g(0) \neq 0$, and $0 \leq f(0) < 1$. Under the transformation

$$u = t/g(t) \tag{3.1}$$

and for some value of t within the circle of convergence of $f(t)$, the power series expansion of $f(t)$, in powers of u, is given by Lagrange's formula (Whittaker and Watson, 1963) as

$$f(t) = \sum_{x \in T} \frac{u^x}{x!} D^{x-1}\left[(g(t))^x D\, f(t)\right]\big|_{t=0}, \tag{3.2}$$

where T is a subset of the set of non-negative integers and $D = \partial/\partial t$.

Taking $D^{x-1}[(g(t))^x Df(t)] \mid_{t=0} \geq 0$ for all $x \in T$, Consul and Shenton (1972) and Consul (1981) defined a random variable X with a probability distribution of the form

$$P(X = x) = \begin{cases} (x!)^{-1} D^{x-1} \left[(g(t))^x \, Df(t)\right] \mid_{t=0}, & x \in T \\ 0, & \text{otherwise.} \end{cases} \tag{3.3}$$

Consul and Shenton (1973) discussed some interesting properties of this class of distributions, which they called "Lagrangian Probability Distributions" (LPD).

Taking $f(t) = \exp[\theta(t-1)]$ and $g(t) = \exp[\lambda(t-1)]$ in (3.3), and on simplifying the LPD, it reduces to the GPD given by (1.1). Also, the probability generating function (pgf) of the GPD becomes

$$H(u) = e^{\theta(t-1)}, \tag{3.4}$$

where $te^{-\lambda(t-1)} = u$. It may be noted that in this pgf of the GPD both the parameters θ and λ are positive.

4. THE GPD MODEL AS THE DISTRIBUTION OF THE TOTAL PROGENY IN A GALTON-WATSON BRANCHING PROCESS

Historically, the study of branching processes began with the Galton-Watson process formulated by Galton to describe a possible mechanism of extinction of distinguished English families. The model of Galton and Watson was used by R. A. Fisher (1930) to study the survival of the progeny of a mutant gene. In ecological and biological applications, we consider a species of animal, insect, bacteria or gene that reproduces asexually. Each individual, before dying, reproduces a certain number of new individuals. If the reproduction process starts with a single individual, we are interested in (a) determining whether or not the species eventually dies out, (b) the distribution of the total progeny at the moment of extinction. While both questions are frequently answered in many standard textbooks on stochastic processes, explicit expression for the distribution of the total progeny was lacking, expecially when the reproduction process starts with a random number of ancestors. To illustrate, let us consider a problem which is of prime interest to government health officials.

In many countries, tourism constitutes a fundamental source of income. In a particular country the tourists arrive at different destinations by sea, air and road through many checkpoints. Although international health cards show vaccination has been completed, there is always a chance that a tourist may carry the bacteria of an infectious disease which spreads to other individuals upon contact. Since the number of tourists arriving is very large and the probability of a tourist carrying the bacteria is very small, it is reasonable to assume that the number of infected tourists is a Poisson random variable X_0 with pgf

$$f(t) = E\left[t^{X_0}\right] = \exp[\theta(t-1)].$$

Here X_0 represents the 0th generation of a branching process or the generation of ancestors. For the validity of the model, we assume Reed and Frost's hypothesis on the theory of epidemics, to be satisfied (Neyman, 1963). Their hypothesis requires that every infected individual undergoes a fixed incubation period, becomes infectious for a short period of time, and then upon contact with a large number of people (drivers, guides, sales people, other uninfected tourists, hotel and restaurant employees, moviegoers, etc.) may infect a random number of susceptibles. We shall assume that the probability distribution of the number of persons Z, who get infected among the susceptibles, is also a Poisson random variable with pgf $g(t) = E(t^Z) = \exp[\lambda(t-1)]$. The newly infected persons will then form the first generation, and each one of them will generate offspring of infected individuals under the same conditions as stated above, and so on. The main objective is to find the probability distribution of the infected individuals up to the moment when the epidemic dies out.

Let X_0, $X_1, \ldots, X_n, \ldots,$ denote the total number of infected individuals in the 0th, 1st, ..., nth, ..., generations, i.e., they denote the sizes of the infected population at a sequence of points in time. One of the basic assumptions here is that the probability distribution of the number of infected individuals is the same. Although this assumption may be valid at the beginning of the epidemic, the situation may change if an anti-epidemic immunization campaign is launched. However, we shall assume the validity of the assumption throughout. Let $g_n(s) = E(s^{X_n})$.

If we assume that $X_0 = 1$ with probability one, so that the process starts with a single infected tourist, then $g_0(s) = s$, and $g_1(s) = g(s)$. Thus, for $n = 2, 3, \ldots,$

$$\begin{aligned} g_{n+1}(s) &= \sum_{k=0}^{\infty} p\left[X_{n+1} = k\right] s^k \\ &= \sum_{k=0}^{\infty} s^k \sum_{j=0}^{\infty} P\left[X_{n+1} = k \mid X_n = j\right] P\left[X_n = j\right] \\ &= \sum_{j=0} P\left[X_n = j\right] (g(s))^j = g_n\left(g(s)\right). \end{aligned}$$

Also

$$g_2(s) = g_1(g(s)) = g(g(s)) = g(g_1(s))$$

and similarly

$$g_{n+1}(s) = g(g_n(s)).$$

The above branching process will stop as soon as $P[X_n = 0] = 1$ for some positive integer n. As is well known, the necessary and sufficient condition for this happening is that $g'(1) = \lambda < 1$.

Now let us assume that the process of spreading the infection stops after the nth generation so that $Y_n = X_0 + X_1 + \cdots + X_n$ (where $X_0 = 1$). Also, let $G_n(s)$ be the pgf of $Z_n = X_1 + \cdots + X_n$. Thus

$$G_1(s) = g_1(s) = g(s),$$

and the pgf of Y_1 is

$$E\left[t^{Y_1}\right] = sg(s) = R_1(s) \text{ (say)}.$$

Since each of the X_1 infected individuals will start a new generation, the pgf of Z_2 becomes

$$G_2(s) = g[sG_1(s)] = g[R_1(s)],$$

and similarly

$$G_n(s) = g[R_{n-1}(s)], \qquad n = 2, 3, 4, \ldots.$$

Thus, the pgf of $Y_n = 1 + Z_n$ is

$$R_n(s) = sG_n(s) = sg[R_{n-1}(s)].$$

Taking the limit, as n increases without limit,

$$G(s) = \lim_{n\to\infty} R_n(s) = sg\left[\lim_{n\to\infty} R_{n-1}(s)\right]$$

and

$$G(s) = sg(G(s)). \tag{4.1}$$

Putting $G(s) = t$ in (4.1), we get $t = sg(t)$.

Thus, the Lagrange expansion of t as a function of s may then be obtained using (3.3) for $f(t) = t$, to give the probability distribution of the total number of infected individuals starting with one infectious individual, as

$$P[Y = j \mid X_0 = 1] = \frac{1}{j!}(j\lambda)^{j-1}e^{-j\lambda}, \quad j = 1, 2, \ldots, \tag{4.2}$$

which is known as the Borel distribution. If $X_0 = M$ with probability one, then $f(t) = t^M$, and

$$P[Y = j \mid X_0 = M] = \frac{m}{(j-M)!} j^{j-M-1} \lambda^{j-M} e^{-\lambda j},$$
$$j = M, M+1, \ldots, \tag{4.3}$$

which is known as the Borel-Tanner distribution.

In the case when X_0 is a random variable with the pgf $f(t) = \exp[\theta(t-1)]$, as assumed and shown by us earlier, the probability distribution of the total number of infected individuals is obtained either by direct substitution for $f(t)$ and $g(t)$ in (3.3), or by multiplying (4.3) by the Poisson probability

$$P[X_0 = M] = e^{-\theta}\theta^M / M!$$

and summing over all values of M. In either case the probability distribution of the total number of infected individuals becomes

$$P(Y = j) = \theta(\theta + \lambda j)^{j-1}(j!)^{-1} e^{-\theta - j\lambda}, \quad j = 0, 1, 2, \ldots, \tag{4.4}$$

which is the GPD model given in (1.1) with the parametric values $\theta > 0$ and $0 \leq \lambda < 1$.

Some Remarks:

It may also be noted that there are two other possible choices of $g(t)$, the pgf of the probability distribution of offspring, which provide other forms of the GPD model. When

$$g(t) = \exp[\alpha\theta(t-1)],$$

the GPD model (4.4) becomes

$$P(X = x) = (1 + \alpha x)^{x-1}\theta^x e^{-\theta(1+\alpha x)}/x!, \quad x = 0, 1, 2, \ldots, \tag{4.5}$$

and when

$$g(t) = \exp[\theta(t-1)],$$

the GPD model (4.4) becomes

$$P(X = x) = (1 + x)^{x-1}\theta^x e^{-\theta(1+x)}/x!, \quad x = 0, 1, 2, \ldots. \tag{4.6}$$

The model given by (1.1) or (4.4) reflects the fact that the generation of ancestors is quite different from that of the immediate offspring, which seems quite reasonable in our example, especially when the 0th generation of the epidemic consists of alien tourists. The probability of regeneration among the local population may be different because of a change in conditions, such as the resistance capacity of the individuals.

Model (4.5) reflects partial similarity between ancestors and progeny and this may be explained by factors such as common inheritance. Model (4.6) will be applicable when the distributions of the ancestors and of their offspring are identical, as in the case of hybrid populations of certain species.

5. APPLICATIONS OF THE GPD MODEL IN INDUSTRY, PSYCHOLOGY, SOCIOLOGY AND BIOLOGY

By considering the problem of the spread of an epidemic, it has been shown in the last section that, under certain conditions, the GPD model can be based upon branching processes. The general hypothesis is that a particular characteristic is transmitted to some members of a group from a source either at an initial point in time or continuously. Then the individuals who have acquired the characteristic spread it to members of other groups. The new generation spreads the same characteristic again and the process continues again and again until either it dies out or else the whole population gets the particular characteristic. Accordingly, the GPD model can be used in all situations where the following conditions can be justified:

(i) The total number of units or individuals in a group is large.

(ii) The probability of acquiring the characteristic by a unit or individual is small.

(iii) Each one of the units (or individuals) having the particular characteristic becomes a spreader of the characteristic for a short period of time.

(iv) The number of members in the group, where each spreader (infectious individual) having the characteristic is likely to spread (infect) the characteristic, is also large.

In each case, the number of individuals having the characteristic (i.e., the total progeny in the branching process) will be modelled by the GPD.

Such conditions may exist in numerous aspects of human society and among other life forms. We shall consider a few important applications.

(A) *Spread of ideas and rumours.* One individual transmits an idea or rumour, either by word of mouth or by some demonstration. The idea or rumour is picked up by a small number of individuals. Each one of them, including the originator, become transmitters like the first one and the process continues. Ultimately, the idea or rumour either dies down, i.e., becomes extinct, or spreads among everybody.

(B) *Spread of fashion and sales.* The effectiveness of an advertising campaign is measured by the total number of adoptions of the advertised product or fashion. For example, on viewing a television commercial advertising a new product, a random number of customers are generated whose pgf is $f(t) = e^{\theta(t-1)}$. When other people see these customers buying the product or using it, some of them become customers. Thus, each original customer generates a further random number of customers, and the branching process continues until the process becomes extinct. Thus, each display of the commercial initiates a regenerative branching process phenomena for the adoption of the product.

(C) *Distributions of salespersons in "pyramid" dealerships, such as AVON, Amway, Tupperware and cleaning chemicals.* Each distributor (infectious individual) initially collects a group of persons (susceptibles) and gives a demonstration or a show and encourages them to become salespersons. Some of them enrol as salespersons (get infected). Then they repeat the same kind of demonstration or show and try to enrol others (infect them). The distributorships mushroom for a while and then die down.

(D) *Population counts of particular species of animals, such as insects, fish and frogs, in different areas.* The conditions for the regeneration of all of these are very similar to those stated in (i) to (iv). The number of eggs laid by each individual is very large and the probability of an egg becoming an adult is very small since a high proportion of eggs, and immature offspring are eaten by other animals or get killed by other means. Again, the process of regeneration is a branching process.

(E) *Distribution of burnt trees in forest fires.* A single spark starts a fire which burns a tree. As the tree burns, numerous sparks are produced which fly out around that tree. Most of these sparks die out, but some of them develop into independent fires, burning other trees. Then the process continues as a regenerative process. Thus, all the conditions (i) to (iv) of the process exist which will provide the GPD model.

ACKNOWLEDGMENT

The financial support for this research by the Natural Sciences and Engineering Research Council of Canada is gratefully acknowledged. We would like to thank the editor for all the suggestions that improved the work.

REFERENCES

Charalambides, C. A. (1974), "The distribution of sufficient statistics of truncated generalized logarithmic series, Poisson, and negative binomial distributions." *Canadian Journal of Statistics* **2**, 261–275.

Consul, P. C. (1981), "Relation of modified power series distributions to Lagrangian probability distributions." *Communications in Statistics A, Theory and Methods* **10**, 2039–2046.

Consul, P. C., and H. C. Gupta (1980), "The generalized negative binomial distribution and its characterization by zero regression." *Siam Journal on Applied Mathematics* **39**, 231–237.

Consul, P. C., and G. C. Jain (1970), "A generalization of the Poisson distribution." Technical report #114. Department of Mathematics, Statistics, and Computing Science, University of Calgary, Calgary, Alberta, Canada.

Consul, P. C., and G. C. Jain (1973), "A generalization of the Poisson distribution." *Technometrics* **15**, 791–799.

Consul, P. C., and L. R. Shenton (1972), "Use of the Lagrange expansion for generating discrete generalized probability distributions." *Siam Journal on Applied Mathematics* **23**, 239–248.

Consul, P. C., and L. R. Shenton (1973), "Some interesting properties of Lagrange distributions." *Communications in Statistics* **2**, 263–272.

Consul, P. C., and M. M. Shoukri (1984), "Maximum likelihood estimation for the generalized Poisson distribution." *Communications in Statistics A, Theory and Methods* **13**, 1533–1547.

Consul, P. C., and M. M. Shoukri (1985), "The generalized Poisson distribution when the sample mean is larger than the sample variance." *Communications in Statistics B, Simulation and Computation* **14**, 667–681.

Fisher, R. A. (1930), *The Genetical Theory of Natural Selection*. London and New York: Oxford University Press.

Hill, J. M., and C. A. Gulati (1981), "The random walk associated with the game of roulette." *Journal of Applied Probability* **18**, 931–936.

Jain, G. C., and P. C. Consul (1971), "A generalization of the negative binomial distribution." *Siam Journal on Applied Mathematics* **21**, 501–513.

Janardan, K. G., W. Kerster, and D. J. Schaeffer (1979), "Biological applications of the Lagrangian Poisson distribution." *BioSciences* **29**, 599–602.

Janardan, K. G., and D. J. Schaeffer (1977), "Models for the analysis of chromosomal aberrations in human leukocytes." *Biometrical Journal* **19**, 599–612.

Kumar, A., and P. C. Consul (1979), "Minimum variance unbiased estimator for modified power series distribution." *Communications in Statistics A, Theory and Methods* **9**, 1261–1275.

Mohanty, S. G. (1966), "On a generalized two-coin tossing problem." *Biometrische Zeitschrift* **8**, 266–272.

Neyman, J. (1963), "Certain chance mechanisms involving discrete distributions." *Classical and Contagious Discrete Distributions*, Calcutta: Statistical Publishing Society.

Takács, L. (1962), "A generalization of the ballot problem and its applications in the theory of queues." *Journal of the American Statistical Association* **57**, 327–337.

Whittaker, E. T., and G. N. Watson (1963), *A Course of Modern Analysis*. Cambridge, MA: Cambridge University Press.

W. T. Federer [1] and B. R. Murty [2]

USES, LIMITATIONS, AND REQUIREMENTS OF MULTIVARIATE ANALYSES FOR INTERCROPPING EXPERIMENTS

ABSTRACT

Since summarization of data from intercropping experiments involves a linear combination of crop yields, it would appear at first glance that multivariate analyses would be ideally suited for this situation. Linear combinations of values of crops, of total calories of crops, of total protein contents of crops, and of land utilization of crops are among some of the linear combinations which have considerable utility in summarizing data from intercropping experiments. From multivariate analyses, canonical variables based upon the criterion of maximum discriminating ability can be obtained when *all* crops are present in a mixture of crops. For v crops taken k at a time, such canonical variables are not obtainable using presently available theory. Even if they were, it is not certain that they would have any general utility for interpretational purposes. Areas of further research in multivariate analysis are discussed. These results are necessary in order to utilize multivariate analyses for intercropping investigations.

1. INTRODUCTION

Intercropping investigations involve the growing of two or more crops simultaneously or sequentially on the same plot of ground. These mixtures of crops have been found to be beneficial in several respects when compared to yield responses from crops grown alone, i.e., sole crops. One problem is how to combine the yield responses from several crops, which may have considerably different means and variances. For a farmer who considers crop values in monetary terms, one could put a monetary value on each crop and

[1] Biometrics Unit, Cornell University, Ithaca, New York 14853

[2] International Atomic Energy Agency, United Nations, Maracaibo, Venezuela

I. B. MacNeill and G. J. Umphrey (eds.), Biostatistics, 269–283.

obtain a total monetary value per hectare or per acre for each of the mixtures and for the sole crops. The form would be:

$$v_1C_1 + v_2C_2 + v_3C_3 + \cdots + v_kC_k, \tag{1}$$

where v_i is the value of the ith crop, C_i is the yield of the ith crop in kilograms/hectare, say, and $i = 1, 2, \ldots, k$, where k is the number of crops in the mixture. Such a linear combination as (1) would provide a useful statistic for a grower of crops. Likewise, if the farmer were interested in the total production of calories for his family needs, again we could use a linear combination of the form:

$$c_1C_1 + c_2C_2 + c_3C_3 + \cdots + c_kC_k, \tag{2}$$

where c_i is the calorie conversion coefficient for the ith crop. A similar linear combination could be used for protein conversion.

If the farmer were interested in land use efficiency, he could use a linear combination of yield responses for each crop of the form:

$$C_1/C_{1s} + C_2/C_{2s} + C_3/C_{3s} + \cdots + C_k/C_{ks}, \tag{3}$$

where C_{is} is the yield of the ith crop when grown as a sole crop. The coefficients v_i and c_i would normally be taken as constants with the C_i being the only random variables. Then if $C_1, \ldots, C_k$ have a k-variate multivariate normal distribution, (1) and (2) each have a univariate normal distribution and no problems of statistical analyses are encountered over those found for sole cropping. If the C_{is} are also random variables, (3) would give a distribution which is a sum of Cauchy variables. However, if the C_{is} are constants, which might be the average yields of crops in that region for the past y years, then (3) also has a univariate normal distribution.

Now crop values and yields fluctuate from year to year but the ratios of prices, v_i/v_1 and of yields, C_{1s}/C_{is}, where v_1 is the value for a base crop and C_{1s} is the yield for a base crop, are relatively stable. Hence one could use $\sum v_iC_i/v_1$ in place of (1), $\sum c_iC_i/c_1$ in place of (2), and $C_{1s}\sum C_i/C_{is}$ in place of (3). Let us call these "relative crop values", "relative calories", and "relative land use". In this form then, all linear combinations have a similar form,

$$C_1 + b_2C_2 + b_3C_3 + \cdots + b_kC_k. \tag{4}$$

The above is also the form for a canonical variable obtained from a multivariate analysis. As we shall see this form is useful when comparing the various forms with each other and especially with the canonical variable obtained from a multivariate analysis.

We first consider a multivariate analysis for p characteristics on crop i. There are no difficulties encountered here which are due to intercropping. In Section 3 the case of c_1 lines of crop one and c_2 lines of crop two in all possible combinations is discussed. The yields of crop one are considered to be the first variable X_1 and the yields of the second crop are considered to be the second variable X_2. A series of interpretational problems are encountered for this situation. The present state of multivariate theory does not allow for solution of these problems and the basic utility of discriminant function analysis of variance (MANOVA), which is to obtain a canonical variable which has maximum discriminating ability, is questioned. In Section 4, the case of p crops in mixtures of k crops, $k = 1, 2, \ldots, p$, is considered. Even if present multivariate theory were applicable in Section 3, it is not sufficient to handle the problems encountered in Section 4 even if k is not allowed to vary. Section 5 considers the situation for which there are c_i lines for the ith crop and these are mixtures of k of p crops. Even if each of the line combinations from the $\prod_{i=1}^{p} c_i = v =$ total number of line combinations are considered separately, we are left with the difficulties encountered in Sections 3 and 4. Finally, some areas of research to extend present multivariate theory and analyses are discussed in the last section.

2. MULTIVARIATE ANALYSIS FOR p CHARACTERISTICS OF ONE CROP

When there are p characteristics or variables for a single crop, and provided that the necessary normality assumptions are satisfied, standard multivariate procedures may be utilized. The purpose here would be to obtain linear combinations (canonical variables) of the p variables which discriminate among the treatments in the experiment. It would be hoped that one or two canonical variables would summarize the whole of the information for the p variables. Also, the relative importance of the p variables in discriminating among the treatments could be assessed.

Thus, when using multivariate procedures for the purpose described here, there appear to be no problems in summarizing information from data derived from intercropping investigations. For example, suppose that an intercropping experiment on maize-bean mixtures in combination and with maize and beans grown alone (sole crops) was conducted. Suppose that there were two lines of maize, X and Y, and four lines of beans, A, B, C, and D, then the 14 treatments in the experiment are given in Table 1.

There are eight mixtures plus two maize sole crops plus four bean sole crops. If M_1 is the total weight of maize grain for an experimental unit and M_2 is the number of ears per maize plant, there would be two characteristics

Table 1. *Treatment Design for a Maize-Bean Experiment*

	Beans				
Maize	A	B	C	D	Sole
X	x	x	x	x	x
Y	x	x	x	x	x
Sole	x	x	x	x	

for maize and a bivariate analysis could be performed for the ten treatments involving maize, i.e., the eight mixtures and the two sole crops for maize. Likewise, for the four variables measured on beans (B_1 is the number of beans per pod, B_2 is the number of pods per plant, B_3 is the grain weight of 100 beans, and B_4 is the grain weight per experimental unit), one could perform a multivariate analysis using the 12 treatments involving beans. All four variables or some subset of the four variables could be included in the multivariate analysis.

3. MULTIVARIATE ANALYSIS FOR MIXTURES OF c_1 LINES OF CROP ONE WITH c_2 LINES OF CROP TWO

As was stated in the introduction, a goal of statistical analyses for intercropping investigations is to obtain some linear combination of the yields from both crops in a mixture. If we let crop one yields be variable one, say X_1, and crop two yields be variable two, say X_2, a bivariate analysis of variance may be performed (see Pearce and Gilliver, 1978, 1979; Mead and Riley, 1981). As mentioned by the above authors, one serious problem here is heterogeneity of covariance matrices. The variance and covariance of a line of one crop may vary from line to line of the second crop. Thus, the lines of any one crop may, and often do, have heterogeneous covariance structures. To overcome this, it is suggested that single degree of freedom contrasts be made from the treatment sum of squares. (Also see Singh and Gilliver, 1983.) It may be necessary to compute an individual error variance for each single degree of freedom contrast. This can easily be accomplished for completely randomized and randomized complete blocks designs.

One problem that appears to be unsolvable using presently available multivariate theory is how to compare sole crop yields for maize, say M_s, with a linear combination of maize and beans in a mixture, say $M_m + bB_m$. If b is a ratio of crop values, for example, this can easily be done. If b is a ratio of yields of sole crop yields in farmers fields, say $b = M_s/B_s$, then again one can compare sole crop and mixture yields. However, if b is a coefficient obtained from a bivariate analysis to form a canonical variable, discriminant analysis is not sufficiently advanced to allow comparisons between M_s and $M_m + bB_m$. This is a serious deficiency in the theory as it is usually desired to compare sole cropping with mixture yields in intercropping investigations.

A second problem arising from the use of discriminant analyses for data from intercropping investigations is the usefulness and interpretation of a canonical variable of the form $M_m + bB_m$. It appears that this canonical variable obtained from a bivariate analysis is useless for interpretation of intercropping data on mixtures of two crops. The criterion used to obtain the canonical variable, i.e., the linear combination producing maximum discriminating ability among treatments, has no meaning. To illustrate consider a set of data obtained from an experiment designed as a randomized complete block with four blocks and including the eight maize and bean mixtures described in Table 1. A bivariate analysis of variance is given in Table 2. For the discussion, ignore the fact that data were missing for beans in one of the bean-maize mixtures. Carrying through the computations, it was found that b in $M_m + bB_m$ was 38. Now a meaningful b in terms of prices of maize and beans is between three and seven. A meaningful b in terms of land use, is in the same range. These values are not even close to the one obtained in the bivariate analysis. To put this into perspective, Figure 1 has been prepared where the canonical correlations are computed for various values of b. The canonical correlation is computed by dividing the treatment sum of squares for the canonical variable by the treatment plus error sums of squares. The formula is

$$S = \frac{M_{tt} + 2b\,C_{tt} + b^2 B_{tt}}{M_{tt} + M_{ee} + 2b(C_{tt} + C_{ee}) + b^2(B_{tt} + B_{ee})}, \tag{4}$$

where M_{tt} and M_{ee} are the treatment and error sums of squares, respectively, for maize yields in the mixtures and similarly for B_{tt} and B_{ee} for beans, and C_{tt} and C_{ee} are the cross products of maize and bean yields for treatments and for error, respectively. The canonical correlation measures discriminating ability of the canonical variable. S has been computed for values of b in the range $0 \leq b \leq 100$. The range of $3 \leq b \leq 7$, or even $2 \leq b \leq 12$, has a practical interpretation, whereas values of b in the range of $b > 12$ have no meaning, throwing considerable doubt on the criterion of "maximum discriminating ability". One other point of interest is the relative flatness of the curve for S when $b > 20$, with a limiting value close in value to the maximum at 38. Because of this problem, it would appear that this type of multivariate analysis is inappropriate for analyzing data from intercropping experiments. One should also note that "maximum discriminating ability" is not invariant with respect to differences among maize or

Table 2. *Bivariate Analysis of Variance for Mixtures of Maize and Beans with Crop Yields as Variables*

Source of variation	Degrees of freedom	Sums of products	Covariance matrix
Total	32[1]	$\begin{bmatrix} 60,936.45 & 628,076 \\ - & 7,097,198 \end{bmatrix}$	–
Correction for mean	1	$\begin{bmatrix} 57,231.903 & 625,055.44 \\ - & 6,826,512.5 \end{bmatrix}$	–
Blocks	3	$\begin{bmatrix} 702.003 & 1,966.65 \\ - & 45,284.75 \end{bmatrix}$	$\begin{bmatrix} 234.00 & 655.55 \\ - & 15,094.92 \end{bmatrix}$
Treatments	7	$\begin{bmatrix} 1,540.980 & 1,810.06 \\ - & 66,340.00 \end{bmatrix}$	$\begin{bmatrix} 220.14 & 258.58 \\ - & 9,777.14 \end{bmatrix}$
Remainder	21[1]	$\begin{bmatrix} 1,461.564 & -756.15 \\ - & 159,060.75 \end{bmatrix}$	$\begin{bmatrix} 73.078 & -37.808 \\ - & 7,574.32 \end{bmatrix}$

[1]less one for first variable bean yields

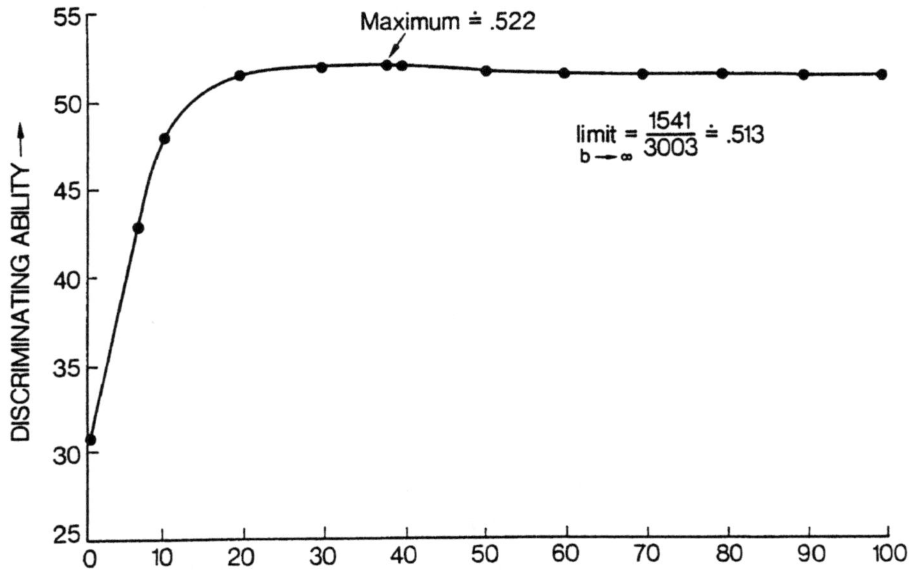

Figure 1. *Treatment/(treatment + error) sums of squares for* $0 \leq b \leq 100$.

among bean lines. The multivariate approach taken by Pearce and Gilliver (1978, 1979) appears to be a fruitful direction for multivariate analyses in the future.

4. MULTIVARIATE ANALYSES FOR p CROPS IN MIXTURES OF k CROPS WITH ONE LINE PER CROP

In the previous section, two crops were considered. There were c_1 lines of crop one and c_2 lines of crop two. Here we consider the case where there are mixtures of k crops from among p crops, $k \leq p$. Also, here we consider that there is only one line per crop. In the next section the case of c_i lines on the ith crop is discussed. To illustrate, consider that four crops are to be included in an investigation wherein mixtures of three of the four crops are studied. All possible mixtures of three crops, for A being crop one, B crop two, C crop three, and D crop four, are given in Table 3.

There also could be four sole crops, six mixtures of two crops, i.e., AB, AC, AD, BC, BD, and CD, and one mixture of four crops, $ABCD$. For the above discussed set of mixtures of three crops, let us designate the yield response on crop i as X_i, $i = A,\ B,\ C,\ D$. The measurements on the four variables (crops) in one of the r complete blocks for mixtures of three

Table 3. *Mixtures of Three of Four Crops*

Mixture			
1	2	3	4
A	A	A	B
B	B	C	C
C	D	D	D

Table 4. *Individual Crop Responses For Mixtures in Table 3*

	Variable (crop)			
Mixture	X_A	X_B	X_C	X_D
1	x	x	x	
2	x	x		x
3	x		x	x
4		x	x	x

are given in Table 4.

In Table 4, "x" denotes that an observation for the variable is available and a blank means that no value for the variable was obtained. The design for the responses in this case would be a balanced incomplete block design with design parameters $v^* = 4$, $b^* = 4$, $r^* = 3$, $k^* = 4$, and $\lambda^* = 2$. In computing a sum of squares for a variable X_i, rr^* measurements would be used. To compute a sum of cross products $r\lambda^*$ observations would be used for any pair of crops. Let T_{ii} and E_{ii} represent a treatment and error sum of squares for variables X_i, respectively, and let T_{ij} and E_{ij}, $i \neq j = 1, 2, \ldots, p$, represent the cross products between crops i and j. Testing in a multivariate analysis involves determinants of matrices composed of sums of squares and cross products, of the form:

$$\frac{\begin{vmatrix} E_{11} & \cdots & E_{1p} \\ \vdots & \ddots & \vdots \\ E_{1p} & \cdots & E_{pp} \end{vmatrix}}{\begin{vmatrix} E_{11}+T_{11} & \cdots & E_{1p}+T_{1p} \\ \vdots & \ddots & \vdots \\ E_{1p}+T_{1p} & \cdots & E_{pp}+T_{pp} \end{vmatrix}} \tag{5}$$

Since the E_{ii} are computed from rr^* responses and the E_{ij} from $r\lambda^*$ responses, one could use a conservative approach and multiply each E_{ii} by λ^*/r^*. This would put sums of squares and sums of cross products on the same basis in order that a test could be used. Such a procedure would not utilize all the information in the data but would allow use of standard multivariate analyses. It should be noted that rr^* and $r\lambda^*$ can be quite different in value. For example, let $v^* = 25$, $k^* = 5$, $r^* = 6$, $b^* = 30$, and $\lambda^* = 1$. For this case, $\lambda^*/r^* = 1/6$, or $rr^* = 6r\lambda^*$. Even for $v^* = 7 = b^*$, $k^* = 3 = r^*$, and $\lambda^* = 1$, $rr^* = 3r\lambda^*$. For $v^* = 27$, $r^* = 13$, $k^* = 3$, $b^* = 117$, and $\lambda^* = 1$, $rr^* = 13\ r\lambda^*$. The number of crops and the size of the mixture k^* of crops determines the disparity in rr^* and $r\lambda^*$ values.

At first glance, it might appear that the paper by Srivastava (1968) would be useful in analyzing incomplete responses such as described above. A study of the paper shows this not to be the case, and it is not known how to adapt the results for situations encountered in intercropping investigations. Srivastava (1968) considers the following situation. For t treatments in a block design on which p responses are to be studied, some or all of the responses are obtained for a subset of the blocks. To illustrate, let $t = 4$, $r = 4$, $k = 3$, $b = 4$, and $\lambda = 2$ be the design on which a set of responses is measured. Let the design on the responses also be a blocked design for the $p = 3$ response variables. For this design let $S_1 = (X_1, X_2)$, $S_2 = (X_1, X_3)$, $S_3 = (X_2, X_3)$, and $S_4 = (X_1, X_2, X_3)$, that is, two of the three responses are obtained in three sets of b blocks and all three responses are obtained in one set of b blocks. This would require a total of $4(4) = 16$ blocks of size $k = 3$ experimental units. If each S_h, $h = 1, 2, 3, 4$, is included twice as in Srivastava's (1968) example, the plan in Table 5 would result. There would be a total of 32 blocks of size $k = 3$ experimental units each for a total of 96 experimental units. The total number of observations would be $6(2)(12) + 2(3)(12) = 216$.

Purportedly the reason for not observing responses on all three variables on every one of the 96 experimental units was cost or lack of time. It would appear that a more reasonable and desirable procedure would be to use

Table 5. *Layout for Srivastava (1968) Example*

Set	Variable	Block 1 treatment 1	2	3	Block 2 treatment 1	2	4	Block 3 treatment 1	3	4	Block 4 treatment 2	3	4
S_1	V_1												
	V_2												
S_2	V_1												
	V_2												
S_3	V_1												
	V_3												
S_4	V_1												
	V_3												
S_5	V_2												
	V_3												
S_6	V_2												
	V_3												
S_7	V_1												
	V_2												
	V_3												
S_8	V_1												
	V_2												
	V_3												

one third fewer blocks and measure all variables on each experimental unit. Then one could use standard multivariate procedures and not run into the difficulties discussed in the Srivastava (1968) paper.

Srivastava (1968) was interested in the treatments in the design. In intercropping the sets S_h are the treatments and are the items of interest in the investigation. Thus interest was on the column totals of Table 5, whereas the row totals of Table 5 represent the treatments in an intercropping investigation. A further study of the procedures described threw no light on how to adapt the procedures of the paper for the situation encountered.

5. MULTIVARIATE ANALYSES FOR p CROPS IN MIXTURES OF k CROPS WITH c_i LINES ON CROP i

Consider the situation wherein there are p crops which will be grown in mixtures of k crops and that there are c_i lines on crop i, $i = 1, 2, \ldots, p$, $k \leq p$, for the $\binom{p}{k}$ combinations or some subset thereof, one could conduct k-

variate analysis in the manner described in Section 3 for one pair of crops. The extension would be straightforward. The difficulties discussed for the bivariate case would carry over to the k-variate case.

If one wished to use a combined analysis for all $\binom{p}{k}$ combinations then all the difficulties described in Section 4 arise when one attempts a discriminant analysis. If the values v_i in (1), c_i in (2), or C_{si} in (3) are given or if the relative values b_i in (4) are known, then there is no difficulty in forming such a canonical variable or a relative canonical variable. The only difficulty that would arise would be in obtaining a range of these values. For two crops, it is easy to vary b in X_1+bX_2. For three crops it is again not too difficult to vary b and c in X_1, X_1+bX_2, X_1+cX_3, bX_2+cX_3, bX_2, cX_3, and $X_1+bX_2+cX_3$ where there would be sole crops, mixtures of two crops, and a mixture of all three crops to compare. One could look at the various levels of b, say ℓ_2, at various levels of c, say ℓ_3. There would be $\ell_2\ell_3$ levels to compare the various treatments (mixtures) in the experiment. With p crops there would be $\ell_2\ell_3\cdots\ell_p$ total levels at which to compare the treatments in the experiment, where ℓ_i is the relative level (crop value, calories, and/or land use) for crop i. This many computations makes presentation and interpretation difficult. Hence, ℓ_i should be as small as possible. For Figure 1, we used $0 \leq \ell_2 \leq 100$, but such a range of values would be extremely tedious if one used $0 \leq \ell_2 \leq 100$, $0 \leq \ell_3 \leq 100$, $0 \leq \ell_4 \leq 100$, etc. Therefore, it is recommended that at most three values for each ℓ_i be used, i.e., a low value, an average value, and a high value. For calorie or protein conversions, it may not be necessary to vary the ℓ_i. For land use and for monetary value, these values should be varied.

6. SOME UNRESOLVED PROBLEMS IN MULTIVARIATE ANALYSES

One problem raised in the preceding discussion was the comparison of sole crops with mixtures of two, three, ..., the comparison of mixtures of two with mixtures of three, four, etc. One method of doing this might be the following. Let α_i be the proportion of the variate in a multivariate distribution with $\sum_{i=1}^{p} \alpha_i = 1$. When the $\alpha_i = \alpha$, we have present multivariate theory. If one were able to generalize present multivariate theory, this would appear to be one way of making the above comparisons, though it must be realized that they may not be useful in a practical interpretation of data from intercropping experiments. It would solve the problem of comparing canonical variables with different numbers of variables.

The use of weight α_i for variable X_i in a multivariate normal distribution could be useful in analyzing replacement series data such as depicted in Figure 2. Here the proportion of one crop goes from zero to one while a second crop goes from one to zero. If the effect of a mixture was simply substitution of crops with no beneficial or no detrimental effect of the mixture, then the yields of the mixture would follow the solid line in Figure 2. If one developed the generalized multivariate normal as described above, it would be for this situation.

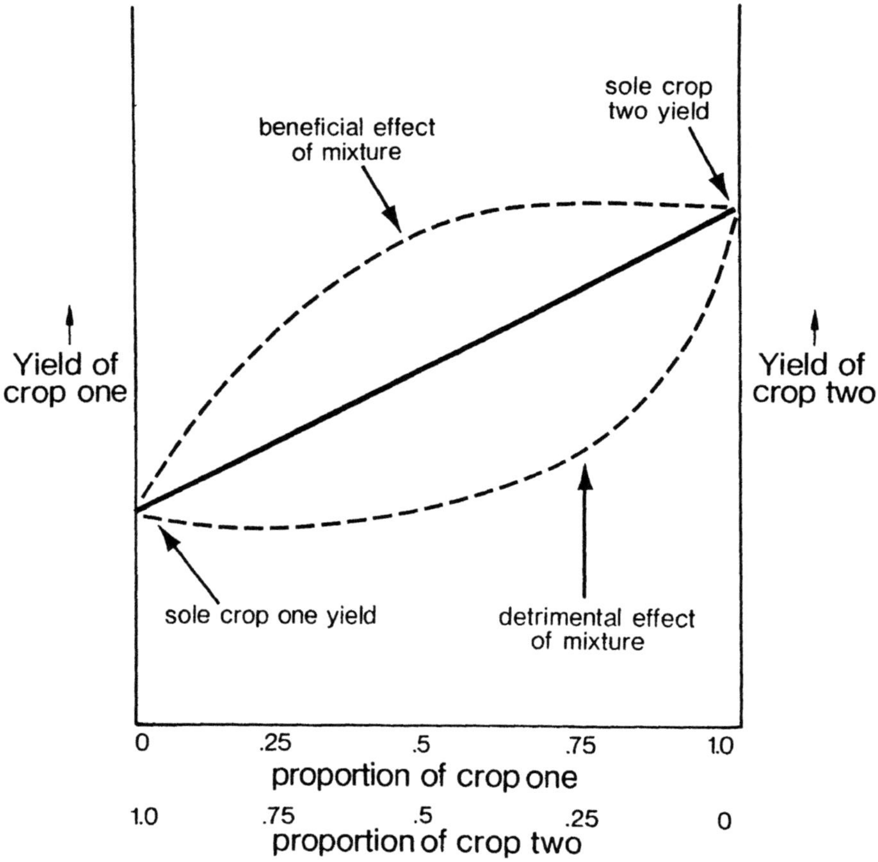

Figure 2. *Replacement series for two crops.*

On the other hand, suppose that the mixture yields followed the dashed line above the solid line. This would indicate a beneficial effect of mixing two crops, which is a usual result in intercropping if the crops are judiciously selected (Okigbo, 1981). If the curve were a smooth function, perhaps only one additional parameter in the generalized multivariate normal distribution would be required to handle the situation. A development of the theory for this situation would be useful.

A second line of development required in multivariate analysis is the one discussed in Sections 4 and 5. This has to do with the computation of sums of squares and cross products from unequal numbers of observations. A simple problem in this area is the distribution of the variance of the linear regression coefficient computed as β = covariance (x, y)/variance (x), where the covariance (x, y) has N_1 degrees of freedom and variance (x) has N_2 degrees of freedom. For the multivariate situation let us consider a specific case for $p = 3$, for sole crop yields X_{1s}, X_{2s}, and X_{3s}, for a mixture of two crop yields X_1 and X_2, X_1 and X_3, X_2 and X_3 and for a mixture of all crops X_1, X_2, and X_3. In a completely randomized design one may compute the following matrix of sums of squares and cross products:

$$\left[\begin{array}{c|c|c} \mathbf{S}_1 & \mathbf{0} & \mathbf{0} \\ \hline \mathbf{0} & \mathbf{S}_2 & \mathbf{0} \\ \hline \mathbf{0} & \mathbf{0} & \mathbf{S}_3 \end{array}\right],$$

where

$\mathbf{0}$ is a 3×3 matrix of zeros,

$$\mathbf{S}_1 = \mathrm{diag}[S_{11}\ (\text{sole}), S_{22}\ (\text{sole}), S_{33}\ (\text{sole})],$$

$$\mathbf{S}_j = \begin{bmatrix} S_{11}(j) & S_{12}(j) & S_{13}(j) \\ S_{12}(j) & S_{22}(j) & S_{23}(j) \\ S_{13}(j) & S_{23}(j) & S_{33}(j) \end{bmatrix} \quad \text{for } j = 2, 3;$$

$S_{ii}(2)$ is a sum of squares computed from the yields of crop i when it is in mixtures of two and computed from $2r$ observations, $S_{ii'}(2)$, $i \neq i'$, is a sum of cross products for crops i and i' computed from r observations, $S_{ii}(3)$ is a sum of squares for crop i computed from mixtures of three and obtained from r observations, $S_{ii'}(3)$, $i \neq i'$, is a sum of cross products computed from the r observations, from mixtures of three, and $S_{ii}(\text{sole})$ is a sum of squares computed from r sole crop yields in the r replicates.

There are only three variates but there is a nine by nine matrix with the population variances for the nine diagonals being different. Also, there are six different parameters for the covariances. The mean for crop X_i changes from sole to mixtures of two to mixtures of three. For this situation, comparisons of sole crop yields with mixtures of two and of three crops and comparisons of mixtures of two crops with a mixture of three crops are desired. One has only a three variate problem which turns into a situation resembling a nine variate problem.

In the above, if we considered only mixtures of two, the sums of squares would be computed from $2r$ observations and the sums of cross products from r observations. Now what is the distribution of error over treatment plus error sums of squares and cross products? One would be considering a quotient of the form given by equation (5). In general if there are $\binom{v^*}{k^*}$ combinations, a balanced incomplete block design can be created with parameters v^*, k^*, $b^* = v^*!/k^*!(v^*-k^*)!$, $r^* = (v^*-1)!/(k^*-1)!(v^*-k^*)!$, and $\lambda^* = (v^*-2)!/(k^*-2)!(v^*-k^*)!$. The sums of squares from crops in mixtures of k would be computed from rr^* observations and the sums of cross products are computed from $r\lambda^*$ observations. How does the estimation and testing proceed for this situation?

A third problem that arises is the interpretation of results from an intercropping experiment when equation (1), (2), (3), or (4) is used as the first canonical variable. Then, one computes a multivariate analysis and obtains canonical variables after (4), say, the computational procedure needs to be detailed as well as determining whether or not this procedure produces any useful results (see, e.g., Burnaby, 1966; Hanley and Parnes, 1983).

In connection with the preceding, the criterion of maximum discriminating ability arises. From the example given it would appear that a canonical variable arising from the criterion of maximum discriminating ability has no practical interpretation in intercropping investigations. The question arises as to what other criteria should be used in analyzing data from intercropping experiments. Should we continue to cling solely to the criteron set forth by Fisher (1936, 1938, 1940) and Smith (1936) or should multivariate theory proceed using other criteria?

Another question that arises is the following. Suppose one has k canonical variables to summarize the information from p variables, $k > p$. Now should one do a second-stage multivariate analysis and treat the k canonical variables as a k-variate problem? Can one now obtain a linear combination of linear combinations and have a stage 2 canonical variable? Should one proceed to an s stage multivariate analysis? This problem arises in intercropping. Should one treat equation (1) as variate one, equation (2) as variate two, and equation (3) as variate three and conduct a three-variate multivariate analysis?

REFERENCES

Burnaby, T. (1966), "Growth invariant discriminant functions and generalized functions". *Biometrics* **22**, 96–110.

Fisher, R. A. (1936), "The use of multiple measurements in taxonomic problems". *Annals of Eugenics* **7**, 179–188.

Fisher, R. A. (1938), "The statistical utilization of multiple measurements". *Annals of Eugenics* **8**, 376–386.

Fisher, R. A. (1940), "The precision of discriminant functions". *Annals of Eugenics* **10**, 422–429.

Hanley, J. A., and M. N. Parnes (1983), "Nonparametric estimation of a multivariate distribution in the presence of censoring". *Biometrics* **39**, 129–140.

Mead, R., and J. Riley (1981), "A review of statistical ideas relevant to intercropping research" (with discussion). *Journal of the Royal Statistical Society,* Series A **144**, 462–509.

Okigbo, B. N. (1981), "Evaluation of plant interactions and productivity in complex mixtures as a basis for improved cropping systems design". *ICRISAT, Proceedings, International Workshop on Intercropping,* pp. 155–179.

Pearce, S. C., and B. Gilliver (1978), "The statistical analysis of data from intercropping experiments." *Journal of Agricultural Science, Cambridge* **91**, 625–632.

Pearce, S. C., and B. Gilliver (1979), "Graphical assessment of intercropping methods". *Journal of Agricultural Science, Cambridge* **93**, 51–58.

Singh, M., and B. Gilliver (1983), "Statistical analysis of intercropping data using a correlated error structure". *Proceedings of the 44th Session of the ISI, Madrid* **1**, 158–163.

Smith, H. F. (1936), "A discriminant function for plant selection". *Annals of Eugenics* **7**, 240–250.

Srivastava, J. N. (1968), "On a general class of designs for multiresponse experiments". *Annals of Mathematical Statistics* **39**, 1825–1843.

THE UNIVERSITY OF WESTERN ONTARIO SERIES IN PHILOSOPHY OF SCIENCE

A Series of Books in Philosophy of Science, Methodology, Epistemology, Logic, History of Science, and Related Fields

1. J. Leach, R. Butts, and G. Pearche (eds.), *Science, Decision and Value.* 1973, vii + 219 pp.
2. C. A. Hooker (ed.), *Contemporary Research in the Foundations and Philosophy of Quantum Theory.* 1973, xx + 385 pp.
3. J. Bub, *The Interpretation of Quantum Mechanics.* 1974, ix + 155 pp.
4. D. Hockney, W. Harper, and B. Freed (eds.), *Contemporary Research in Philosophical Logic and Linguistic Semantics.* 1975, vii + 332 pp.
5. C. A. Hooker (ed.), *The Logico-Algebraic Approach to Quantum Mechanics.* 1975, xv + 607 pp.
6. W. L. Harper and C. A. Hooker (eds.), *Foundations of Probability Theory, Statistical Inference, and Statistical Theories of Science.* 3 Volumes. Vol. I: *Foundations and Philosophy of Epistemic Applications of Probability Theory.* 1976, xi + 308 pp. Vol. II: *Foundations and Philosophy of Statistical Inference.* 1976, xi + 455 pp. Vol. III: *Foundations and Philosophy of Statistical Theories in the Physical Sciences.* 1976, xii + 241 pp.
7. C. A. Hooker (ed.), *Physical Theory as Logico-Operational Strúcture.* 1979, xvii + 334 pp.
8. J. M. Nicholas (ed.), *Images, Perception, and Knowledge.* 1977, ix + 309 pp.
9. R. E. Butts and J. Hintikka (eds.), *Logic, Foundations of Mathematics, and Computability Theory.* 1977, x + 406 pp.
10. R. E. Butts and J. Hintikka (eds.), *Foundational Problems in the Special Sciences.* 1977, x + 427 pp.
11. R. E. Butts and J. Hintikka (eds.), *Basic Problems in Methodology and Linguistics.* 1977, x + 321 pp.
12. R. E. Butts and J. Hintikka (eds.), *Historical and Philosophical Dimensions of Logic, Methodology and Philosophy of Science.* 1977, x + 336 pp.
13. C. A. Hooker (ed.), *Foundations and Applications of Decision Theory.* 2 volumes. Vol. I: *Theoretical Foundations.* 1978, xxiii + 442 pp. Vol. II: *Epistemic and Social Applications.* 1978, xxiii + 206 pp.
14. R. E. Butts and J. C. Pitt (eds.), *New Perspectives on Galileo.* 1978, xvi + 262 pp.
15. W. L. Harper, R. Stalnaker, and G. Pearce (eds.), *Ifs. Conditionals, Belief, Decision, Chance, and Time.* 1980, ix + 345 pp.
16. J. C. Pitt (ed.), *Philosophy in Economics.* 1981, vii + 210 pp.
17. Michael Ruse, *Is Science Sexist?* 1981, xix + 299 pp.

18. Nicholas Rescher, *Leibniz's Metaphysics of Nature.* 1981, xiv + 126 pp.
19. Larry Laudan, *Science and Hypothesis.* 1981, x + 258 pp.
20. William R. Shea, *Nature Mathematized.* Vol. I, 1983, xiii + 325 pp.
21. Michael Ruse, *Nature Animated.* Vol. II, 1983, xiii + 274 pp.
22. William R. Shea (ed.), *Otto Hahn and the Rise of Nuclear Physics.* 1983, x + 252 pp.
23. H. F. Cohen, *Quantifying Music.* 1984, xvii + 308 pp.
24. Robert E. Butts, *Kant and the Double Government Methodology.* 1984, xvi + 339 pp.
25. James Robert Brown (ed.), *Scientific Rationality: The Sociological Turn.* 1984, viii + 330 pp.
26. Fred Wilson, *Explanation, Causation and Deduction.* 1985, xviii + 385 pp.
27. Joseph C. Pitt (ed.), *Change and Progress in Modern Science.* 1985, viii + 398 pp.
28. Henry B. Hollinger and Michael John Zenzen, *The Nature of Irreversibility.* 1985, xi + 340 pp.
29 Kathleen Okruhlik and James Robert Brown (eds.), *The Natural Philosophy of Leibniz.* 1985, viii + 342 pp.
30. Graham Oddie, *Likeness to Truth.* 1986, xv + 218 pp.
31. Fred Wilson, *Laws and Other Worlds.* 1986, xv + 328 pp.
32. John Earman, *A Primer on Determinism.* 1986, xiv + 273 pp.
33. Robert E. Butts (ed.), *Kant's Philosophy of Physical Science.* 1986, xii + 363 pp.
34. Ian B. MacNeill and Gary J. Umphrey (eds.), Vol. I, *Applied Probability, Stochastic Processes, and Sampling Theory.* 1987, xxv + 329 pp.
35. Ian B. MacNeill and Gary J. Umphrey (eds.), Vol. II, *Foundations of Statistical Inference.* 1987, xvii + 287 pp.
36. Ian B. MacNeill and Gary J. Umphrey (eds.), Vol. III, *Time Series and Econometric Modelling.* 1987, xix + 394 pp.
37. Ian B. MacNeill and Gary J. Umphrey (eds.), Vol. IV, *Stochastic Hydrology.* 1987, xv + 225 pp.
38. Ian B. MacNeill and Gary J. Umphrey (eds.), Vol. V, *Biostatistics.* 1987, xvi + 283 pp.
39. Ian B. MacNeill and Gary J. Umphrey (eds.), Vol. VI, *Actuarial Science.* 1987, xvi + 250 pp.

GPSR Compliance

The European Union's (EU) General Product Safety Regulation (GPSR) is a set of rules that requires consumer products to be safe and our obligations to ensure this.

If you have any concerns about our products, you can contact us on ProductSafety@springernature.com

In case Publisher is established outside the EU, the EU authorized representative is:

Springer Nature Customer Service Center GmbH
Europaplatz 3
69115 Heidelberg, Germany

Batch number: 10607222

Printed by Printforce, the Netherlands